Contributions on the Physical Oceanography of the Gulf of Mexico

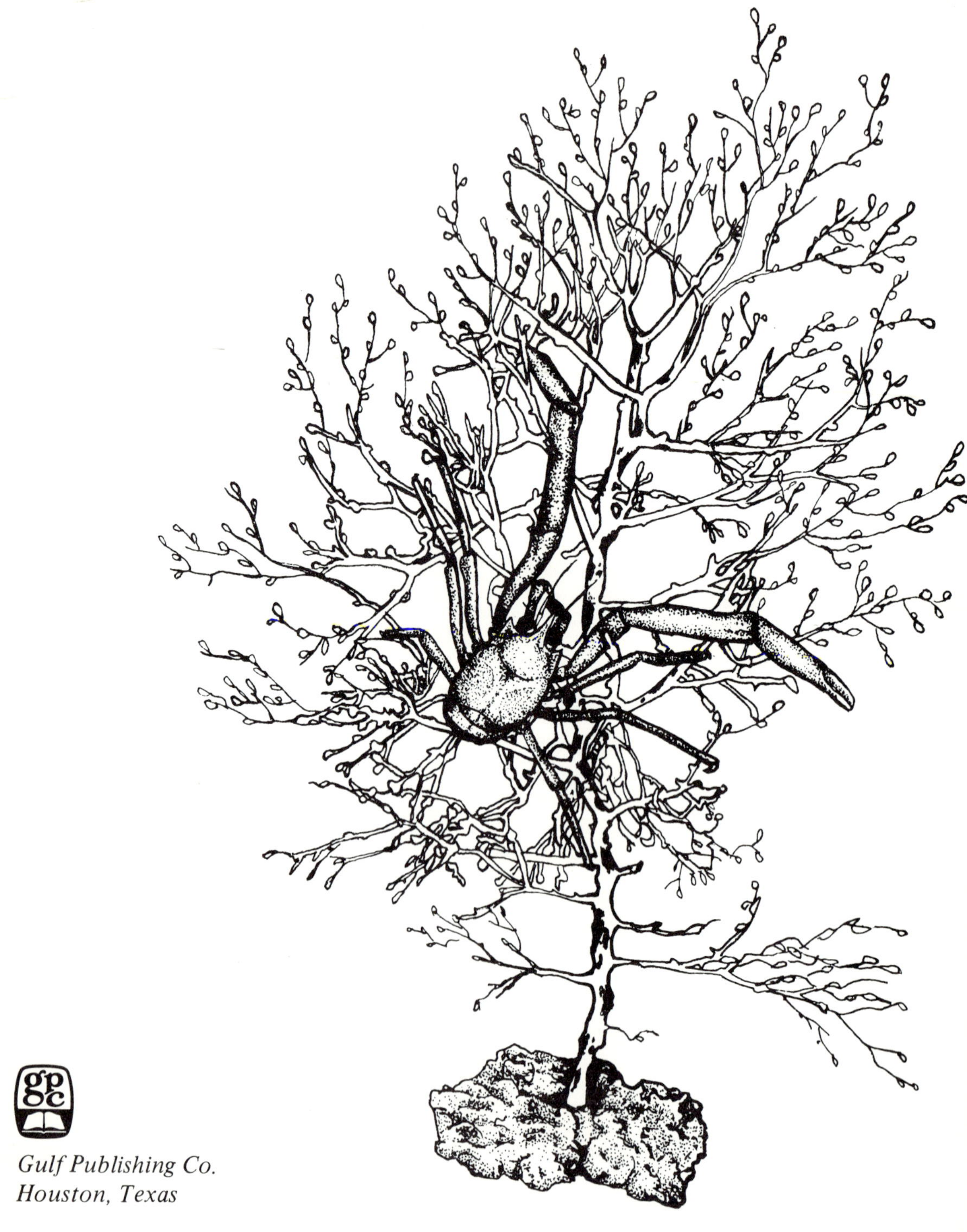

Gulf Publishing Co.
Houston, Texas

Volume 2
Texas A&M University
Oceanographic Studies

Contributions on the Physical Oceanography of the Gulf of Mexico

L. R. A. Capurro

Editorial Director
Department of Oceanography, UNESCO
Paris, France
Lecturer, Texas A&M University, College Station, Texas

Joseph L. Reid

Research Oceanographer, Department of Oceanography
Scripps Institution of Oceanography, La Jolla, California

Publication Committee
Department of Oceanography

Richard A. Geyer, Chairman

John D. Cochrane, Physical Oceanography
Bobby J. Presley, Chemical Oceanography
Willis E. Pequegnat, Biological Oceanography
Richard Rezak, Geological Oceanography

Art facing title page by Ramah Taylor

Library of Congress Catalog Card Number 71-135998
ISBN 0-87201-347-2

Manuscript received at Gulf Publishing Co. in October, 1970

Contents

Preface

The second volume of the Texas A&M University Oceanographic Series contains physical oceanography studies of the Gulf of Mexico. Professor Geyer's introduction to these papers reviews the history of research in the Gulf—from the earlier, principally biologically oriented investigations through the few pre-World War II expeditions that included physical oceanography. With the establishment of a Department of Oceanography at Texas A&M University in 1949, the Gulf received new attention, encompassing a wider range of studies than had been possible before. The department's studies have by no means been limited to the Gulf of Mexico, but it was considered appropriate in the earliest series volume dealing with biological oceanography to take the Gulf as a focus.

This book begins with a description of the Gulf waters and their general circulation, proceeds through a discussion of the major current and its variability, examines the various problems through both numerical and experimental modeling, discusses the effects of hurricanes and concludes with a chapter on tides.

The first three chapters describe the overall situation within the Gulf. In the initial chapter all earlier data as well as more extensive data from the Texas A&M expedition in the winter of 1962 are used in describing the origin, characteristics and stratification of the Gulf waters and of their circulation. The second chapter gives a more detailed study of the distribution of the intermediate water (Wüst's "Subantarctic Intermediate Water") in the Gulf, with a specific examination of the finer scale variation of its salinity. The third chapter discusses direct measurements of a rather strong flow at about 3 km, whose existence was first suggested from the distribution of animals and sediments.

In the second section, six chapters deal with the Loop Current and its variability. In Chapter 4 the breakoff of an anticyclone from the Loop Current is described and discussed on the basis of measurements made on several expeditions in 1969. The fifth chapter describes the first isolated eddy to be observed clearly within the Gulf and its subsequent movements from observations made in August and September, 1965. The passage of a major hurricane during this period added some interesting features to the study. In Chapter 6 the situations of the Loop Current in 1966 and 1967 are contrasted as two different modes, one being more typically a loop in the major flow and the other a somewhat isolated anticyclonic ring. In Chapter 7 the direct current measurements described in Chapter 6 are used to derive a stream function using various boundary conditions. A good fit is obtained and the conclusion is reached that the major flow in the eastern Gulf is nondivergent.

In the eighth chapter the relation between the Yucatan Current and bottom topography is examined, assuming conservation of potential vorticity and using measured velocity in the Yucatan Straits and actual bottom topography. There is a very good fit for some distance downstream from the Straits, but beyond that, in the deep water, conditions other than topography may be important. These are examined in the ninth chapter, where the Loop's dimensions over the deeper water are explained in terms of current speed and variation of the Coriolis parameter. These two chapters present a very interesting explanation of the dynamics of the steady-state aspects of the Loop Current. The topographic waves discussed in Chapter 8 may account for the initial disturbances of the Loop, but their rapid growth and subsequent development (with coalescence in some years) and the detached eddies remain to be explained.

Numerical modeling is taken up in Chapters 10 and 11, with more of the topographic complexities and more complete dynamical investigation. The seasonal variations in the Yucatan Current are considered, and an attempt is made to consider the effect of stratification. The results seem fairly realistic and represent a major step in the numerical model approach.

The results of experimental modeling of the Gulf and the Caribbean are presented in Chapter 12. A tank was used and the rotation and inflow were varied. With appropriately set β-effect, Rossby number and Reynolds number the flow patterns showed similarity to the actual circulation.

In the fifth section Chapters 13 and 14 examine the rapid and extensive effects of hurricanes on circulation and thermal structure. Along the path of the hurricane's eye the depth to the thermocline appears to decrease, and on the edges to increase. This behavior is examined by theoretical models of both a homogeneous and a two-layer ocean. In Chapter 14 the effects are examined from pre-hurricane and post-hurricane bathythermograph sections taken about a month before and a month after the hurricane passage. In the latter case the upper thermal structure had returned to about the pre-hurricane state within a month, but the deeper structure still appeared to be disturbed.

In Chapter 15, the final section, a previous review of tides in the Gulf is brought up to date with a new contribution on energy dissipation within the Gulf.

The studies included in this book involve the methods of both the descriptive oceanographer and the theoretical investigator. These two approaches are not often attempted together. The breadth of the department is reflected by the extent to which the two approaches are used here, both individually and in combination.

Introduction

Expeditions directed primarily toward the exploration, conquest and colonization in the Gulf of Mexico began at the end of the 15th century and continued for several hundred years. However, there were few scientific expeditions in the area even as late as the 19th and throughout the first half of the 20th centuries. For example, Bencker (1930), in his paper summarizing worldwide oceanographic expeditions from 1800-1930, lists 133 expeditions to the Arctic regions, 36 to the Antarctic, 10 to the Indian Ocean and 15 around the world. But, during this same period only three were to the Gulf of Mexico.

The era of comprehensive, systematic and continuing study in all branches of oceanography for the Gulf of Mexico did not begin until the Department of Oceanography of Texas A&M University was established in 1949. During the past 20 years, the department has grown significantly. Its faculty and senior research associates now number 26. It is currently operating its third major oceanographic research vessel, the 850 ton R/V *Alaminos*, and a coastal research vessel, R/V *Orca*, of 100 tons displacement. Significant advances have been made during this period in understanding many basic problems of the Gulf of Mexico in all fundamental branches of oceanography.

To make many of these results available in more detail and continuity than can generally be presented in diverse professional journals, it was decided to establish an occasional paper series. The early volumes will provide an overview of the current status of various results of basic oceanographic research on the Gulf of Mexico and contiguous areas. These will contain papers prepared by a broad cross section of the department's faculty and senior research associates. They will deal with different aspects of biological, chemical, geological, geophysical, meteorological and physical oceanography. Later volumes will present results of other diversified oceanographic research activities conducted by members of the department. It is also planned to publish pertinent papers prepared by other scientists investigating the Gulf of Mexico and contiguous areas, as well as papers by the A&M staff on oceanographic research in other geographic areas.

Those familiar with the literature on the Gulf of Mexico are aware of Bulletin No. 89 of the Bureau of Commercial Fisheries (Galtsoff, 1954). This was the first attempt to present within a single volume a relatively comprehensive treatment of the current status of oceanographic knowledge in the Gulf. The occasional paper series intends to give to the scientific community a comparable publication describing the most recent advances in oceanography for this region. The tremendous increase in interest in the Gulf during the past few years will continue to accelerate. Hence, it is hoped that information in these publications will serve as the springboard on which to base the ever increasing scientific studies to be conducted in the future, beginning with the "International Decade of Oceanographic Exploration."

Volume 1 presents a series of related papers directed toward a better understanding of selected aspects of the biological oceanographic characteristics of the Gulf. These include not only taxonomic but systematic and environmental aspects. Descriptions of many species heretofore unavailable are provided. Volume 2 comprises a collection of papers dealing with the physical oceanography of the Gulf. It includes a presentation of the results of theoretical and modeling experiments, as well as descriptive treatments of basic circulation and tidal studies. Volume 3 contains papers describing the geological, geophysical and tectonic characteristics of the Gulf of Mexico, as well as its chemical properties, including results of theoretical and descriptive studies.

Scientific research in the Gulf of Mexico has not had the emphasis given to other comparable oceanic areas, but it would be remiss to infer that it has been completely neglected. There was some scientific interest in the Gulf toward the end of the 19th century. But, from then until the late forties, only a few scientific expeditions were active in these waters. The earliest as well as relatively comprehensive studies of the Gulf included the cruises of the U.S. Coast and Geodetic Survey's steamer *Blake* and of the Fish Commission's steamer *Albatross*. Scientific results were presented by Alexander Agassiz (1888) in his two-volume treatise describing three cruises of *Blake* in the Gulf of Mexico.

The expeditions of *Pawnee* reported on by Boone (1927) and of *Mabel Taylor* of the Bingham Oceanographic Institute of Yale University headed by Parr mark the next organized and relatively detailed effort to study the Gulf after a lapse of more than a quarter of a century. Parr (1935) reported on hydrographic observations in the Gulf and adjacent straits. This expedition was followed a few years later by a cruise of the Woods Hole Oceanographic Institution's *Atlantis*. Riley (1937) discussed the significance of the Mississippi River drainage for biological conditions in the northern Gulf of Mexico from data taken during this cruise. He also included information on the phosphate content and other nutrients for this area.

Shortly after World War II, the *Atlantis* again operated in the Gulf for several months, emphasizing geological and paleontological research. Several expeditions to the Gulf were also conducted under the auspices of the Lamont Geological Observatory, stressing geophysical and geological studies. The recent advent of the research drilling activities of *Glomar Challenger* in the Gulf marks a new era designed to better define the geological and geophysical history and structure of this area. When analyzed, the data obtained from these cruises will significantly improve our basic understanding in these two fields of oceanography.

Numerous additional references are available in the literature describing the results of past oceanographic research in this region. Hence, it would be appropriate to briefly mention some of the earliest references available in the various fields of oceanography. Two major sources exist for this purpose: one in the bibliographies for each chapter in Bulletin No. 89 of the Bureau of Commercial Fisheries (Galtsoff, 1954), the other in a special annotated bibliography prepared by Geyer (1950) of oceanography, marine biology, geology, geophysics and meteorology of the Gulf of Mexico. The latter also includes a small section containing early references, dating back more than 100 years, of occurrences of mass mortality in the Gulf, as well as an appendix listing early navigation charts prepared by various sources, including hydrographic offices of the United States and other countries.

Kohl (1863) described the earliest history of the discovery and exploration of the Gulf of Mexico and contiguous areas by Spaniards from 1492 through 1543. Lindenkohl (1896) presented a chart showing the specific gravity of the surface waters of the Gulf of Mexico and the Gulf Stream compiled from data from *Blake* and *Albatross* expeditions. A plate was presented in the same publication, summarizing temperatures in the Gulf of Mexico and the Gulf Stream at a depth of 460 meters. Lindenkohl described the results of the temperature and density observations of the Gulf Stream and Gulf of Mexico waters made by the U.S. Coast and Geodetic Survey. Sweitzer (1898) speculated on the origin of the Gulf Stream and on the general circulation in the Gulf of Mexico, but Pillsbury's (1890) classical current measurements in the Yucatan and Florida Straits and the southeastern Gulf laid the foundation for quantitative studies of the circulation of the Gulf of Mexico. Haupt et al. (1898) described the origin of the Gulf Stream and circulation of waters in the Gulf of Mexico with special reference to jetty construction. Dietrich (1936) summarized and evaluated much of the available information on the Gulf up to that time, as well as including some new material on physical oceanography.

Hilgard (1871) presented one of the first discussions of the geological history of the Gulf with additional information in the form of a generalized description of the bathymetric character of the area, including a map. Macgree (1892) presented detailed discussions on the Gulf of Mexico as a measure of isostasy, and Lawson (1942) described the Mississippi Delta as an example of a study in isostasy. Murray (1885) described bottom sediments on parts of the continental shelf of western Florida and summarized the results on a map. Wells (1919) reported on carbon dioxide determinations in the Gulf. Hayes and Kennedy (1903) mentioned the presence of oil ponds on the Gulf about two miles offshore near the mouth of the Sabine. Turner (1903) gave a geologic and biologic analysis of mud and sea wax obtained along the shores of the Gulf from the Sabine River to Corpus Christi. Some of these occurred in cakes six to eight feet long and one to two inches thick. Oil slicks have been known to occur for a long time in the Gulf of Mexico, and their locations have been summarized on maps (U.S. Hydrographic Office, 1900, 1905, 1906). Numerous early references on the mud lumps off the mouth of the Mississippi Delta include those of Delafield (1829), Thomassy (1860), Anonymous (1868), Hilgard (1871, 1906), Corthell (1884) and Harris (1902).

The phenomenon of mass mortality due to excessive temperature changes and upwelling has also been reported by Bartlett (1856) and Smith (1899). Pierce (1884) and Carlson (1908) speculated on the cause of mortality of fishes in the Gulf of Mexico. Another mass mortality reference is that of Willcox (1887), who recorded fish killed by cold along the Gulf and the coast of Florida. Glazier (1882) recorded fish killed from polluted waters. Other early general references to mass mortality in the Gulf are those of Jefferson (1878), Anonymous (1881), Moore (1882), Porter (1882), Ingersoll (1882), Walker (1884) and Weber (1887).

Evermann and Kendall (1894) described the fishes of Texas and the Rio Grande Basin with reference to their geographical distribution. Stevenson (1893) and Rathburn (1892) discussed the fisheries of Texas, and Collins (1887) reported on *Albatross* cruises. Tauner (1887), in his report on *Albatross* cruises, described experiments using torpedoes to study effects of explosions on fish in the Gulf, as well as recording data on hydrography and bottom conditions.

Emphasis in the initial volumes of this series is on investigations in the Gulf of Mexico. However, it should not be interpreted that the research activities of the Department of Oceanography have been or are confined entirely to this region. A substantial and continuing effort is also being maintained in many other parts of the world oceans. These include such diverse areas as the Antarctic, south and east Pacific and south and central Atlantic Oceans. Here research has been and is being conducted to solve basic oceanographic problems, particularly in biological, chemical and physical oceanography. Many of these investigations have been and are being conducted in cooperation with other oceanographic institutions and government agencies. Examples include the Equatorial Atlantic (*Equalant*) and Eastern Tropical Pacific (*Eastropac*) expeditions, as well as research conducted aboard other vessels, such as USNS *Eltanin*, USCGC *Glacier*, R/V *Thompson* and ARA *San Martin*. It is perhaps fitting that the publication of the first volume of this series, together with the newly established Gulf of Mexico Folio Series to be published by the American Geographical Society, should commemorate the twentieth anniversary of the founding of the Department of Oceanography at Texas A&M University.

Richard A. Geyer

Head, Department of Oceanography
Texas A&M University
April 15, 1970

References

Agassiz, A. (1888) Three cruises of the U.S. Coast and Geodetic Survey steamer *Blake* in the Gulf of Mexico, in the Caribbean Sea, and along the Atlantic Coast of the United States from 1877 to 1880. *Bull. Mus. Comp. Zool. Harv.* Cambridge, Mass. vol. 14, 15.

_____. (1888) *Three cruises of the* Blake, *Gulf of Mexico*. Houghton Mifflin Co., 2 vol.

_____. (1897) North American starfishes. *Mem. Mus. Comp. Zool. Harv.* 5 pt. 1.

Anonymous. (1868) The Mississippi River. *DeBow's Review* 5, pp. 454-471.

Anonymous. (1881) Mortality of fish in the Gulf of Mexico. *Ann. Mag. Nat. Hist.*, ser. 5, vol. 8, pp. 238-240.

Bartlett, J.R. (1856) *Personal narrative of explorations and incidents in Texas, etc.* Appleton, N. Y.

Bencker, H. (1930) The bathymetric soundings of the oceans (with chronological list of ocean expeditions from 1800-1930). *Hydro. Rev.*

Carlson, Y.A. (1908) Brilliant Gulf waters. *Monthly Weather Review*, 36, pp. 371-372.

Boone, L. (1927) Scientific results of the first oceanographic expedition of the *Pawnee*, 1925, Crustacea from tropical east American seas. *Bull. Bingham Oceanogr. Collect.* 1(2):1-147.

Collins, J.W. (1887) *Report on the discovery and investigation of fishing grounds made by the Fish Commission steamer* Albatross *during a cruise along the Atlantic Coast and in the Gulf of Mexico, with notes on the Gulf fisheries.* Rept. U.S. Comm. Fisheries 1885, pp. 217-311.

Corthell, E.L. (1884) The South Pass jetties, etc. *Trans. Amer. Soc. Civil Engrs.* 13, pp. 313-330.

Dall, W.H. (1886) Reports on the results of dredging. . .by the U.S. Coast Survey Steamer *Blake*. . .Report on the Mollusca. Part I. Brachiopoda and Pelecypoda. *Bull. Mus. Comp. Zool.*, 12(6): 171-318.

_____. (1890) Preliminary report on the collection of Mollusca and Brachipoda obtained in 1887-88. Scientific results of explorations by the U.S. Fish Commission Steamer *Albatross*, No. 7. *Proc. U.S. Nat. Mus.*, 12: 219-392.

Delafield, R. (1829) *Report on the survey of the passes of the Mississippi.* 21st Cong., 1st Sess., House Doc. 7, no. 1, pp. 7-14.

Dietrich, Guenter. (1936) Das "ozeanische Nivellement" und seine Anwendung auf die Golfkueste und die atlantische Kueste der Vereinigten Staaten von Amerika, *Zeitschr. f. Geophysik, Jahrg.* 12, Heft 7/8, p. 287-298.

____. (1939) *Das Amerikanische Mittelmeer,* (The Gulf of Mexico). Gesellsch. f. Erdkunde zu Berlin, Zeitschr., pp. 108-130, photo, pp. 115, 120, 121, 127.

Evermann, B.W. and Kendall, W.C. (1894) The fishes of Texas and the Rio Grande Basin considered chiefly with reference to their geographic distribution. *Bull. U.S. Fish Comm.* 12, pp. 57-126.

Galtsoff, Paul (1954) *Gulf of Mexico–its origin, waters, and marine life.* U.S. Dept. of Interior, Fish and Wildlife Service, Fishery Bulletin 89 of Fish and Wildlife Service vol. 55.

Glazier, W.C.W. (1882) On the destruction of fish by polluted waters in the Gulf of Mexico. *Proc. U.S. Nat. Mus.* (1881), vol. 4, pp. 126-127.

Geyer, Richard A. (1950) A Bibliography on the Gulf of Mexico. *The Texas Journal of Science,* vol. 2, no. 1, pp. 44-93.

Haupt, E.M. et. al. (1898) Discussion on Paper 875. *Trans. Amer. Soc. Civil Engrs.* 40, pp. 99-112.

Harris, G.D. (1902) *The geology of the Mississippi embayment.* Rept. of Geol. Survey of La., Spec. Rept. 1, pp. 1-39.

Hayes and Kennedy. (1903) Oil fields of the Texas-Louisiana Gulf Coastal plains. *Bull. U.S. Geol. Survey 212.*

Hilgard, E.W. (1871) On the geology of the delta and the mud lumps of the passes of the Mississippi. *Amer. Jour. Sci.* 2 S. 238-246, 356-364, 425-435.

____. (1871) The basin of the Gulf of Mexico. *Amer. Jour. Sci.* 3, vol. 21, pp. 283-291.

____. (1906) The exceptional nature and genesis of the Mississippi delta. *Science n.s.,* 24, pp. 861-866.

Ingersoll, E. (1882) On the fish mortality in the Gulf of Mexico. *Proc. U.S. Nat. Mus,*, vol. 4 (1881), pp. 74-80.

Jefferson, J.P. (1878) On the mortality in the Gulf of Mexico in 1878. *Proc. U.S. Nat. Mus.*, vol. 1 (1881), pp. 244-245.

Kohl, J.G. (1863) Aelteste Geschichte die Entdeckung und Erforschung des Golfs von Mexico und der ihn umgehenden Kuesten durch die Spanien von 1492 bis 1543. *Zeits. fuer allge.* Erdkunde N. F., Bd 15, pp. 1-40, 169-194.

Lawson, A.C. (1942) Mississippi Delta–a study in isostasy. *Geol. Soc. Amer.,* vol. 53, pp. 1231-1254.

Lindenkohl, A. (1896) Temperature im Golf von Mexico und im Golfstrom in den tiefe von 460 meters. *Petermanns Mitt.,* vol. 42, 3 pl. at end.

____. (1896) Spezifisches gewicht des oberflaechenwassers im Golf von Mexico und im Golfstrom. *Petermanns Mitt.* vol. 42, pl. 3 at end.

____. (1896) Resultate der Temperatur und Dichtigskeitbeobachtungen in den Gewaessern des Gulfstroms und der Golf von Mexico durch das Bureau des USCGS. *Petermanns Geogr. Mitt.,* Heft 2, pp. 25-29, Map.

McGee, W. (1892) The Gulf of Mexico as a measure of isostasy. *Amer. Jour. Sci.* (3), 44, pp. 177-192.

Milne Edwards, A. and Bouvier, E.L. (1897) Results of dredging, under the supervision of Alexander Agassiz, in the Gulf of Mexico (1877-78) in the Caribbean Sea (1878-79), and along the Atlantic coast of the United States (1880), by the U.S. Coast Survey Steamer *Blake*. 35. Description des Crustaces de la Famille des Galatheides recueillis pendant l' expedition. *Mem. Mus. Comp. Zool. Harv.,* 19(2): 1-141, pls. 1-12.

____. (1909) Reports on the results of dredging, under the supervision of Alexander Agassiz, in the Gulf of Mexico (1877-78), in the Caribbean Sea (1878-79), and along the Atlantic coast of the United States (1880), by the U.S. Coast Survey Steamer *Blake*. 44. Les Peneides et Stonopides. *Mem. Mus. Comp. Zool.*, 27 (art. 8): 177-274.

Moore, M.A. (1882) Fish mortality in the Gulf of Mexico. *Proc. U.S. Nat. Mus.*, vol. 4, (1881), pp. 125-126.

Murray, J. (1885) Report on the specimens of bottom deposits. Reports on the results of dredging. . .by the U.S. Coast Survey Steamer *Blake. . .Bull. Mus. of Comp. Zool. Harv.* Cambridge, Mass., no. 27 vol. 12, S. 37-61.

——. (1899) On the survey by the SS *Britannia* of the cable route between Bermuda, Turk's Islands and Jamaica. *Proc. Roy. Soc. Edinburgh* 22, S. 409-429.

Parr, A.E. (1935) Report on hydrographic observations in the Gulf of Mexico and the adjacent straits made during the Yale Oceanographic Expedition on the *Mabel Taylor* in 1932, *Bull. Bingham Ocean, Collect.* vol. 5, Art. 1, pp. 1-93.

Phillips, B. (1884) Notes on a trip in the Gulf of Mexico. *Bull. U.S. Fish.* 4, p. 144.

Pillsbury, J.E. (1890) The Gulf Stream. A description of the methods employed in the investigation, and the results of research. Annual Report of the Supt. U.S. Coast and Geodetic Survey for 1890, appendix no. 10.

Pierce, H.D. (1884) Notes on the bluefish, mortality of Florida fishes, etc. *Bull. U.S. Fish Comm.*, vol. 4, pp. 263, 266.

Porter, J.Y. (1882) On the destruction of fish by poisonous water in the Gulf of Mexico. *Proc. U.S. Nat. Mus.*, vol. 4 (1881), pp. 121-123.

Rathburn, R. (1892) *The fisheries of the Gulf of Mexico.* Rept. U.S. Comm. of Fish. 1888-1889, pp. 54-59.

Riley, G.A. (1937) The significance of the Mississippi River drainage for biological conditions in the northern Gulf of Mexico. *Jour. Marine Res. 1,* no. 1, pp. 60-74.

Shaw, E.W. (1913) Gas from mud lumps at the mouths of the Mississippi. *U.S. Geol. Survey Bull.* 541A, pp. 12-15.

Smith, H.M. (1899) *Peridinium as a possible cause of red oysters.* Rept. U.S. Comm. Fish 1898, pp. 127-129.

Smith, S.I. (1882) Report on the results of dredging, under the supervision of Alexander Agassiz, on the East coast of the United States, during the summer of 1880 by the U.S. Coast Survey Steamer *Blake.* 17. Report on the Crustacea. Part I. Decapoda. *Bull. Mus. Comp. Zool.*, 10(1): 1-108.

Stevenson, C.H. (1893) *Report on the coast fisheries of Texas.* Rept. U.S. Comm. Fish 1889-1891, pp. 373-420, Doc. 218.

Sweitzer, N.B., Jr. (1898) Origin of the Gulf Stream and circulation of waters in the Gulf of Mexico, with special reference to the effect of jetty construction. *Trans. Amer. Soc. Civil Engrs.* 40, pp. 86-98.

Tauner, L.J. (1887) *Report on the work of the U.S. Fish. Commission steamer* Albatross *for year ending December 31, 1885.* Rept. of U.S. Comm. Fish. 1885. Also as Document 118.

Theel, Hjalmar. (1886) Report of the Holothurioidea. *Bull. Mus. Comp. Zool. Har.*, vol. 13.

Thomassy, V. (1860) *Geologique practique de La Louisiane* (Practical Geology of Louisiana). New Orleans.

Turner, H.J. (1903) Examination of mud from the Gulf of Mexico. *Bull. U.S. Geol. Survey* 212, pp. 107-112.

United States Hydrographic Office (1900, 1905, 1906) Maps showing oil slicks in the Gulf of Mexico.

Walker, S.T. (1884) Fish mortality in the Gulf of Mexico. *Proc. U.S. Nat. Mus.*, vol. 6 (1883), pp. 105-109.

Weber, J.G. (1887) The mortality of fish in the Gulf of Mexico. *Bull. U.S. Fish Comm.*, vol. 6 (1886), pp. 11-13.

Wells, R.C. (1919) New determinations of carbon dioxide in water of the Gulf of Mexico. USGS Prof. Paper #120a.

Willcox, J. (1887) Fish killed by cold along the Gulf of Mexico and coast of Florida. *Bull. U.S. Fish Comm.* 6,p. 123.

CHARTS

Alruie chart of the Gulf of Mexico, London, R. H. Laurie, 1862, 25x36.

Copley, C., The north coast of the Gulf of Mexico from St. Marks to Galveston. Pub. by E. and G. W. Blunt, New York, 1842, 25x39.

Villiers du Terrage, M.L'expedition de Cavalier de LaSalle dans le golfe du Mexico 1684-1687 par le baron Marc de Villiers, Paris, A. Maisonneure, 1931.

Contributions on the Physical Oceanography of the Gulf of Mexico

Section 1
Circulation and Water Masses

1
Winter Circulation Patterns and Property Distributions

Worth D. Nowlin, Jr.

Abstract

Based on their characteristic properties, the water masses of the Gulf of Mexico and their vertical stratification are discussed. The *T-S* relationships specific to the region are presented. For the basin waters, below a sill depth of about 2000 m, the potential temperature, salinity and dissolved-oxygen concentrations show no measurable horizontal variation, although weak vertical density gradients evidence slight positive stability. Variations in the characteristics of the water in the following layers are shown, and the likely origins of these water masses are identified: North Atlantic Deep Water, Subantarctic Intermediate Water, oxygen minimum layer and Subtropical Underwater. For the winter season, the property distributions in the mixed surface layers are described.

On the basis of dynamic computations and GEK measurements, the general winter circulation patterns within the Gulf are examined. The mode most often observed in the eastern Gulf is one dominated by the Loop Current; water enters through Yucatan Strait as the Yucatan Current and flows in a clockwise loop which extends well into the Gulf and exits via Florida Strait. The extent of penetration and location of this loop is quite variable. In other seasons, large current rings are known to separate from the Loop Current. In contrast, the winter circulation in the western Gulf seems more predictable; it consists primarily of a clockwise cell centered over the western central Gulf, having broad westward flow for its southern limb, a narrow east northeastward flow for its northern limb and flanked to the north by a west-southwestward current along the outer Texas-Louisiana shelf.

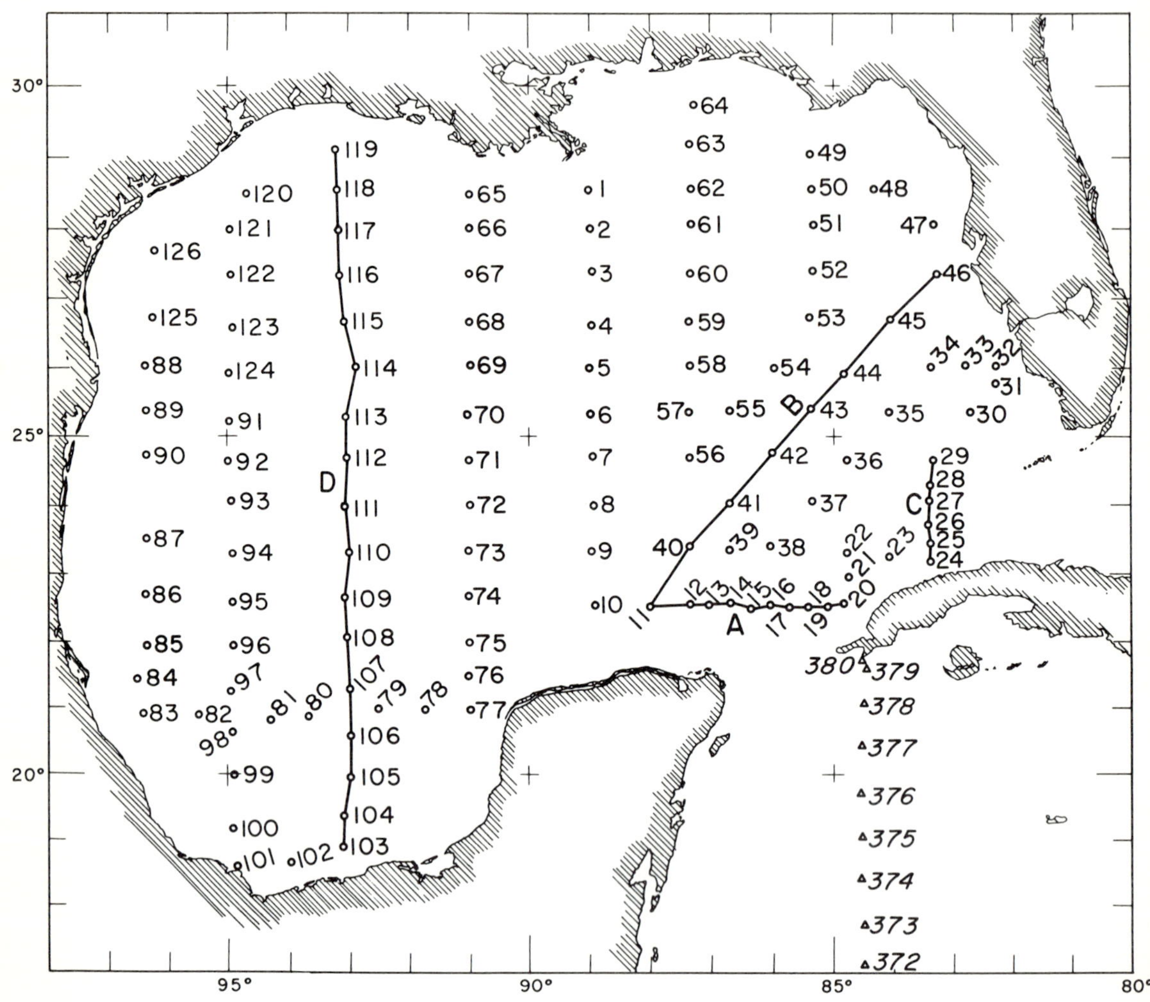

Figure 1-1. Stations occupied during Hidalgo *62-H-3, February-March, 1962 (circles), and* Crawford *Cruise 17, February-March, 1958 (triangles). Lines indicate vertical sections discussed in text.*

Introduction

This paper presents a characterization of the Gulf of Mexico waters for the winter season. There have been three major exploratory cruises within the Gulf during winter. During 25 January through 27 April 1932, the *Mabel Taylor*, under the direction of A.E. Parr from Yale, occupied 69 hydrographic stations in the Gulf of Mexico; Parr presented the results in 1935. Unfortunately, there is understandable uncertainty attached to the reported depths for these *Mabel Taylor* samples because Parr did not have unprotected reversing thermometers aboard during the cruise. The *Atlantis* produced reliable hydrographic data from 98 Gulf stations occupied between 15 February and 13 April of 1935. Neither of these cruises occupied a sampling pattern adequately uniform or dense to

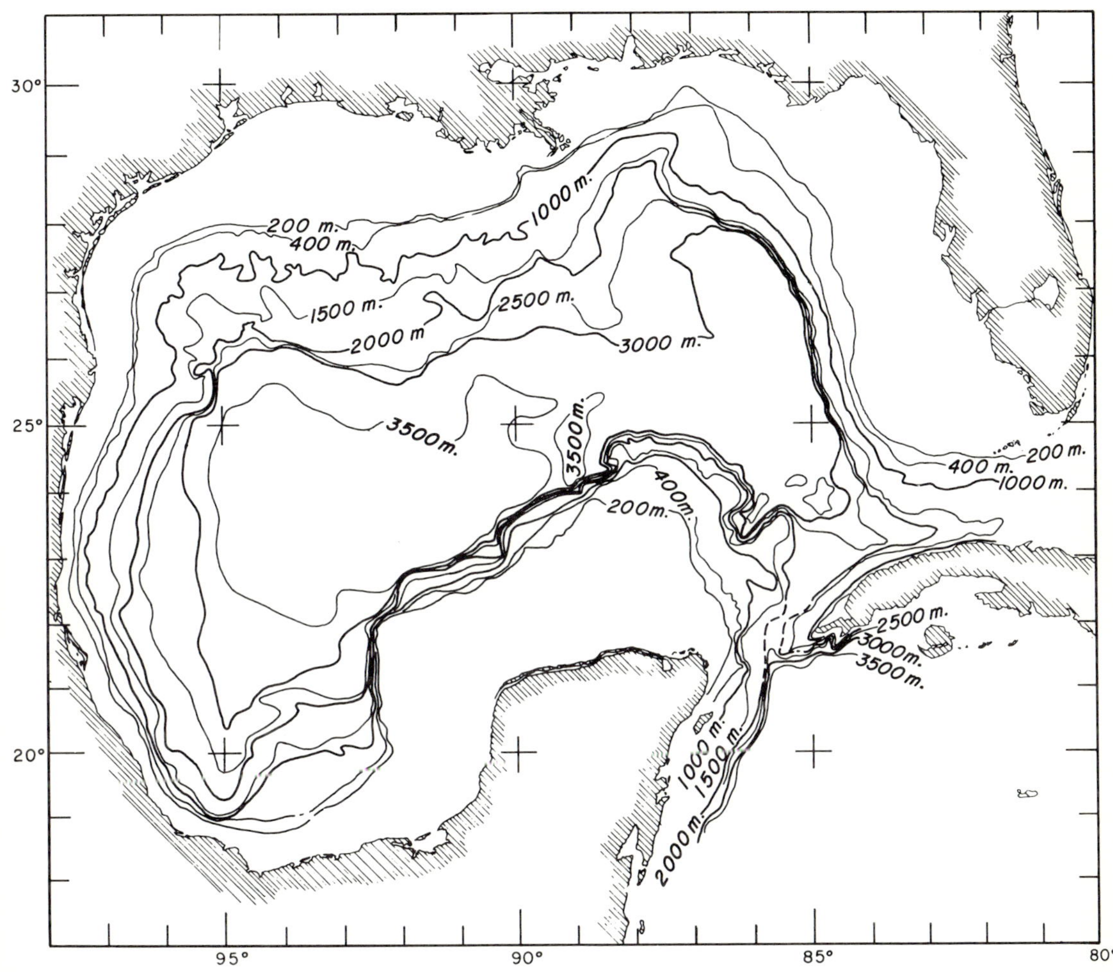

Figure 1-2. Bathymetry of Gulf of Mexico based on U.S. Coast and Geodetic Survey Chart 1007 and soundings on file at Department of Oceanography, Texas A&M University.

describe the entire Gulf. In 1962, the *Hidalgo* of Texas A&M University was used by H. J. McLellan and W.D. Nowlin, Jr. in a hydrographic survey to obtain data that might for the first time fully characterize the Gulf of Mexico waters during the winter. This cruise (62-H-3), which was executed as rapidly as the vessel's capability permitted—from 12 February through 31 March—systematically covered the entire Gulf of Mexico. Figure 1-1 shows the station positions, numbered serially in the sequence of their occupation. The maximum sampling depth at most stations was near the bottom; Figure 1-2 shows the general bathymetry of the Gulf.

Although data from many cruises are utilized, the cornerstone for this description is the winter characterization of the Gulf of Mexico based on the findings from Cruise 62-H-3. Unless otherwise stated, the data presented and inferences therefrom are from Cruise 62-H-3 and are meant to represent conditions during only February and March of 1962. This point is emphasized because of the temporal variability in the strength and distribution of circulation features in the Gulf shown by other studies, e.g., the work of Cochrane (1963, 1965) dealing with the Yucatan Current in and near Yucatan Strait. Because of this temporal variability, circulation patterns arrived at by combining data collected over a period of many years and seasons (such as the very complicated averaged dynamic topographies presented by Duxbury in 1962) may not represent conditions typical of the region during any season.

When planning Cruise 62-H-3 a careful evaluation of existing data, including those collected in other seasons of the year, and interpretations of data were made. The sequence of observations that ranks second to the 1962 survey in approaching a synoptic survey of the entire Gulf of Mexico is the series of 124 hydrographic stations occupied between 22 April and 21 August 1951 by the *Alaska* of the U. S. Fish and Wildlife Bureau in cooperation with Texas A&M University. Combining data from these three cruises, Austin (1955) presented a pattern for the dynamic topography of the sea surface relative to the 1000-db surface (Figure 1-3); Collier et al. (1958) presented a similar pattern for the 500-db dynamic topography relative to 1000 db. These geopotential anomalies may be interpreted as indicating a very complicated circulation pattern within the central and western Gulf. The circulation within the eastern Gulf, except over continental shelves, seems dominated by the Loop Current, which is the Yucatan Current's downstream continuation through the Gulf and into the straits of Florida.

Every "high" and "low" of the Gulf circulation as inferred by Austin (Figure 1-3) was intersected by at least one line of stations on the 62-H-3 cruise plan. In regions where existing data indicated large horizontal current shear, an effort was made to space stations rather closely on transects perpendicular to the indicated current directions. Nansen bottles and paired reversing thermometers were used at each hydrographic station to collect water samples and to measure the *in situ* temperatures and depths of the samples. The water samples were analyzed aboard ship by using a conductivity-type salinometer (University of Washington, No. 12) for salinity and the Winkler method for dissolved oxygen. Because of the apparently limited accuracy of the dissolved oxygen measurements, data from other cruises have been used to establish the oxygen distribution within the deep waters of the Gulf. The observed hydrographic data as well as meteorological observations are available from the National Oceanographic Data Center or in an unpublished data report (Ref. 62-16D, Department of Oceanography and Meteorology, The A.&M. College of Texas). Bathythermograms and geomagnetic electrokinetograph (GEK) current measurements were made at one-hour intervals while underway between stations.

Characteristic *T-S* Relationship

Figure 1-4 shows temperature versus salinity for all observations at temperatures lower than 17°C and for the near-surface layers at each of 17 stations in the regions of inflow and outflow. For the realm below 17°C (representing 849 data points) the plot shows a remarkable uniformity, indicating that the waters constitute essentially a single system. Comparison with historical data indicates no departures from this characteristic curve that could not be attributed to observational error.

If the data are examined on a regional basis the interagreement between stations is even more remarkable. Based on the 62-H-3 data, Caruthers (1969) made a careful comparative analysis of the potential temperature (θ) versus salinity (S) relation for the intermediate depth waters. Potential temperatures were obtained using the tables of Helland-Hansen (1930). The analysis was restricted to depths between 100 m and 1200 m and to potential temperatures between 4° and 20°C. A "standard" $\theta - S$ relation was selected, and regional deviations from this standard were quantified and discussed.

Stations 16-22 and 38, within the water of the

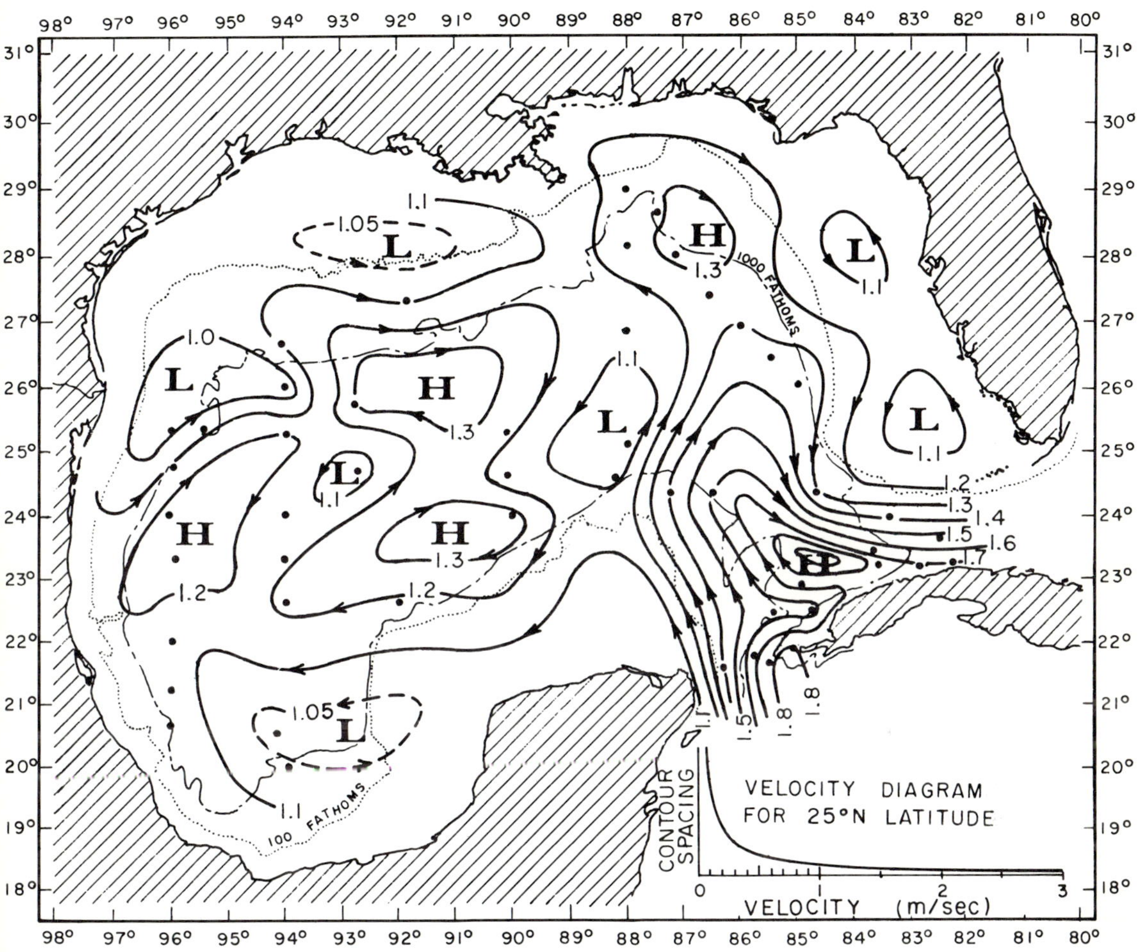

Figure 1-3. Dynamic topography of surface relative to 1000-db surface; Alaska *Cruises 1-1A, 2-1B and 3-1C, 22 April-21 August, 1951. (After Austin, 1955: Figure 2.)*

eastern Gulf bounded by the Loop Current and Cuba, were used to define the standard relation. Caruthers found that according to the F-test the best fit to these 33 temperature-salinity pairs is a seventh degree polynomial in potential temperature representing salinity by

$$S = \sum_{n=0}^{7} A_n (\theta/10)^n.$$

The standard error is 0.0045 per mil if the coefficient $A_0, ..., A_7$ have the values: 32.58314, 23.11518, −82.50041, 143.1953, −135.2648, 71.99103, −20.27993, 2.34921

For the 62-H-3 data Caruthers computed the differences between the observed salinity and the salinity computed from the standard relation for the corresponding temperature. For each station, polynomials were fitted to these differences. It was noted that "the water type found at 14°C is very constant and represents the zero crossing" of these graphs. Caruthers (1969) also confirmed in quantitative form the observations that there is a tendency for the salinities near the minimum, which occurs at temperatures below 14°C, to uniformly increase and a tendency for salinities at temperatures above 14°C to decrease as one moves into the northern or western Gulf.

Figure 1-4. Temperature versus salinity, Hidalgo *62-H-3, for all observations with temperatures less than 17°C and for the near-surface waters at 17 stations in the regions of inflow and outflow. For temperatures greater than 17°C,* T-S *relations are ploted for 10* Crawford *stations occupied in the Caribbean. See Figure 1-1 for station locations.*

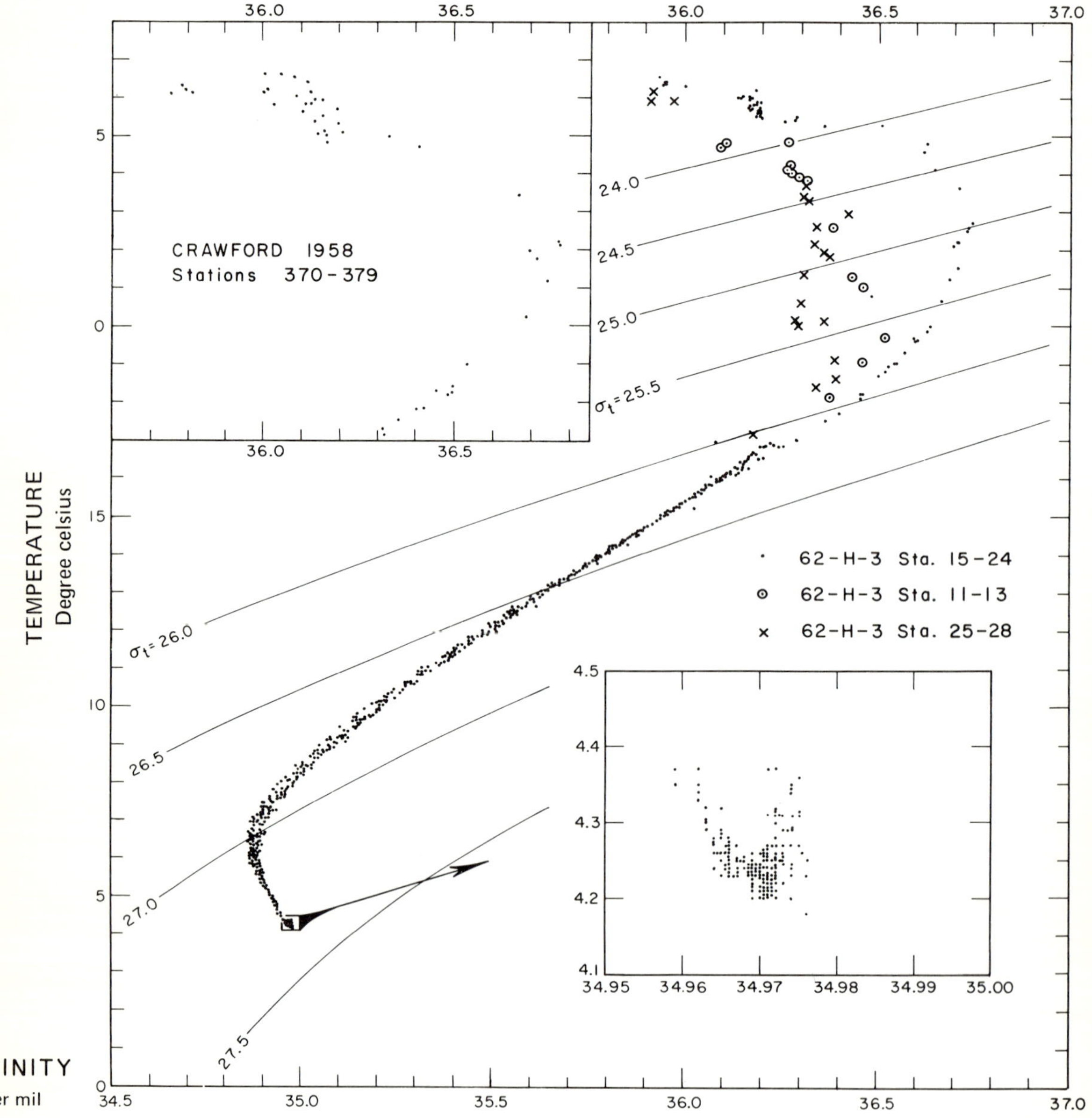

The salinity minima, which occur in the eastern Gulf at temperatures of approximately 6.3°C, are specific to the remnant of Subantarctic Intermediate Water. For comparison, within the Antillean sector of the North American Basin of the North Atlantic, the core of this Subantarctic Water layer (Wüst, 1964) is found at depths of 700-850 m, with salinities as low as 34.60 per mil and corresponding temperatures as low as 5.2°-5.6°C. On entering the eastern Caribbean this core layer has typical minimum salinities (Wüst, 1964: plate XVI) just under 34.70 per mil, with temperatures around 5.7°C. There is a consistent spatial increase in salinity and temperature within this core layer as one progresses eastward through the Caribbean Sea.

Within the western Gulf of Mexico this feature appears to be eroded to an extent that increases the minimum salinity some 0.02-0.03 per mil and decreases the associated temperature at the minimum some 0.1-0.2°C relative to the characteristics as this water enters the eastern Gulf. The width of the characteristic plot (Figure 1-4) from 5-14°C is occasioned by this mixing activity. For the waters in this potential temperature range, the maximum departure of salinity from the standard θ–S relation was obtained at each station from Caruthers' (1969) fitted polynomials. These values were plotted, and the contoured number field is presented as Figure 1-5. (Remember that the standard relation is based on data from stations within the Loop Current regime.) Salinity increases seem to become regularly greater with increasing distance into the Gulf; the maximum anomaly values of 0.04 per mil occur in the Bay of Campeche, which is separated from the Yucatan Strait inflow by the broad shelf of the Yucatan Peninsula, a barrier to waters at intermediate depths. Caruthers (1969) notes that the maximum increase of salinity occurs around 9°C instead of at the salinity minimum.

In Figure 1-4 the characteristics of water below 4.4°C are shown in an insert. The pressure effect leads to an increase in *in situ* temperature with depth below 2000 m, although potential temperature appears to decrease slowly to at least 3000 m, as discussed below. Similarly, the salinity of the deep water is shown to increase slightly, though continuously, with depth.

The relationship between potential temperature and salinity for observations from depths greater than 1400 m is shown in Figure 1-6. Comparison of this θ–S plot for waters of the Gulf Basin with a corresponding plot for waters of the Yucatan Basin (*Crawford* Sts. 372-379; Figure 1-1) will convince one that at any given potential temperature in the range 4.01-4.15°C, significant differences in salinity between the waters of the two basins do not exist, i.e., the salinity differences do not exceed 0.006 per mil, which is roughly twice the error claimed for the conductive measurement of salinity.

The *T-S* curve from 4.5-17°C is similar in form to that found in the Caribbean (Wüst, 1964, Figure 3); but there especially in the east, the salinity minimum is a much more distinct feature. This is consistent with the west-northwestward flow of water through the Caribbean and into the Gulf through Yucatan Strait.

All stations off the shelf display a salinity maximum within the upper 200 m. For Sts. 15 through 24 this feature traces a smooth curve, with a maximum near 22.5°C and 36.75 per mil–rather like Wüst's (1964: Figure 3) case of Subtropical Underwater for the Caribbean. This same relationship shows clearly (insert, Figure 1-4) at *Crawford* Sts. 370 to 379 taken across the western Yucatan Basin in February and March 1958. Within the Gulf, those waters having such a *T-S* relationship (i.e., that shown for Sts. 15 to 24) are commonly bounded by the Loop Current, comprised of the Yucatan Current and its downstream extension through the eastern Gulf into the Florida Straits. Wennekens (1959) gave the name Yucatan Water to that having this set of characteristics. A specific name does not seem apropos, since these characteristics are just the same as for the upper waters of the northwest Caribbean. In addition to these waters bounded by the Loop Current, anticyclonic current rings surrounding

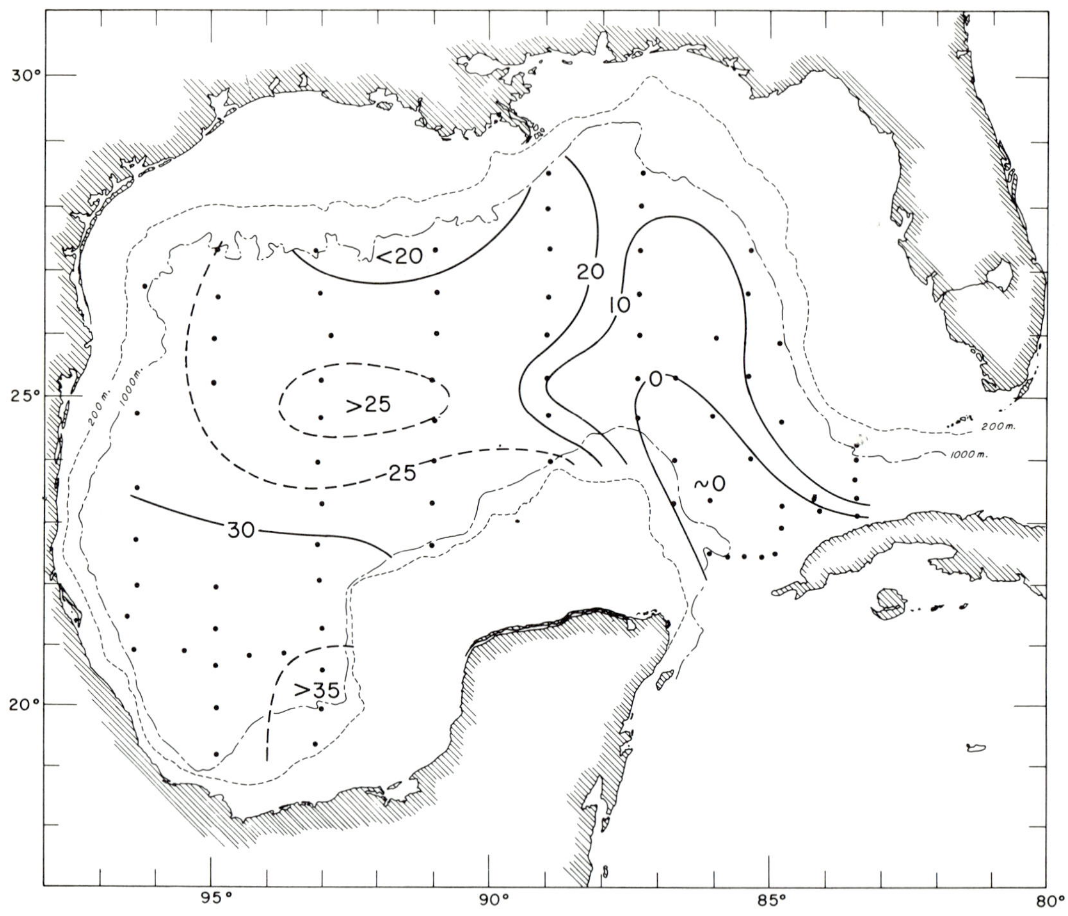

Figure 1-5. For waters with potential temperatures between 5° and 14°C, the maximum departure of salinity parts per million from "standard" potential temperature-salinity relation for Gulf given by Caruthers (1969). Positive values indicate observed salinity at individual station was greater than standard salinity at same potential temperature. Analysis based on Hidalgo *62-H-3 data from the station positions indicated.*

waters of these same characteristics are known (Nowlin et al., 1968) to become detached from this Loop Current in the eastern Gulf. (See Chapter 6 for a detailed description.)

Wennekens (1959) has indicated that the Yucatan Water, at least at times, is found in Florida Straits as far downstream as the Miami-Bimini region. Wenneken's data indicate that the northern limit of this water in the southern Florida Straits is usually about midway between Cuba and the western Florida Keys. On Cruise 62-H-3 definite evidence of this water was found in Section *C* at St. 24 only. Its distribution within the Gulf is discussed further below.

Sts. 11 through 13 in the western part of Yucatan Section *A* and Sts. 25 through 28 in the Florida Straits Section *C* show less regular *T-S* curves and display salinity maxima at lower salinities and temperatures. This feature was noted by Cochrane (1963) for the Yucatan Strait, and his suggestion (Cochrane, 1965) that part of the Yucatan Current flows inside Cozumel Island and Arrowsmith Bank invites speculation that vigorous vertical mixing may occur in this part of the passage.

Based on *Hidalgo* 62-H-3 data, Wilson (1967) estimated quantitatively the volumes of water in the Gulf of Mexico of different classes according to potential temperature-salinity characteristics. Much of the Gulf water was shown to be derived from the Caribbean water by mixing. Wilson concludes that "a dominant factor in the creation of new water types in the Gulf of Mexico appears to be the reduction in strength of the salinity maximum associated with the Subtropical Underwater," and that "the region of the Campeche Bank appears to be a focal point for water mass formation, modification and distribution in the Gulf of Mexico." According to Wilson's volumetric distributions by potential temperature-salinity classes, the occurrence of water in classes which differ from water found also in the Caribbean is expect-

Figure 1-6. Potential temperature-salinity relationships for all observations from deeper than 1400 m on Hidalgo *62-H-3, Gulf of Mexico.*

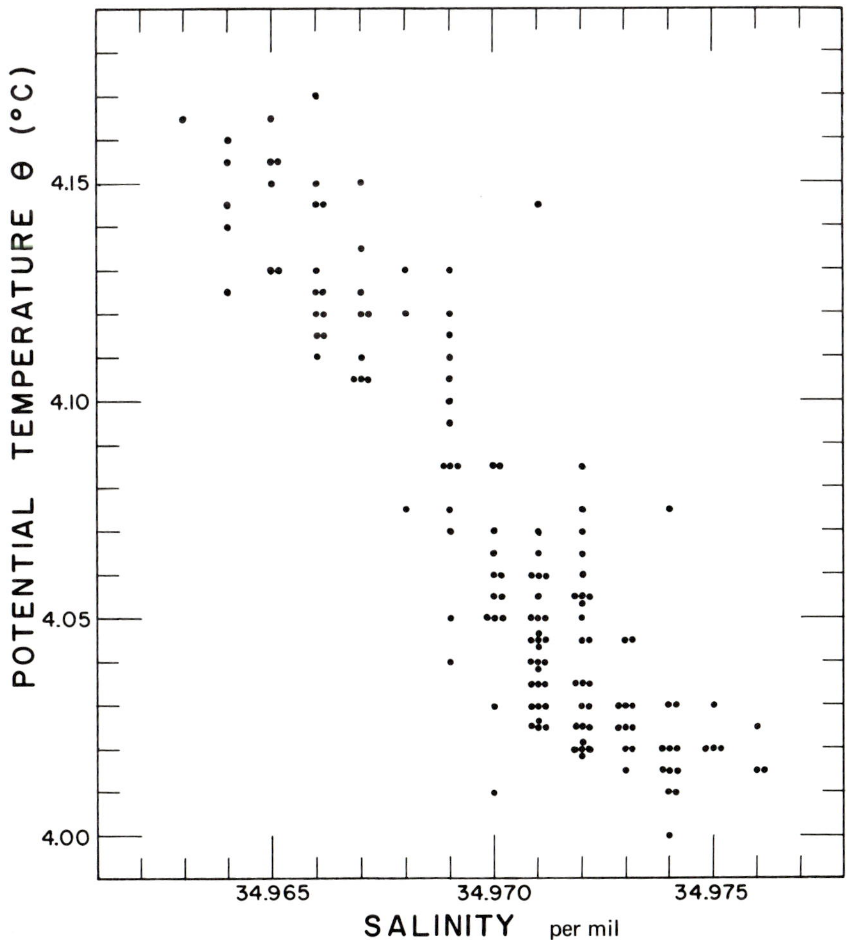

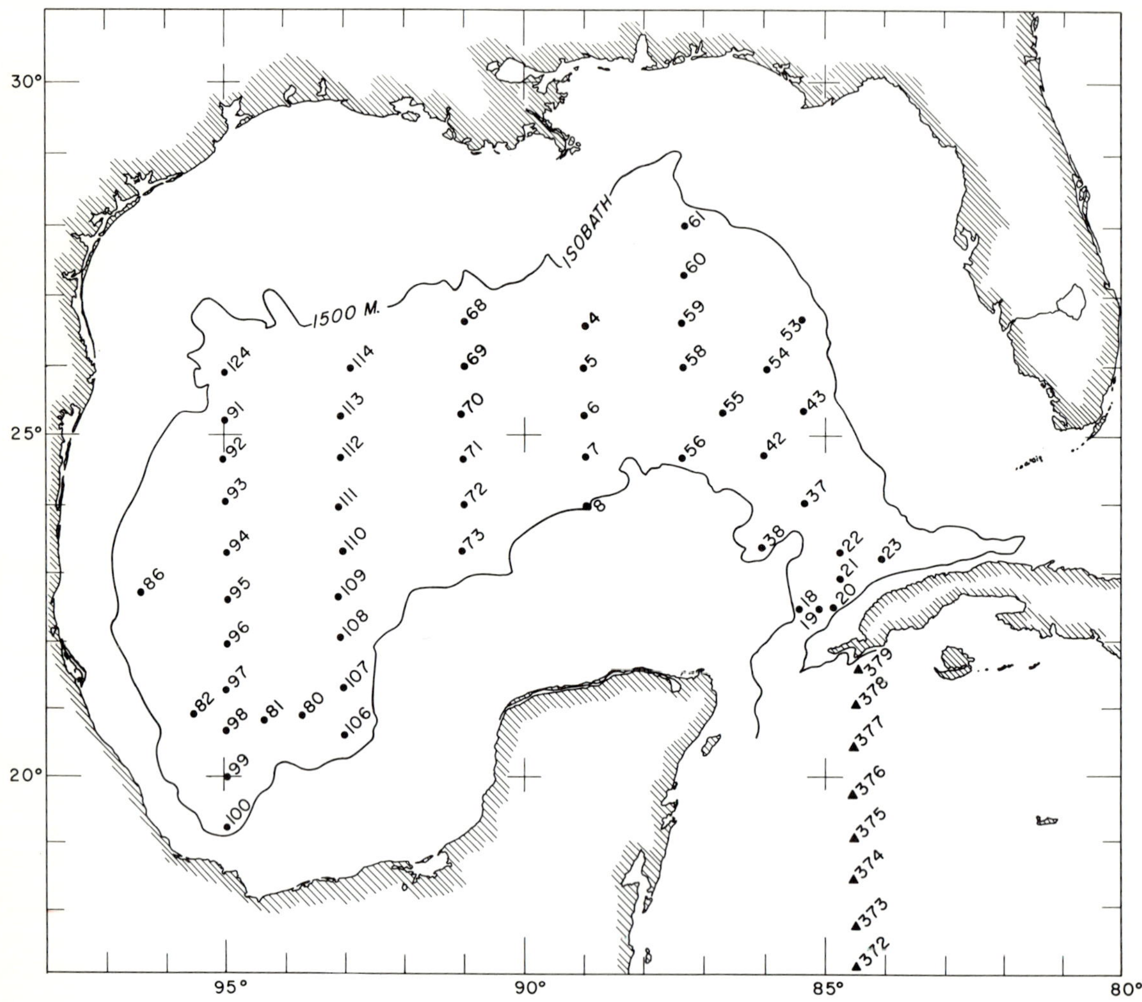

Figure 1-7. Location of stations occupied during Hidalgo *62-H-3 (solid dots) where observations at depths greater than 1500 m were made. Location of* Crawford *Cruise 17, Sts. 372-379 (solid triangles), in the western Caribbean are also shown.*

edly more common in the western than in the eastern Gulf. These conclusions support the previously mentioned suggestions regarding vertical mixing.

Water Masses and Stratification

The Basin Waters

The central Gulf of Mexico, having depths in excess of 3400 m, is a basin isolated from the adjacent Caribbean Sea by a sill with a controlling depth of roughly 2000 m (Figure 1-2); the sill depth in the Yucatan Channel has not been determined with precision. Figure 1-7 shows the 52 stations at which observations from depths greater than 1500 m were obtained. The characteristics and origins of the waters below 1500 m are discussed in this section.

Salinity and Potential Temperature. Figure 1-8 shows a composite plot of salinity versus depth for

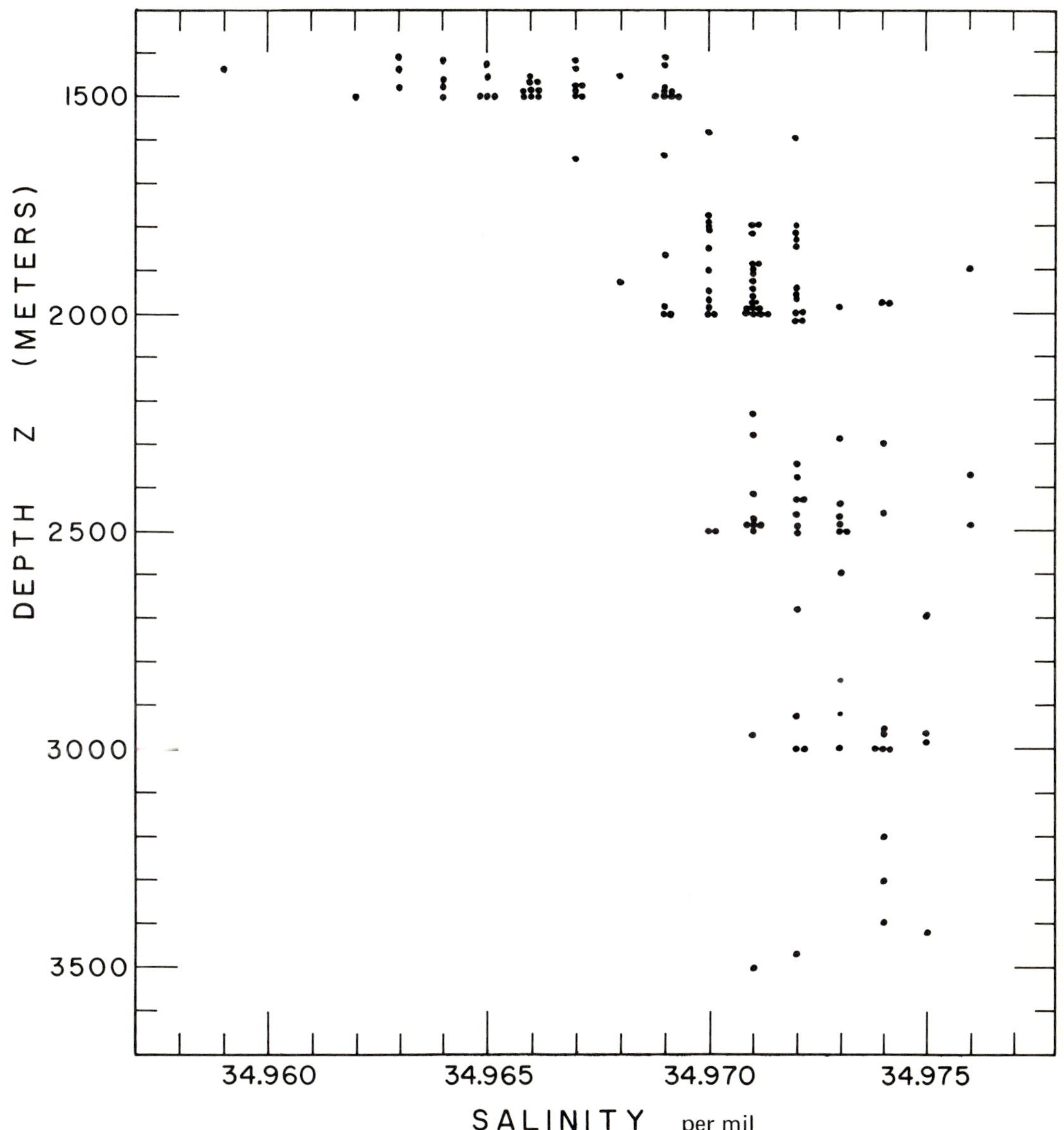

Figure 1-8. Salinity-depth relationships for all observations from deeper than 1400 m on Hidalgo *62-H-3, Gulf of Mexico.*

all observed depths below 1400 m at the 52 deep stations of Figure 1-7. From 2000 m to the maximum depth sampled, the range in observed salinity was only 0.008 per mil. All salinity observations from depths greater than 1500 m were fitted to both first-and second-degree polynomials (in depth) by the method of least squares. The resulting linear and quadratic forms, each with a standard error of 0.0015 per mil, appeared to be equally good. Salinity values and gradients at standard depths were obtained from this linear fit and are presented in Table 1-1. This trend line (omitted

Table 1-1
The Mean Stability Parameter, E,
from Mean Values and Gradients of T and S at Stated Depths

Z (m)	T (°C)	S (per mil)	$dT/dZ \times 10^3$ (°C m^{-1})	$dS/dZ \times 10^3$ (per mil m^{-1})	$E \times 10^8$ (m^{-1})
2000	4.23	34.971	0.034	0.002	1.3
2500	4.26	34.972	0.077	0.002	0.8
3000	4.31	34.973	0.120	0.002	0.2

from Figure 1-8 for clarity) indicates that throughout this deep water salinity continued to increase with depth, with a gradient of some 0.002 per mil per 1000 m. Although the error claimed in the measurement of salinity by the conductive method is ± 0.003 per mil or more, this gradient appears to be real rather than due to scatter. Even if this gradient is real, however, it may be a measurement of something other than salinity as usually defined, e.g., a slight departure from constancy of composition.

In Figure 1-9 it is seen that a general decrease in potential temperature with depth was observed from 1400 to 3000 m; these potential temperatures were computed from tables given by Helland-Hansen (1930). The potential temperatures for all observed depths greater than 1500 m were fitted to several alternate functional forms. The form chosen was the second-degree polynomial in depth,

$$\theta = 4.334 - 2.139 \times 10^{-4}Z + 3.596 \times 10^{-8}Z^2, \quad (1.1)$$

with θ in degrees Celsius and Z in meters; this gave a standard error of 0.012 C degrees. Resulting mean potential temperatures for selected depths are presented in Table 1-2; these values agree to three significant figures with the arithmetic means of the potential temperatures observed within a depth range of 50 m on either side of the tabulated depths. Although the number of observations below 3000 m is small, it may be said that the potential temperature does not decrease further below this depth.

That the vertical gradients of both potential temperature and salinity were observed to be small below the 2000-m depth is consistent with the concept of a basin isolated by a sill whose controlling depth is 2000 m or less. The observed vertical gradients indicate diffusion of salt (upward) and heat (downward at least to 3000 m). No significant pattern of horizontal variations in salinity or potential temperature was discernible below 2000 m.

Table 1-2
Mean Potential Temperatures (θ)

Depth (m)	θ(°C)
2000	4.050
2500	4.024
3000	4.016
3500	4.026

Stability of the Basin Waters. On the basis of the vertical distribution of potential temperature, the waters of the basin appear to be stable down to 3000 m. Moreover, if the salinity gradient is to be accepted, this stability is enhanced, and it may be that the waters are stable to the bottom.

Since the ranges of both potential temperature and salinity in Figures 1-8 and 1-9 are limited, it is easily possible to overestimate the degree of positive stability from a consideration of these figures. It seems worthwhile, therefore, to consider some parameter whereby the vertical stability of these waters may be quantitatively compared with that of waters in other areas. Such a quantitative

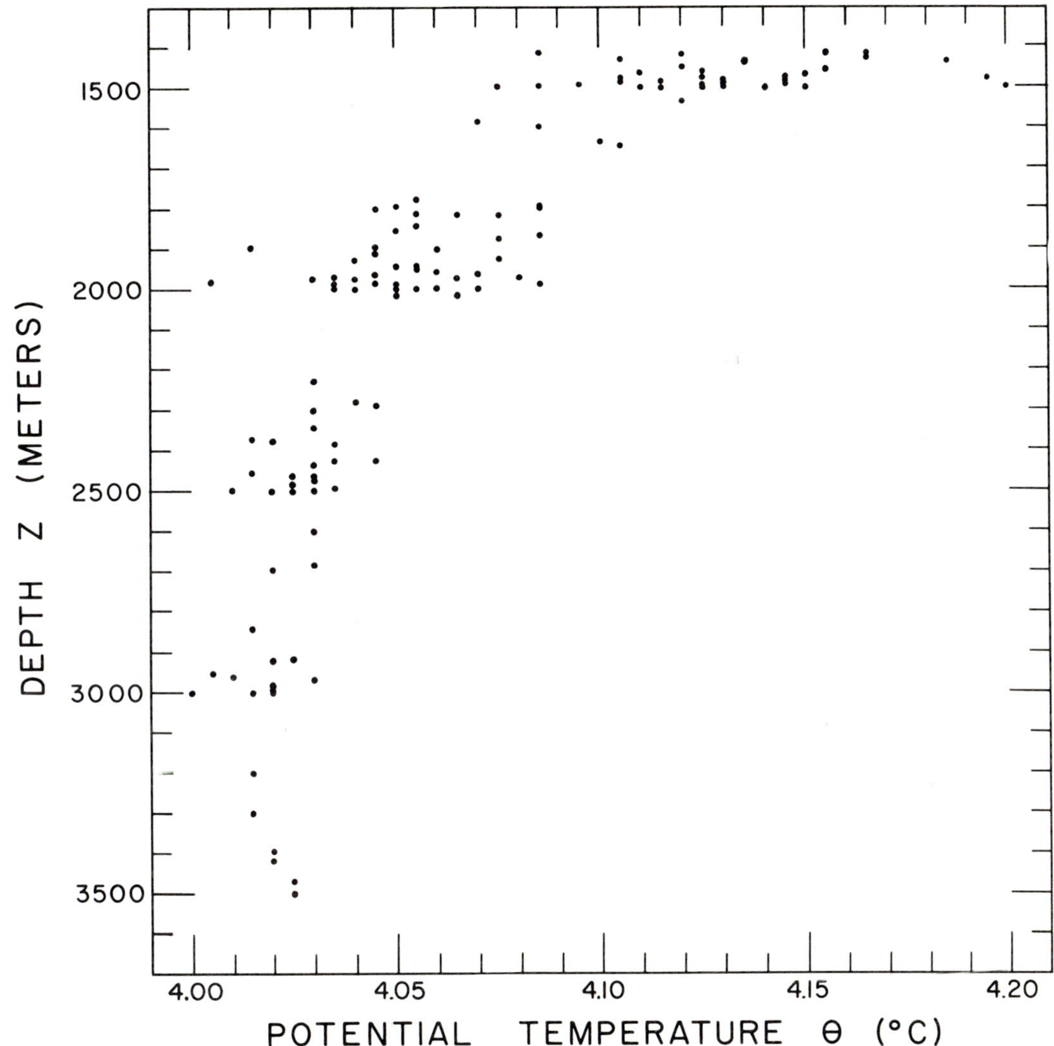

Figure 1-9. Potential temperature-depth relationships for all observations from deeper than 1400 m on Hidalgo *62-H-3, Gulf of Mexico.*

measure is provided by the Hesselberg-Sverdrup (1915) stability parameter, E, an equivalent of which may be defined by

$$E = (1/\rho)[(\partial\rho/\partial S)(dS/dZ) + (\partial\rho/\partial T)(dT/dZ - \gamma)],$$

where: ρ is density, S salinity, T temperature, and Z depth (increasing downward); DS/dZ and dT/dZ denote environmental gradients; $\partial\rho/\partial S$ and $\partial\rho/\partial T$ denote thermodynamic partial derivatives; γ denotes the increase in temperature per unit increase in depth under adiabatic conditions.

The mean values of E and the mean values and mean gradients of salinity and temperature (denoted by superior bars) from which they were computed are presented in Table 1-1. The values and gradients of temperature were obtained from a

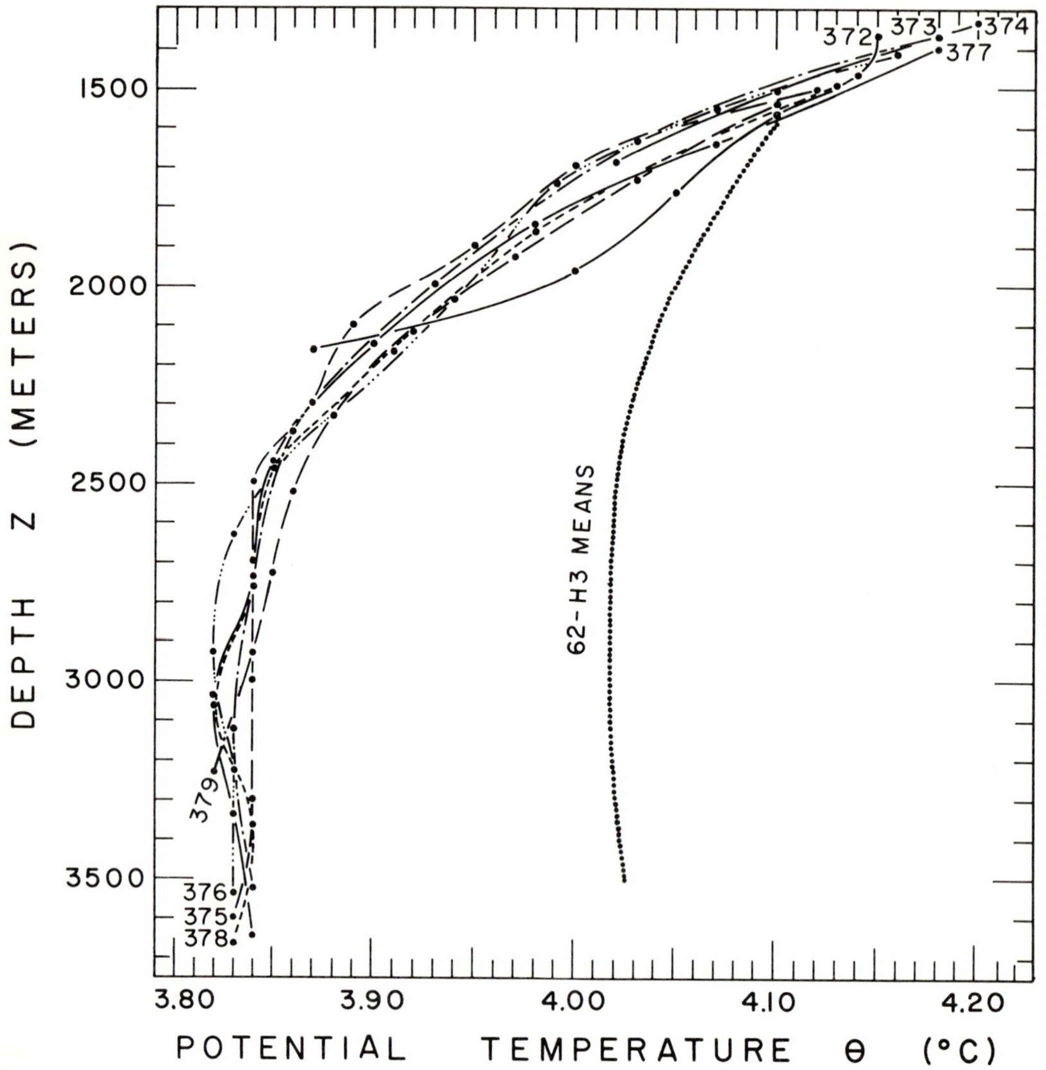

Figure 1-10. Potential temperature-depth relationships for Crawford *Cruise 17, Sts. 372-379. Range of* Hidalgo *62-H-3 observations shown.*

regression analysis of temperature data collected on Cruise 62-H-3 from depths greater than 1500 m. The standard error of the resulting curve fit was 0.014°C. The values of thermodynamic partial derivatives were taken from Hesselberg and Sverdrup (1915).

Although it seems significant that positive stability is indicated as the mean condition, the values of E are so small as to strain the accuracy of the data from which they were computed. If these values of the stability parameter are compared with values available for comparable depths in open ocean areas, where $E \times 10^8$ is of the order of $10\ m^{-1}$, it is concluded that the Gulf of Mexico Basin waters have nearly neutral stability, as expected. Stability computations based on data from consecutive sampling depths at individual stations also indicate stable conditions with few exceptions: 11 stations evidenced neutral conditions; St. 110 indicated unstable conditions, $E \times 10^8 = -1.8 m^{-1}$, in the depth range 2750-3250 m.

Origin of the Basin Waters. The lack of any discernible tendency toward horizontal variations in salinity or potential temperature within the basin has been mentioned. This seems to indicate that either the basin waters had a common source or their residence time is of a magnitude sufficient to insure that horizontal gradients have been destroyed by exchange processes. Since the only major inflow to the Gulf at present is through the Yucatan Channel, it is reasonable to compare the characteristics of the waters of this basin with those of the Cayman Basin.

Figure 1-10 shows θ-Z curves for Sts. 372-379 of *Crawford* Cruise 17, located on a north-south line between the western tip of Cuba and Honduras (Figure 1-1). Also pictured (Figure 1-10) is the graph of the least-squares fit, equation (1-1), relating potential temperature and depth as observed on Cruise 62-H-3. If it is assumed that waters from the Cayman Basin are presently displacing the bottom waters of the Gulf of Mexico Basin, then consideration of potential temperatures leads to an estimated maximum controlling sill depth of between 1650-1900 m (less than 1800 m if *Crawford* St. 372 is ignored). If it is assumed that the bottom waters of the Gulf Basin have a greater potential density than the present Yucatan waters at the controlling sill depth, then Yucatan waters may presently be flooding the Gulf Basin at some depth between 1500 m and the bottom. In this case the controlling sill depth must be less than 1900 m. It is seen, however, that this sill depth must be at least 1400-1500 m, for at these depths the θ-Z curves for the Yucatan and Gulf waters coincide. Moreover, if the controlling sill depth is within this shallower range, the interpretation might be that the Gulf Basin was filled at some past time (when the distribution of temperature within the Yucatan Basin differed from that evidenced by the *Crawford* data).

During June, 1967, a number of deep hydrographic stations were made in the Gulf of Mexico aboard *Alaminos* on Cruise 67-A-4. Two deep stations were also made in the Yucatan Basin. (For station locations, see Figure 1-11.) Although the values of adiabatic cooling that were applied to *in situ* temperatures to determine potential temperatures were based on data presented by Cox and Smith (1959) rather than on the Helland-Hansen (1930) data, the results (Nowlin et al., 1969: figure 2) substantiate Figure 1-10. Curves of potential temperature versus salinity drawn from 67-A-4 data taken on opposing sides of the Yucatan sill intersect at about 1550 m.

It must be emphasized that the controlling sill depth as discussed is only an estimate. The difficulty of accurately relating the controlling sill depth to the available hydrographic data is reflected in the following statements:

1. Unless isothermal surfaces are horizontal within the Yucatan Basin in the 1400-2000-m depth range, the water found some hundreds of kilometers removed from the sill cannot be expected to represent the water available for inflow into the Gulf of Mexico at the same depth.
2. Since the northward currents through the Yucatan Channel are thought to extend practically to the bottom and affect the slope of isosteric surfaces, it is possible that the intensity of this flow (not necessarily stationary) may determine the greatest depth from which Caribbean waters enter the Gulf.
3. Even though no horizontal gradients of conservative properties have been observed within the Gulf Basin, such gradients may exist near the bottom in the area north of the Yucatan sill if Caribbean waters are presently moving northward in a downslope flow and displacing the bottom waters of the Gulf Basin.

In a foregoing section, Yucatan and Gulf Basin waters were compared on the basis of

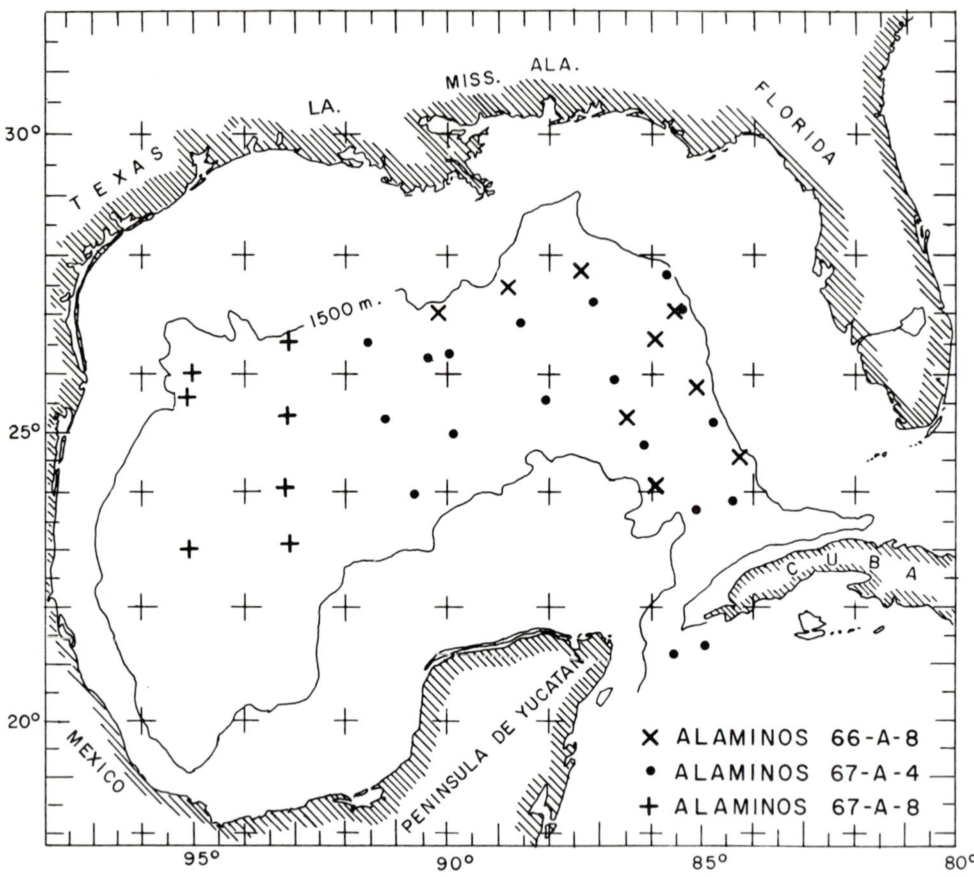

Figure 1-11. Locations of stations on Alaminos *Cruises 66-A-8 (1-24 June 1966), 67-A-4 (10 June-1 July 1967) and 67-A-8 (9-12 September 1967) from which data were utilized.*

salinity. The relationships of potential temperature versus salinity near sill depth in the two basins are quite similar and so are consistent with the idea of present-day displacement of Gulf bottom water by Cayman water.

Distribution of Dissolved Oxygen

In this section, oxygen data obtained in 1966 and 1967 are compared with selected data obtained in 1935, 1958, 1959, 1962 and 1964. It is shown that there is no clearly discernible horizontal variation in dissolved oxygen in the basin waters of the Gulf of Mexico.

The available dissolved oxygen measurements provide areal coverage of the entire Gulf Basin and were obtained on cruises of the *Atlantis,* the *Hidalgo* and the *Alaminos.* The *Atlantis* in 1935 and the *Hidalgo* in 1958 (58-H-4) provided observations principally from the central eastern Gulf; observations were obtained from the western part on the *Hidalgo* in 1958 and 1959 (58-H-1, 59-H-2) and on the *Alaminos* in 1964 (64-A-2, 64-A-3); observations from the entire Gulf Basin were obtained on the *Hidalgo* in 1962 (62-H-3) and on the *Alaminos* in 1966 and 1967 (66-A-8, 67-A-4, 67-A-8); and observations in the Yucatan Basin were obtained on the *Alaminos* in 1967

(67-A-4). The *Atlantis* data are available from the Woods Hole Oceanographic Institution, the other data from the Department of Oceanography at Texas A&M University. Figure 1-11 shows the locations of the 1966 and 1967 *Alaminos* stations from which oxygen data are presented.

Oxygen determination on Cruise 66-A-8 were made by the Winkler method, using potassium bi-iodate as the standard; oxygen data on 67-A-4 and 67-A-8 were determined by the Chesapeake Bay Institute technique for the Winkler method described by Carpenter (1965); and all oxygen determinations for cruises prior to 1966 were made with modifications of the Winkler method.

Figure 1-12 displays plots of dissolved-oxygen concentration versus depth for the basin waters based on the three cruises in 1966 and 1967. Also shown are visually determined best-fit curves. In Figure 1-13, oxygen versus depth is plotted for seven pre-1966 cruises.

For each cruise considered here, Table 1-3 gives the standard deviation and the numerically averaged oxygen concentration, $\overline{O}_2$, for all samples taken at or below 1500 m. There is no evidence that the oxygen concentration changed during the period covered. The oxygen data collected on the 1962 and 1964 cruises show much more scatter at any given depth than do the oxygen data presented from the other cruises. As noted, it is very probable that this scatter resulted from poor sampling or inaccurate analyses. Moreover, horizontally uniform oxygen concentrations are consistent with the horizontal uniformity of salinity and potential temperature observed for the basin waters.

Wüst (1964:45, table 15) has given a value of 5.6 ml/l for the dissolved-oxygen concentration in

Figure 1-12. Oxygen versus depth in the Gulf of Mexico Basin. Plotted are all data from the ≥ 1400-m sampling depths on the indicated cruises. The solid curves were selected by eye to best fit the data. The dashed curve is for the 67-A-4 data from the Yucatan Basin.

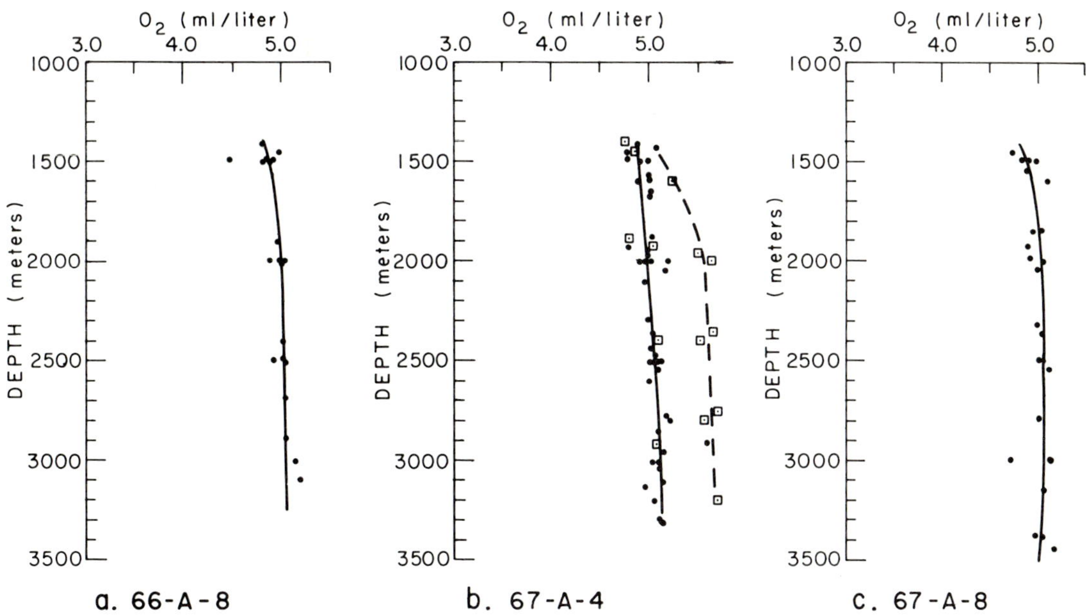

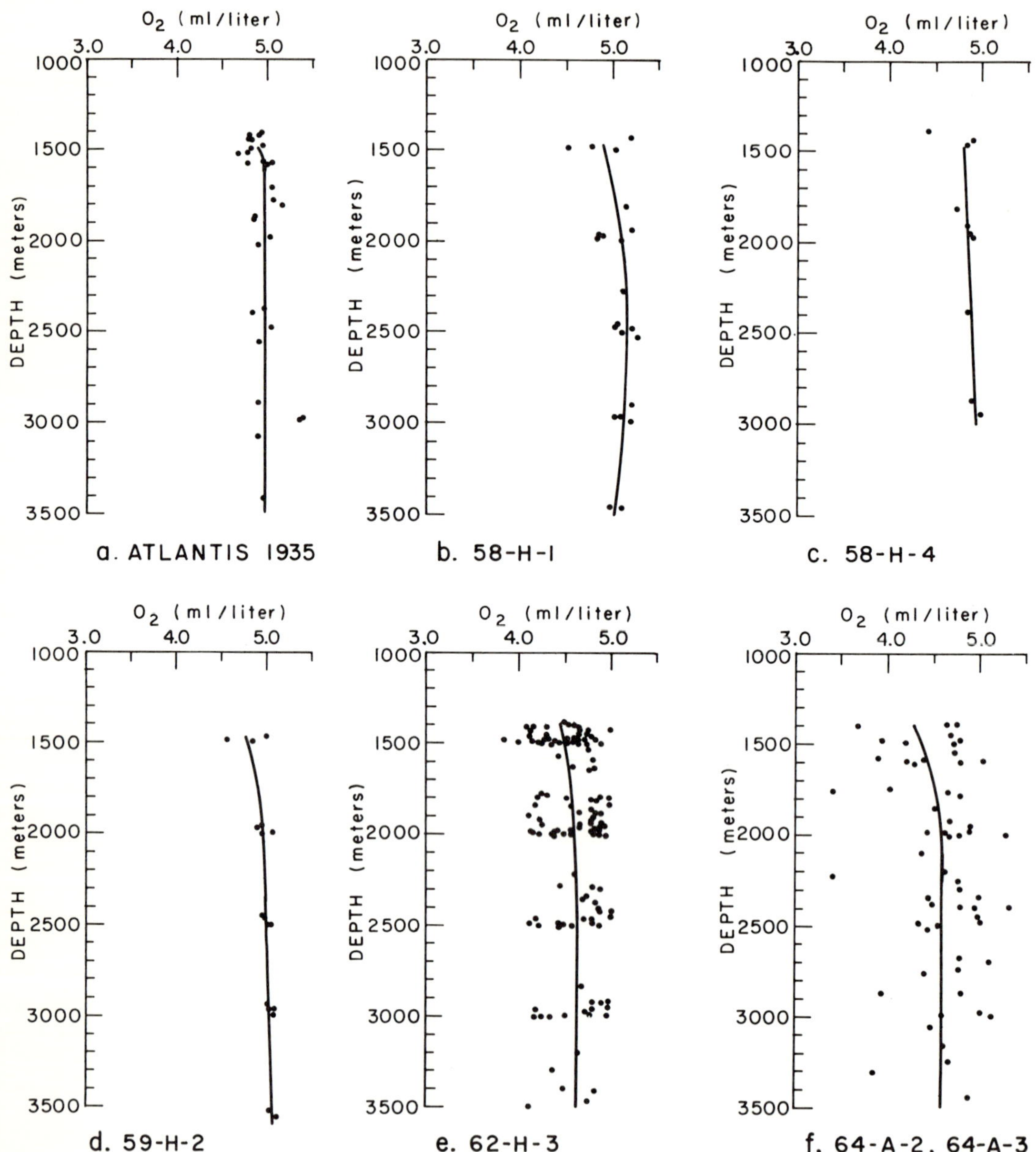

Figure 1-13. Oxygen versus depth in the Gulf of Mexico Basin. Plotted are all the data from the ≥ 1400-m sampling depths on the indicated cruises. The solid curves were selected by eye to best fit the data.

the deep waters of the Yucatan Basin. This agrees with our value of 5.6 ml/l obtained on Cruise 67-A-4 at two deep stations in the Yucatan Basin near Yucatan Strait (see Table 1-3).

Our mean value for oxygen in the Gulf of Mexico Basin waters is 4.91 ml/l when based on data from the 10 cruises listed in Table 1-3, or 4.99 ml/l if the 1962 data are excluded. Wüst shows values near 5 ml/l within the Yucatan Basin at depths of about 1500-2000 m, i.e., depths from which waters could enter and fill the Gulf of Mexico Basin.

Although the uncertainty in dissolved-oxygen values obtained on Cruise 62-H-3 is large enough to prohibit the use of those data in characterizing oxygen distributions within the basin waters, measurements made on that same cruise can be used to indicate distributions within the overlying waters, in which variations in dissolved-oxygen content far exceed the errors. Figure 1-14 shows typical oxygen-depth curves from the surface to 1250 m. Surface values were found to be generally at or near saturation for the particular temperature and salinity. The most persistent feature is a continuous layer of minimum dissolved-oxygen content. This shows up at depths of about 700 m on the right-hand side of the main Yucatan inflow (Sts. 18 and 20), in the center of the Loop Current regime (Sts. 41 and 55) and in the southerly portion of the outflow (St. 24). To the left of the inflow (St. 15) and of the outflow (St. 26) the minima occur at considerably lesser depths, reflecting the mass adjustment associated with the circulation. Extreme values in this oxygen-minimum core were in the range 2.5-2.7 ml/l, consistent with those shown by Wüst (1964: plate XV), and they are clearly continuous with the core that he traces through the Caribbean (plate XIX). This oxygen-minimum core does not coincide with the salinity--minimum core but occurs at a depth less by some 200 m.

A secondary oxygen minimum at 150-300 m, observed throughout the water bounded by the Loop Current, was almost completely suppressed at St. 26. The shallow maximum and minimum observed at St. 26 are illustrative of a number of such regional features observed in the northern and northeastern part of the survey, especially over the continental shelf. These near-surface maxima, which are not necessarily associated with

Table 1-3
Mean Values of Dissolved Oxygen Concentration Observed in the Basin Waters of the Gulf of Mexico

Cruise	Standard deviation (ml/l)	O_2 (ml/l)	Number of stations ($Z \geqslant 1500$ m)	Number of data points ($Z \geqslant 1500$ m)
Atlantis 1935	0.134	4.96	9	23
58-H-1	0.124	5.08	6	19
58-H-4	0.054	4.87	4	7
59-H-2	0.055	5.01	5	17
62-H-3	0.280	4.62	52	114
64-A-2 and 3	0.482	4.58	19	67
66-A-8	0.048	4.99	7	19
67-A-4:				
Mexico Basin	0.077	5.03	16	55
Yucatan Basin	0.099	5.61	2	11
67-A-8	0.111	5.01	9	21

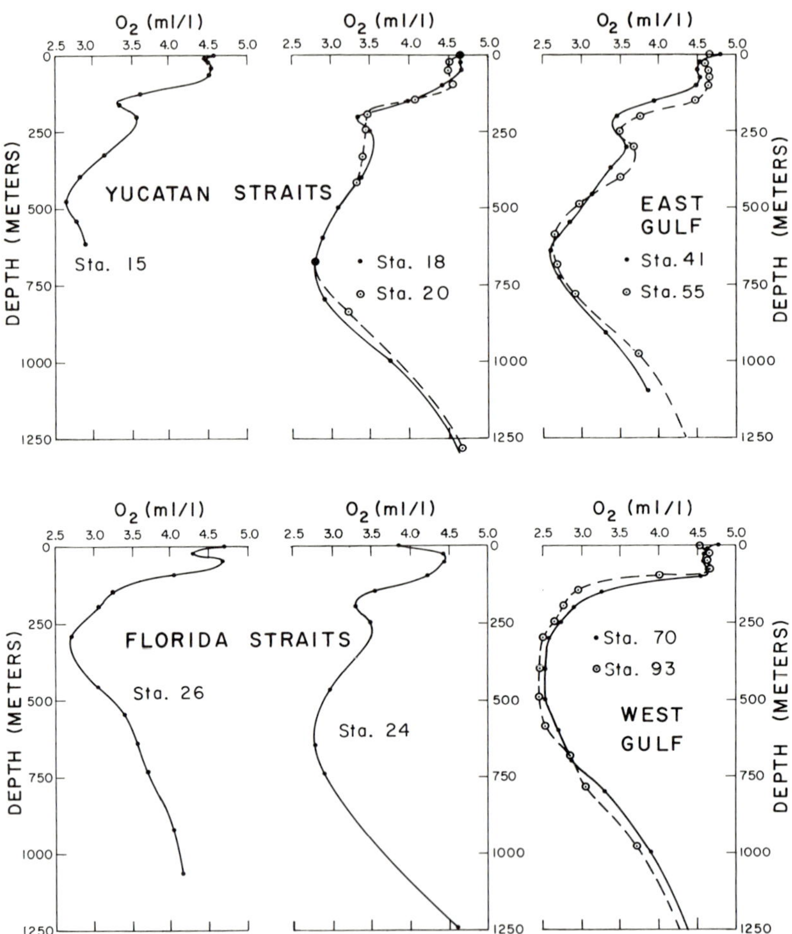

Figure 1-14. Dissolved-oxygen concentration versus depth at selected stations on Hidalgo *62-H-3, Gulf of Mexico.*

corresponding temperature or salinity extrema, are likely associated with biological activity. They are presently unexplained and seem worthy of future investigation.

In the western Gulf (represented by Sts. 70 and 93) the oxygen-minimum layer is considerably thicker, and the lower extreme values (typically 2.4 ml/l or less) indicate a considerable residence time for this water. The oxygen content in vertical profile along Section *D* (approximately 93°W) is shown in Figure 1-15. The maximum vertical gradient of dissolved oxygen occurs between depths of 75-150 m across the section, corresponding to the top of the thermocline (Figure 1-24). The depth variations in the oxygen-minimum layer appear to be related to the circulation, and below this the configuration of oxygen isopleths generally conforms to the isotherms and isosteres (see Figures 1-24 and 1-27). The very low value observed in the minimum layer at St. 105 (1.67 ml/l) suggests the possibility of a permanent eddy in Campeche Bay.

Overlying the basin waters within the Gulf of Mexico are several strata having distinct water mass characteristics indicative of their sources. In addition to the layers (already discussed) in which the concentration of dissolved oxygen has local extrema, there are strata characterized by local maximum values of salinity, minimum values of salinity or (near the surface) vertically uniform concentrations of both salinity and temperature.

Remnant of Subantarctic Intermediate Water

Wüst (1964) has reviewed his earlier (1936) findings concerning the distinctive water mass that is formed near the Polar Front in the South Atlantic (48° -52′S). To this water mass he gave the name Subantarctic Intermediate Water. The mechanism of formation is not well understood, but these waters sink and spread out northward. The core of minimum salinities enters the Caribbean through those passages of the Lesser Antilles that have sufficient sill depth and then spreads out in the Caribbean as shown by Wüst (1964: plates XVI and XVII). Salinities in this core may be less than 34.7 per mil in the eastern Caribbean, but mixing that accompanies the horizontal spreading raises the salinity to 34.85 per mil at the Yucatan Strait.

Figure 1-16 shows the depth of the salinity minimum as observed in the Gulf; it also shows representative salinities and temperatures. Prior to contouring, the depths at each station were selected from smoothed curves of observed salinity versus depth. From the following discussion of currents it will be recognized that the depths are related to the circulation, being greater than 900 m in the center of the water bounded by the Loop Current and close to 500 m in the northern central Gulf.

Figure 1-15. Dissolved-oxygen concentration along vertical Section D, *22-27 March 1962,* Hidalgo *62-H-3. Upper 1250 m at interval of 0.2 ml/l. Observation positions indicated by dots. Vertical exaggeration 555:1.*

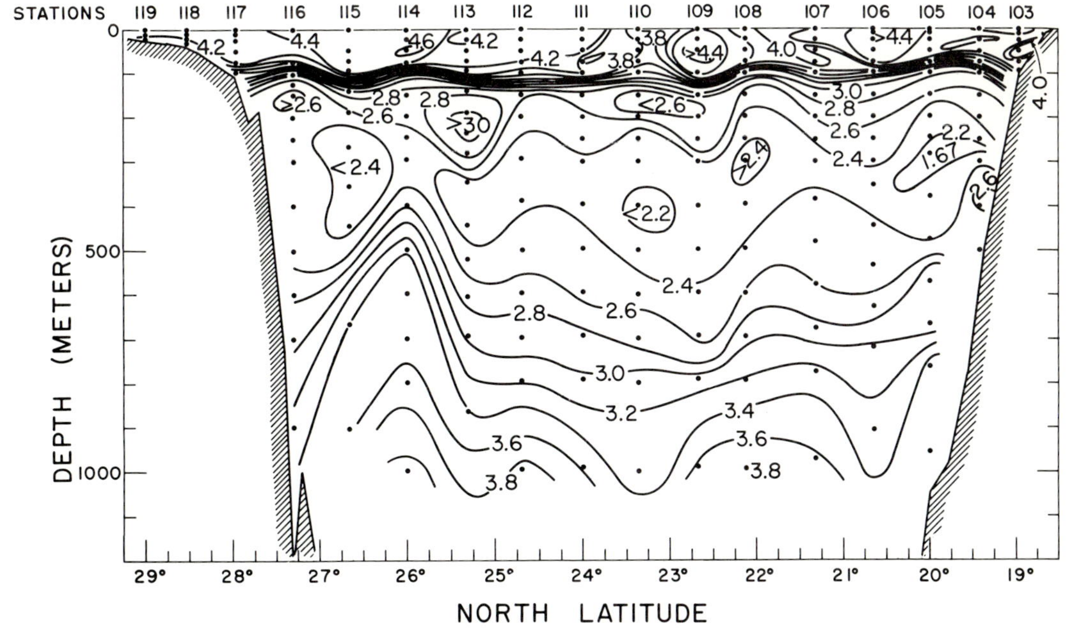

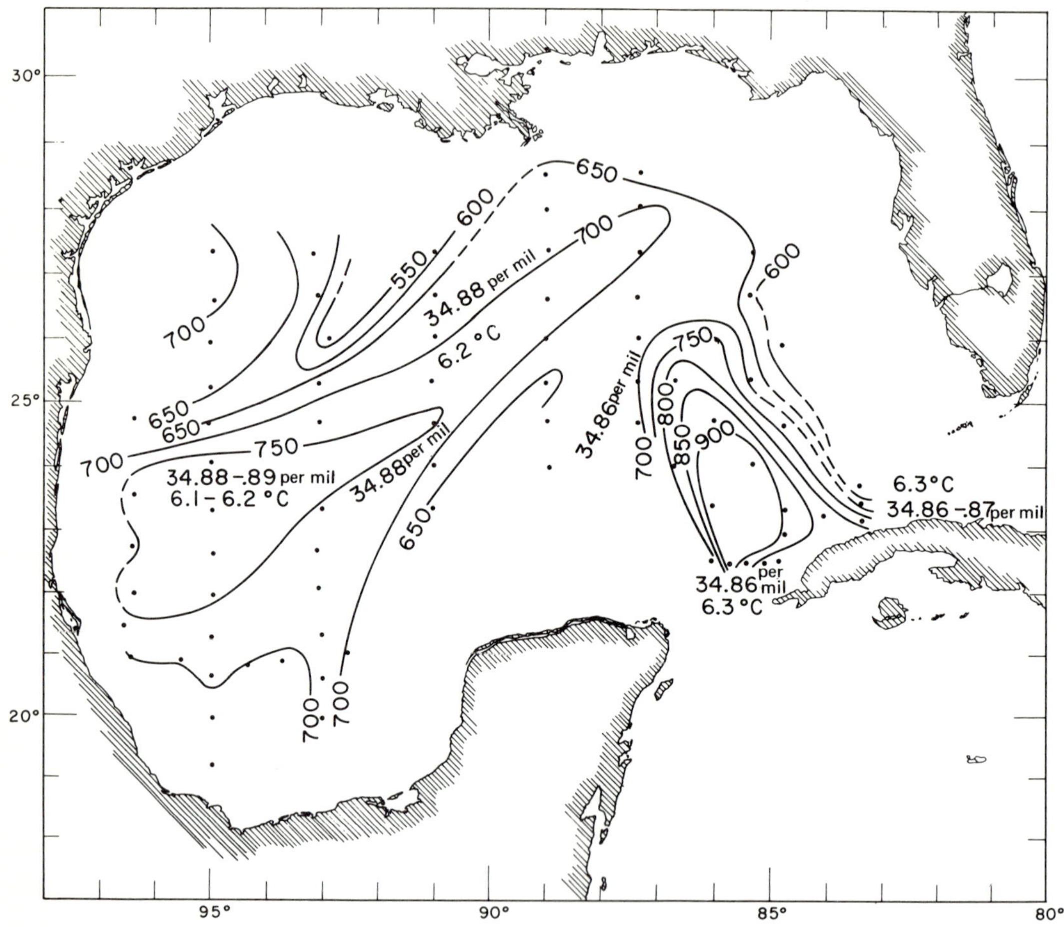

Figure 1-16. Depth (50-m intervals) of core layer in Subantarctic Intermediate Water remnant (salinity minimum), Hidalgo *62-H-3. Selected values of salinity and temperature at the core are shown.*

On the basis of minimum salinity, Wüst (1964) has constructed a table for estimating the percentage composition of the Subantarctic component in the core. This would indicate that at Yucatan (34.86 per mil) there is already less than 5 percent and in the western Gulf (34.88-34.89 per mil) only some 1 or 2 %. Even so, the origin of this distinctive feature of minimum salinity within the Gulf is clear, and it is proper to label it a remnant of the Subantarctic Intermediate Water.

Tropical Salinity Maximum

Another distinct layer in the water structure of the Gulf shows up as a subsurface salinity maximum. Wüst (1964) has traced this layer through the Caribbean and has given it the name Subtropical Underwater. According to him, it originates at the surface-salinity maxima in the North and South Atlantic (cf. Defant, 1961: plate 5), where its maximum surface expressions are found at 20°-25°N, 30°-50°W and at approximately

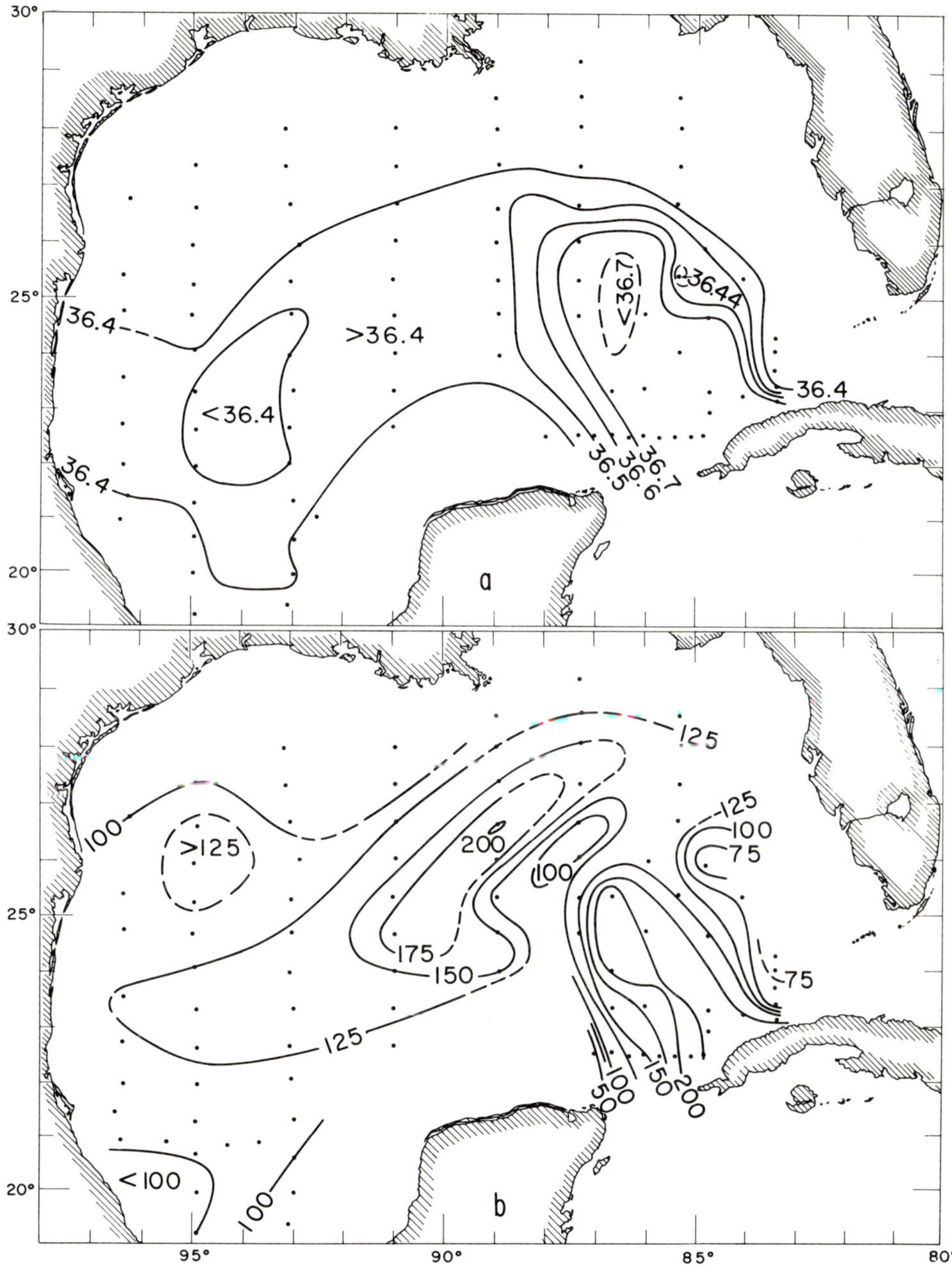

Figure 1-17. Core of salinity maximum, Hidalgo *62-H-3: (a) salinity (0.1 per mil intervals); (b) depth (25-m intervals). Locations of maxima indicated by dots.*

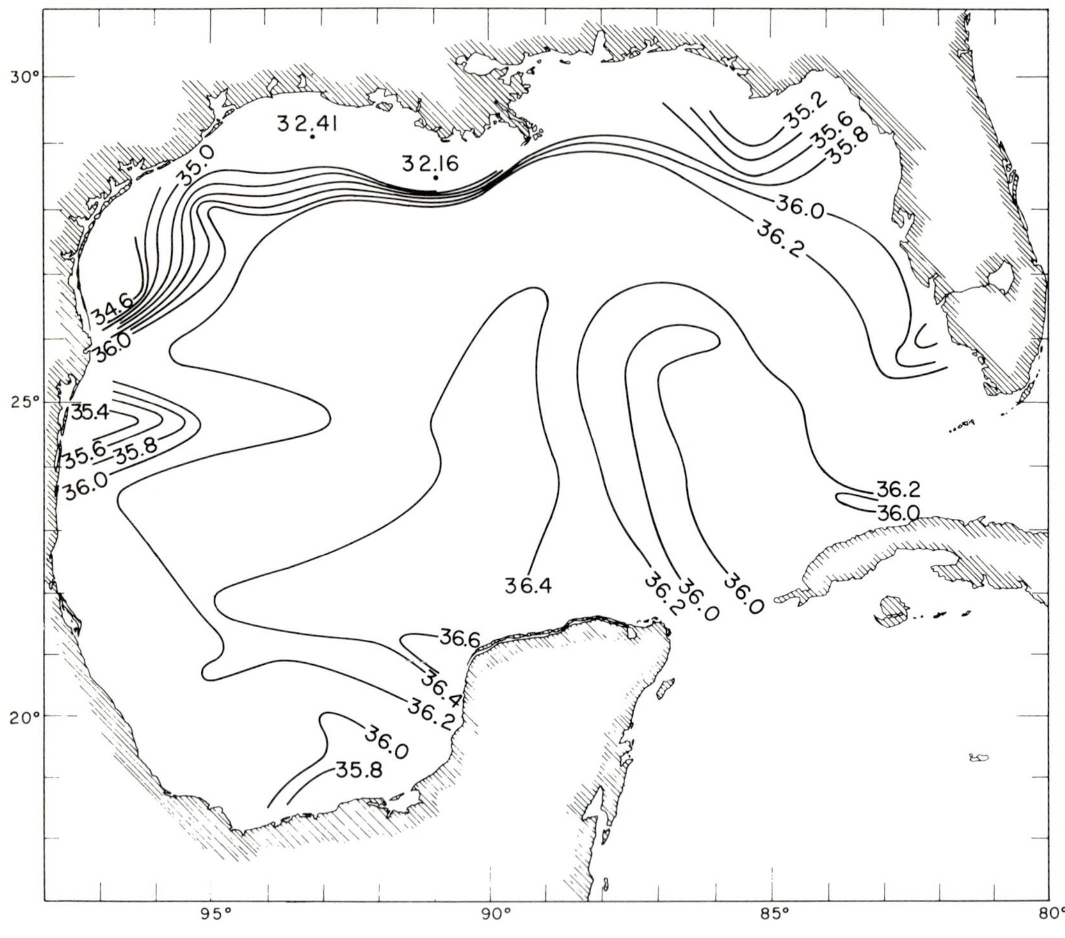

Figure 1-18. Surface salinities (parts per mil), Hidalgo *62-H-3.*

15°S, 30°W; westward these maxima spread out as intermediate layers. Although the core of the South Atlantic component extends along the northern coast of South America, Cochrane's (1965) presentation of data from the EQUA-CHEQUE Expedition makes it clear that the maximum salinites encountered offshore from northern Brazil are already too low for the South Atlantic to be considered a source for the Subtropical Underwater in the Caribbean. We may point then to the tropical North Atlantic as the source region for this layer in the Caribbean and the Gulf.

Salinity values at the subsurface maxima are shown in Figure 1-17. The maximum salinity in this layer is about 36.75 per mil, in agreement with Wüst (1964: plate XIII), and is confined to the water bounded by the Loop Current. In the western part of the Yucatan Current, the salinity appears to be decreased by vertical mixing, as discussed above. As a consequence, the salinity maxima throughout most of the Gulf are lower than those in the water bounded by the Loop Current. As shown in Figure 1-17, the average depth of this layer is about 150 m, which is in general agreement with the depths at which the layer is found in the Caribbean.

The Surface Layers

The observed surface salinities are depicted in Figure 1-18; surface salinities taken underway were used to augment station data in constructing this figure. In the Yucatan area and the water bounded by the Loop, the values 35.9-36.2 per mil are typical of the Caribbean during the winter.

An interesting feature of the surface salinity pattern in Figure 1-18 is the relative high (S>36.4 per mil) covering the western Campeche Bank and extending over the deep water areas to the west and north. Salinities near the bank were reexamined in an effort to understand that distribution. During Cruise 62-H-3, salinities along the eastern bank edge exceeded 36.4 per mil (Figure 1-20*b*). It was suspected that high-salinity waters might be moving from the east up onto the central Campeche Bank in a frictional bottom layer (in a manner described by Cochrane, 1967) and then upwelling to give the surface distribution of high salinity shown over the western bank. Although this cannot positively be denied on evidence from Cruise 62-H-3, the data from Sts. 10 and 11 show salinities over the central bank of approximately 36.32 per mil from top to bottom. At the four other Campeche Bank stations salinity was nearly uniform with depth, having values of 36.48 per mil at Sts. 9 and 75, 36.52 per mil at St. 76 and 36.67 per mil at St. 77. Either during or before Cruise 62-H-3, it is possible that a band of high-salinity water, extending south of St. 10, connected the high-salinity surface feature and the high-salinity intermediate waters of the Yucatan Current along the eastern bank edge. The data were also examined for evidence of a connection between the surface high and underlying waters of the subtropical salinity maximum, found west and north of Campeche Bank at depths of 100-150 m. At each station the high-salinity surface was separated from the high-salinity subsurface waters by a relatively lower salinity layer.

Runoff from the Mississippi and Atchafalaya and from smaller rivers to the east and west gives rise to a band of low-salinity water along the northern shore. Most of this water flows westward over the Texas Shelf and shows its freshening influence as far south as 26°N. The tongue of fresher water protruding eastward from the Mexican coast at about 25°N may have been a continuation of the coastal flow augmented by the Rio Grande River, or it may have been entirely due to local runoff. Station spacing was inadequate for a positive interpretation.

The horizontal distribution of average salinities within the upper 50 m presented by Parr (1935: figure 10) from the *Mabel Taylor* data is strikingly similar to the pattern in Figure 1-18, although there are not enough *Mabel Taylor* data to conclusively establish that relatively high-salinity (S>36.25 per mil) surface waters covered the Campeche Bank, as Parr pictured. Parr found surface salinities less than 36.0 per mil over the northern shelves; values less than 24 per mil were observed near the mouth of the Mississippi River. During the winter of 1932, the surface waters in a band southeast of the Yucatan Current and along the northern coast of Cuba had salinities less than 36.0 per mil, just as did the surface waters bounded by the Loop Current in the winter of 1962.

Franceschini (1961) studied the hydrologic balance of the Gulf of Mexico for the year October, 1958, through September, 1959, using meteorological observations from stations around the region as well as observed river discharge rates. He showed the Gulf to be an area where evaporation generally exceeds precipitation, but he found a distinct seasonal variation in computed values of evaporation minus precipitation. During the winter season of 1958-1959, particularly during February and March, the western Gulf experienced a net addition of fresh surface water. In summertime, salinities in excess of those shown for the winter season might therefore be expected and have been observed.

The surface-temperature field (Figure 1-19), constructed from Cruise 62-H-3 bathythermograph and station data, shows mainly a latitudinal variation, with an intrusion of warm water from the Caribbean in the southeast, the periphery of which is marked by strong gradients. Winter cooling in the Gulf generally proceeds stepwise as cold fronts move across the central United States and out over the Gulf, but few of these outbreaks extend completely across the Gulf. Some of the apparent surface-temperature structure is due to weather changes during the survey period, e.g., the southerly extension of isotherms in the west, which resulted from two outbreaks of cold air that occurred just prior to sampling. The cooling season prior to this cruise had not been severe; a comparison of the 1962 temperature data with those collected during the winter of 1964 is presented in the next section.

Temperature and Salinity in Vertical Sections

Figure 1-20 shows the temperature and salinity distributions in Section *A*, the east-west line just north of Yucatan Strait (see Figure 1-1). In addition to station data, temperature observations from bathythermograph lowerings between sta-

Figure 1-19. Surface temperatures (degrees Celsius), Hidalgo *62-H-3.*

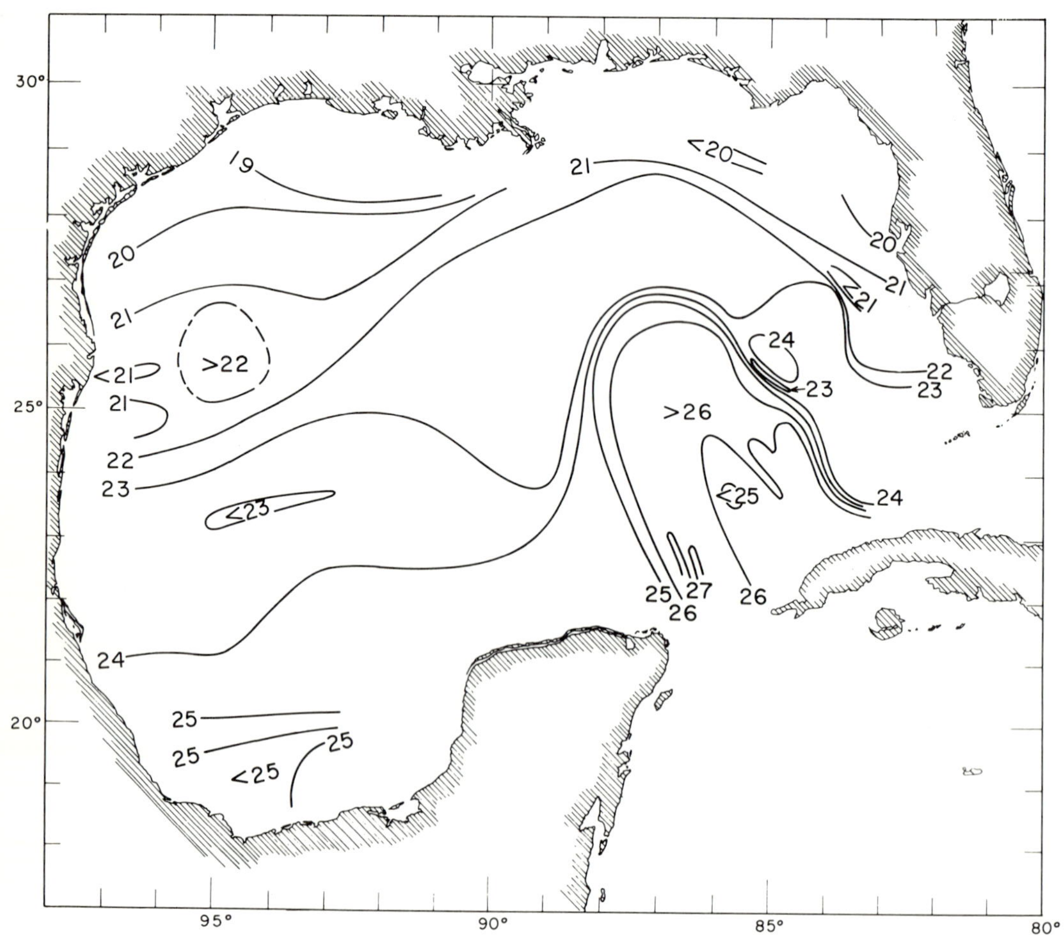

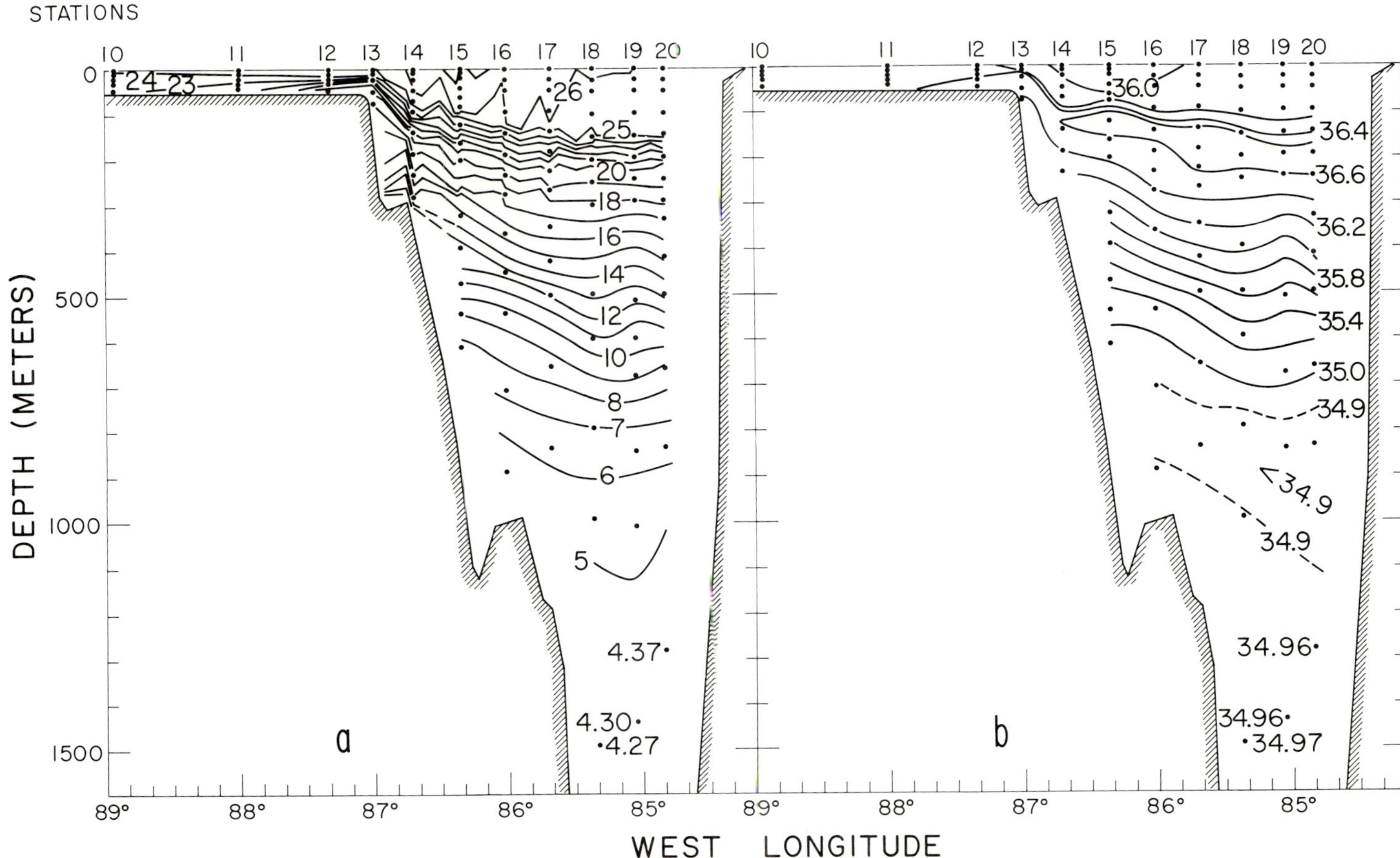

Figure 1-20. (a) Temperature (degrees Celsius) and (b) salinity (parts per mil) along Section A, Hidalgo *62-H-3, 16-18 February 1962. Sampling positions indicated by dots. Vertical exaggeration 340:1.*

tions have been used in constructing Figure 1-20, as for each of the vertical temperature sections. Consequently, the isotherms in the upper 275 m show more structure than do the isolines elsewhere in Figure 1-20. The slopes of the isotherms and isohalines indicate northward flow through most of the section; in the western part some northward flow is indicated to at least 1000-m depth or to the bottom. (Here, as in the discussion of vertical sections to follow, the distributions of temperature and salinity are used to infer roughly the currents within the upper 1000 m, with the tacit implication that such flow is relative to an assumed reference level at greater depth.) Southward flow is suggested between Sts. 18 and 19. The thermal structure indicates that there is a marked intensification of the Yucatan Current along its westward flank, with a maximum just to the west of St. 14. East of St. 14 the flow slackens and again increases to another maximum near St.

Figure 1-21. Temperature (degrees Celsius), Section B, Hidalgo *62-H-3. St. 11 occupied 16 February 1962; Sta. 40 through 46, 25-27 February 1962. Sampling positions indicated by dots. Vertical exaggeration 555:1.*

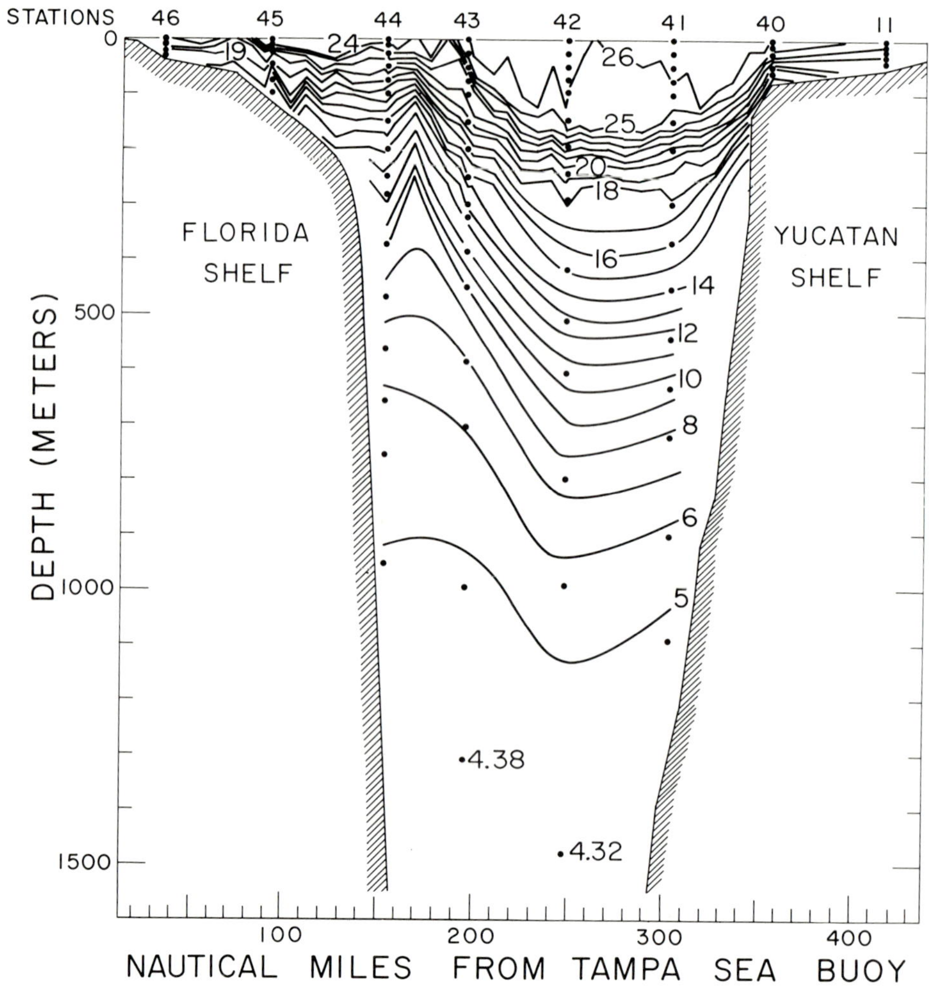

15. Cochrane (1963, 1965) has shown this sort of banded structure to be a consistent feature of the Yucatan Current. Note that, as a consequence of the current structure, waters over the shallow Campeche Bank are of considerably lower temperature and higher salinity than elsewhere at comparable depths.

Figures 1-21 and 1-22 show, respectively, temperature and salinity distributions in Section *B*, from the western Florida Shelf to the Campeche Bank. The subsurface salinity minimum at shallow depths over the Florida Shelf is a feature observed in much of this region. The configuration of isolines shows the two crossings of the major current loop. Bathythermograph data, again used in Figure 1-21, give evidence of a strong flow, with a northwestward component just northeast of St. 40 and a strong southeastward flow in the region of St. 43. Thermal data indicate a region of strong horizontal shear between Sts. 43 and 44, as do the GEK current measurements presented below.

Figure 1-22. Salinity (parts per mil), Section B, Hidalgo *62-H-3. St. 11 occupied 16 February; Sts. 40 through 46, 25-27 February 1962. Sampling positions indicated by dots. Vertical exaggeration 555:1.*

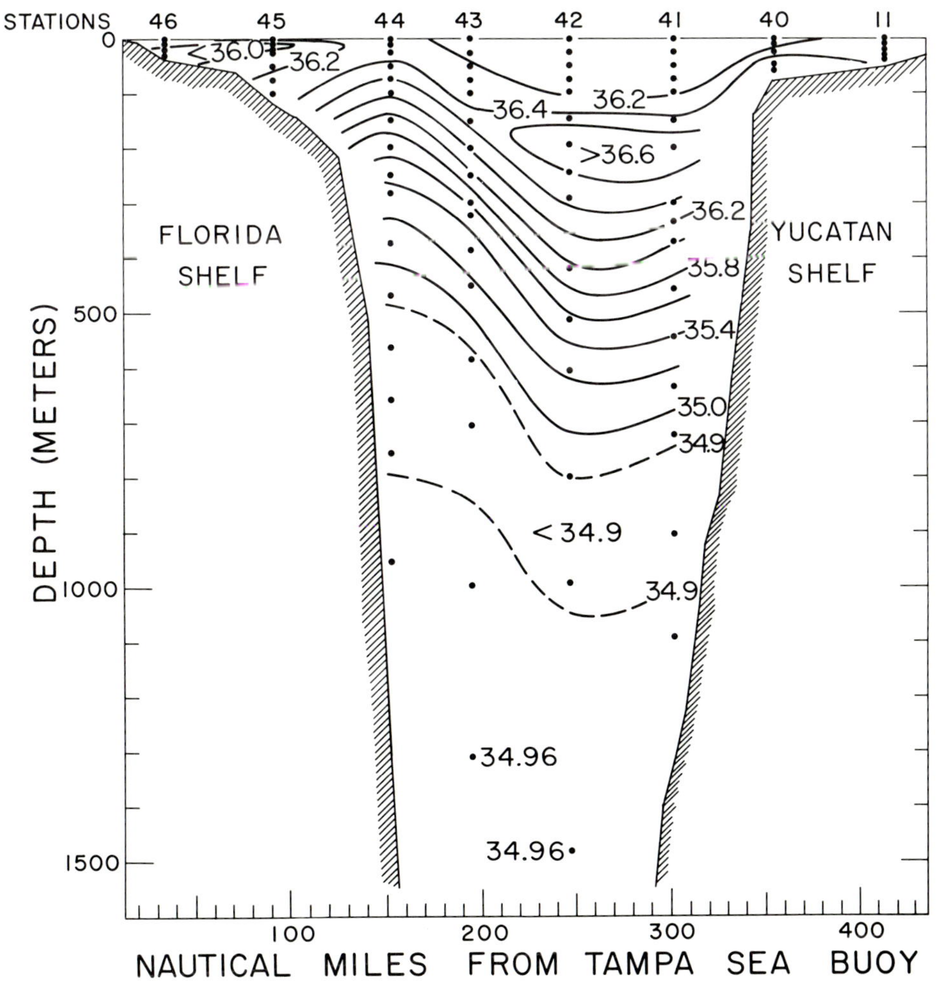

Figure 1-23. (a) Temperature (degrees Celsius) and (b) salinity (parts per mil), Section C, Hidalgo *62-H-3, 19-20 February 1962. Sampling positions indicated by dots. Vertical exaggeration 555:1.*

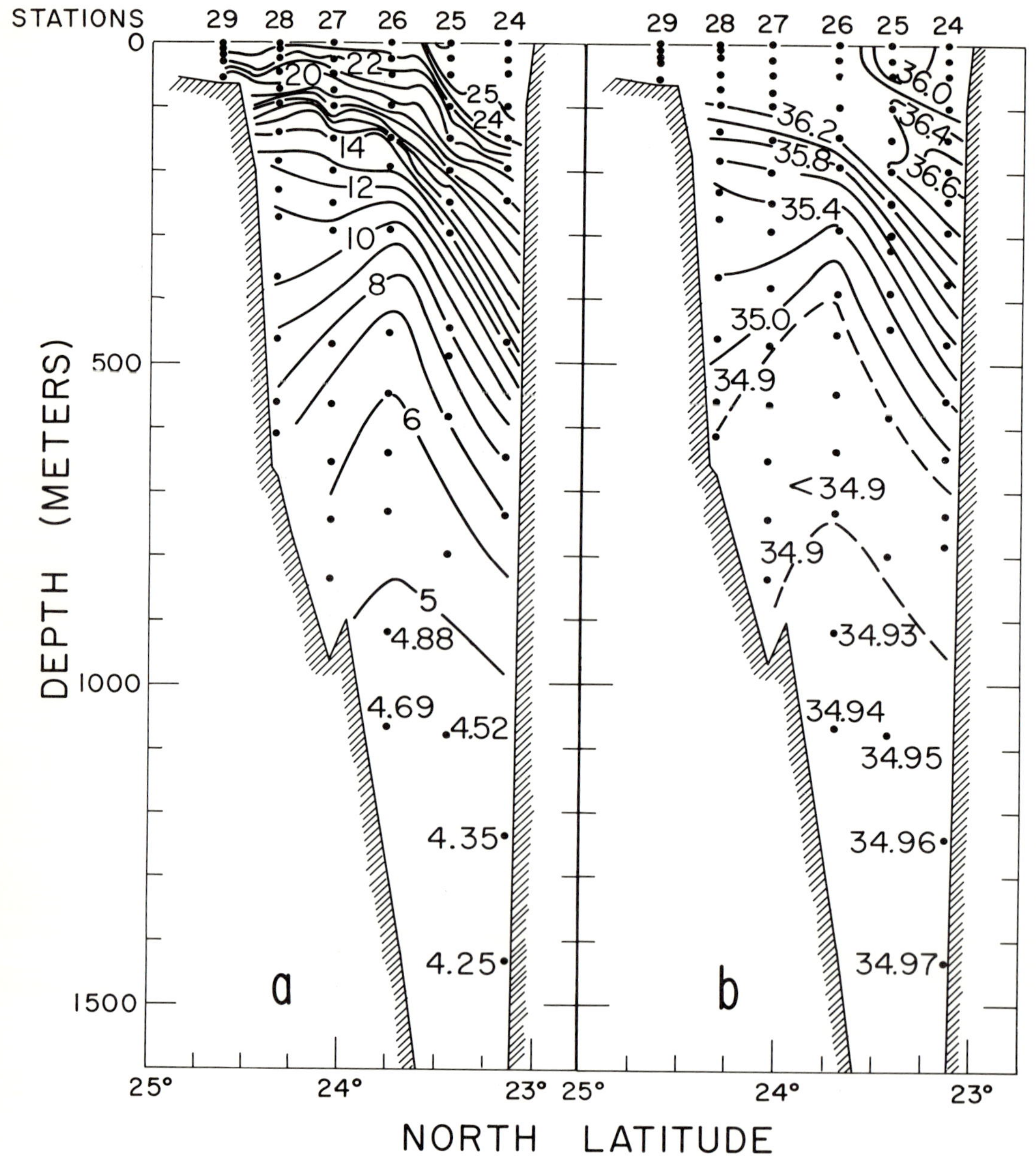

Temperature and salinity for Section *C* are shown in Figure 1-23. The slopes of isolines indicate a strong eastward surface current close to the southern end of the section and some eastward flow reaching to at least 1000-m depth. The flow appears to be reversed (based on the slope of isolines in the deeper waters) in the northern half of the section.

Figures 1-24 and 1-25 show temperature and salinity in Section *D* (93°W). In this section the tropical salinity maximum is only a weak feature. The striking feature of the salinity distribution in the near-surface layers is the occurrence of low-salinity waters, resulting from continental drainage, over the inshore portions of the shelves at each end of the section. The finer scale temperature structure in the upper layers suggests that the flow between the northern shelf and 26°N is not a simple current structure, although a generally westward flow appears to exist to a considerable depth between Sts. 114 and 116. Between Sts. 114 and 112 there is indication of a strong eastward flow based on the configurations of isotherms and isohalines below the thermocline.

The temperature pattern in Figure 1-24 may be compared with that in Figure 1-26, which was taken along approximately the same longitude in late January and February, 1964. The most striking difference is in the extent of development of the mixed surface layer. The thermocline lay almost 100 m deeper in 1964 than in 1962, and the waters above it were cooler and more uniform.

Again in 1964 there was evidence (24°-25°W) of an eastward flow extending to a considerable depth south of the Texas Shelf. However, the current seems to have been centered farther south in 1964 than in 1962. A westward flow near 22°N is indicated for both years. Although these features were less well developed in 1964 than in 1962, and although the eastward flow near 25°N was not in the same location during both years, it is tempting to suggest that there is a regularity in the circulation pattern of the western Gulf. This concept is explored at some length in a succeeding section of this chapter.

Density (σ_t) for Section *D* is shown in Figure 1-27. The features discussed in connection with the temperature and salinity fields are apparent. In the thermocline the maximum stability was typically 1.5-2.0 x $10^{-5}m^{-1}$, with an extreme value of about 3.5 x $10^{-5}m^{-1}$ at St. 105. Below the thermocline, stability decreased monotonically with increasing depth, having values of about $10^{-6}m^{-1}$ at 500 m and 5 x $10^{-8}m^{-1}$ at 1000 m. Below sill depth the stability decreases even further to average values presented in Table 1-1.

Low-Salinity Waters over the Texas-Louisiana Shelf

Figure 1-28 shows the salinity distributions in north-south vertical sections over the outer shelf of the northwestern Gulf. Waters of salinity less than 35 per mil were observed as far west of the Mississippi River as St. 126. Apparently such water, formed by the mixing of river discharge and near-shore waters, is confined to a band along shore some 40-60 nautical miles (75-110 km) wide. Lower salinities were observed at the near-shore locations closest to the Mississippi, which points to this river as the chief source and indicates a general westward spreading. The configurations of isohalines over the outer Texas Shelf, reflected in the density distribution (Figure 1-27), indicate a strong westward flow associated with this low-salinity water.

A similar situation was observed in January, 1966 (Nowlin, unpublished data taken on *Alaminos* Cruise 66-A-1), when a band some 30 nautical miles (55 km) in width containing salinities of less than 35 per mil lay along the coast from Southwest Pass on the Mississippi Delta to at least 95°20′W (the western limit of the survey). During this latter survey the coastal waters (at the 12-m isobath) had salinities of approximately 30 per mil over this region, with the exception of the immediate vicinity of Southwest Pass, where salinities approaching 22 per mil were observed.

In sharp contrast, the January and February, 1964, data taken along 93° and 95°W on *Alaminos* Cruises 64-A-2 and 64-A-3 show no surface salin-

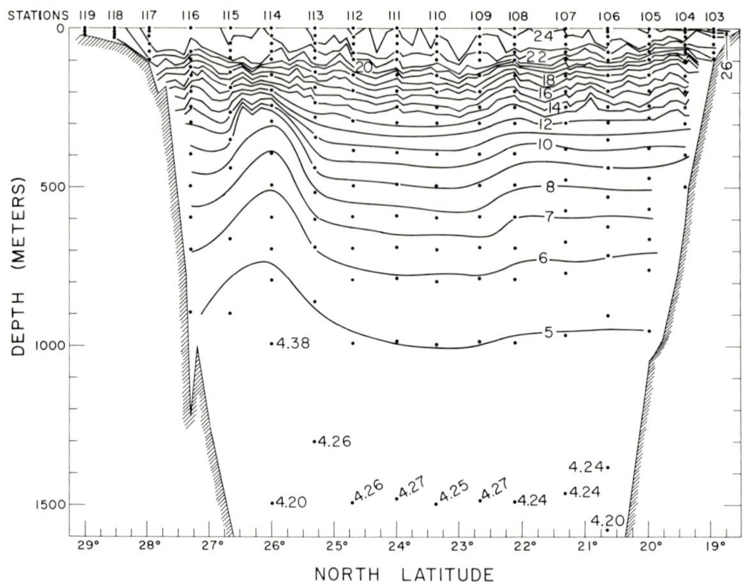

Figure 1-24. Temperature (degrees Celsius), Section D, Hidalgo *62-H-3, 22-27 March 1962. Sampling positions indicated by dots. Vertical exaggeration 555:1.*

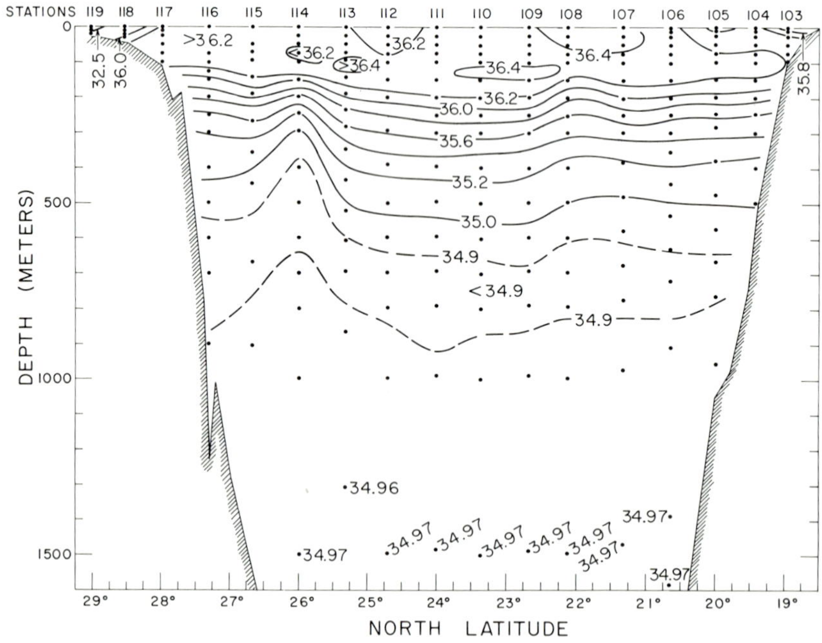

Figure 1-25. Salinity (parts per mil), Section D, Hidalgo *62-H-3, 22-27 March 1962. Sampling positions indicated by dots. Vertical exaggeration 555:1.*

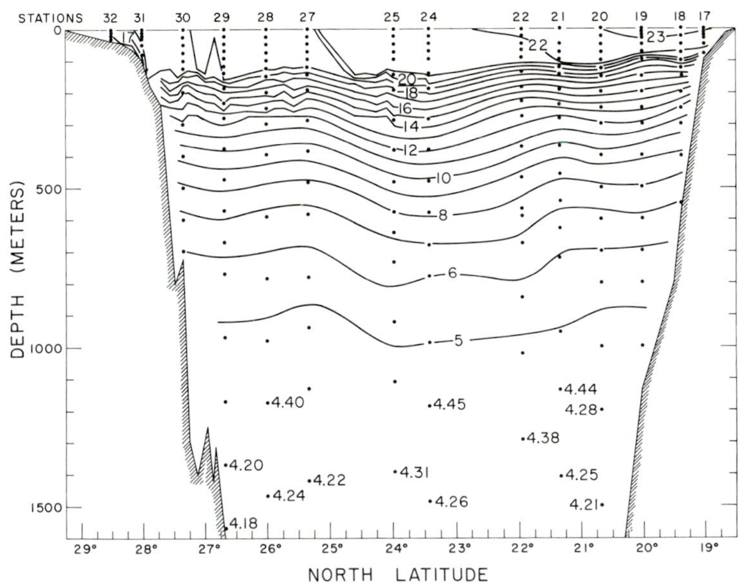

Figure 1-26. Temperature (degrees Celsius) along 93° W (approximately Section D*), Sts. 17 through 21, 24-25 January 1964,* Alaminos *64-A-2; Sts. 22 through 32, 27 February-2 March, 1964,* Alaminos *64-A-3. Sampling positions indicated by dots. Vertical exaggeration 555:1.*

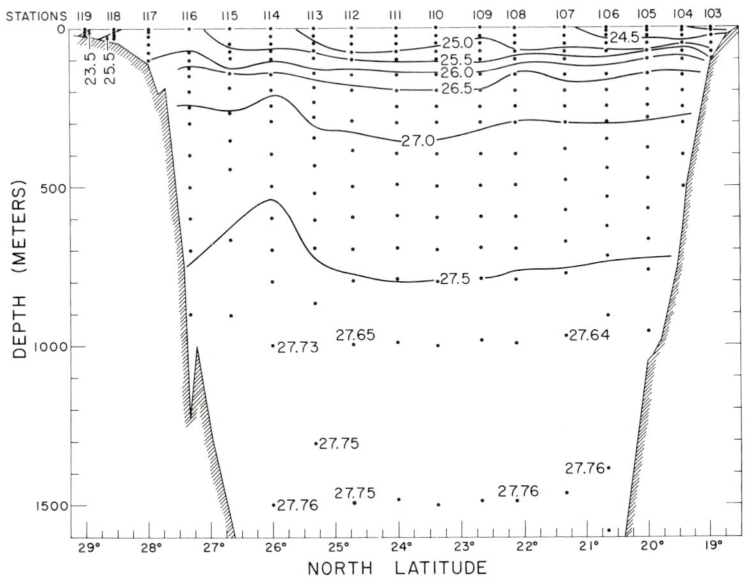

Figure 1-27; σ_t along Section D, Hidalgo *62-H-3, 22-27 March 1962. Sampling positions indicated by dots. Vertical exaggeration 555:1.*

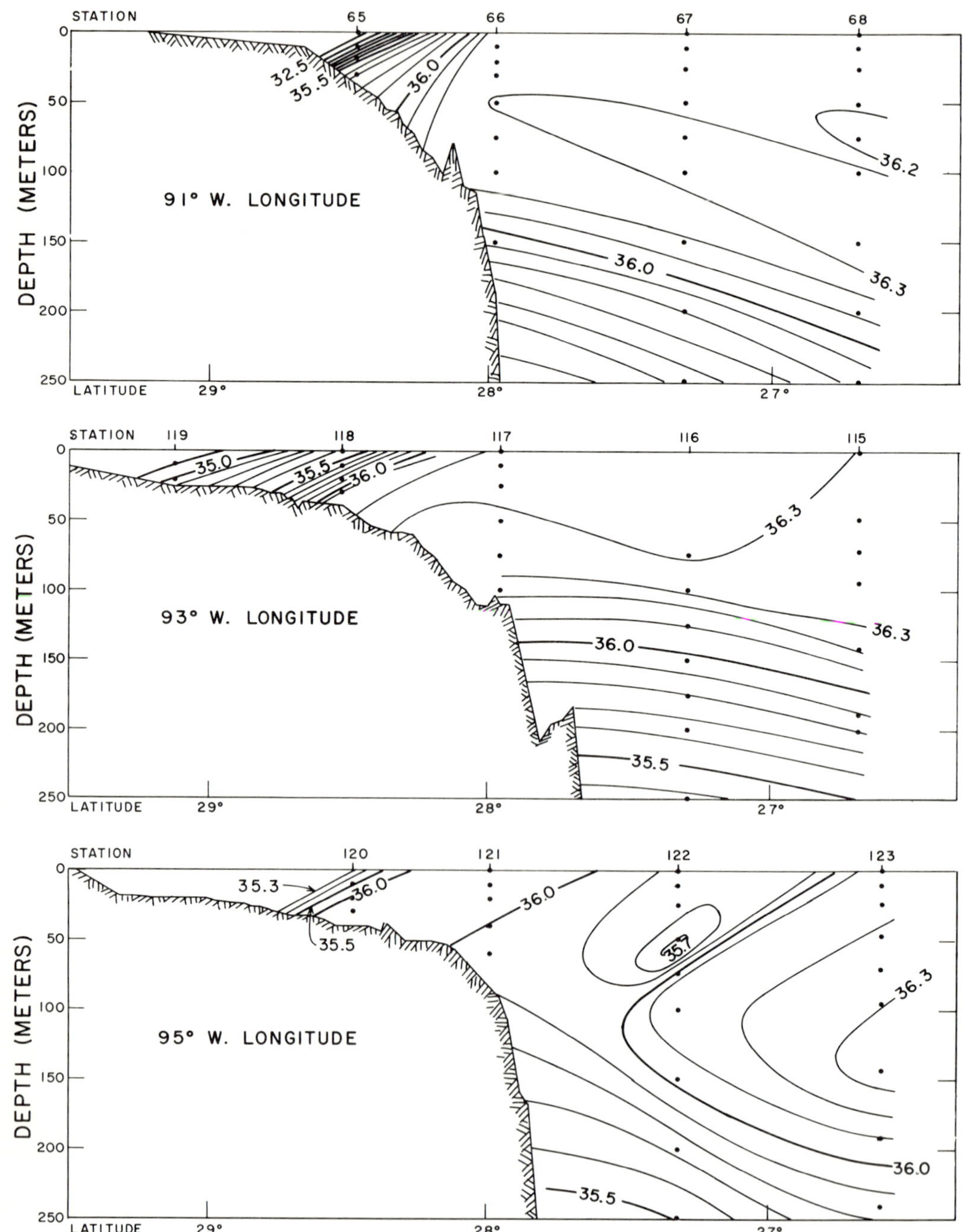

Figure 1-28. Vertical sections of salinity (parts per mil) over outer Texas Shelf, Hidalgo *62-H-3. Data from section along 91°W collected 7-8 March 1962; data along 93° and 95°W collected 26-30 March 1962. Sampling positions indicated by dots. Vertical exaggeration 555:1.*

ities lower than 36.5 per mil beyond the closest station to the shore (35 nautical miles; 65 km).

Currents and Transports

Gross Features

Circulation patterns at the time of the 1962 cruise may be inferred from GEK surface-current measurements and from observed distributions of physical properties. To draw inferences about the current and transport regimes, we have tacitly assumed geostrophic flow and relied on dynamic calculations, even though there may be strong arguments against applying such an assumption to a confined region such as the Gulf.

The dynamic topography of the sea surface relative to the 1000-db surface (Figure 1-29) pictures the Loop Current of the eastern Gulf as the main feature of the surface circulation. The one large eddy centered 60-80 nautical miles (110-150 km) north of the western tip of Cuba, within the confines of the Loop Current, is so often present that some have considered it to be a semipermanent feature, although we now know that such eddies or rings sometimes become detached

Figure 1-29. Dynamic topography of sea surface relative to the 1000-db surface, Hidalgo *62-H-3;* x's *indicate some extrapolation. Contour interval, 0.05 dynamic meters.*

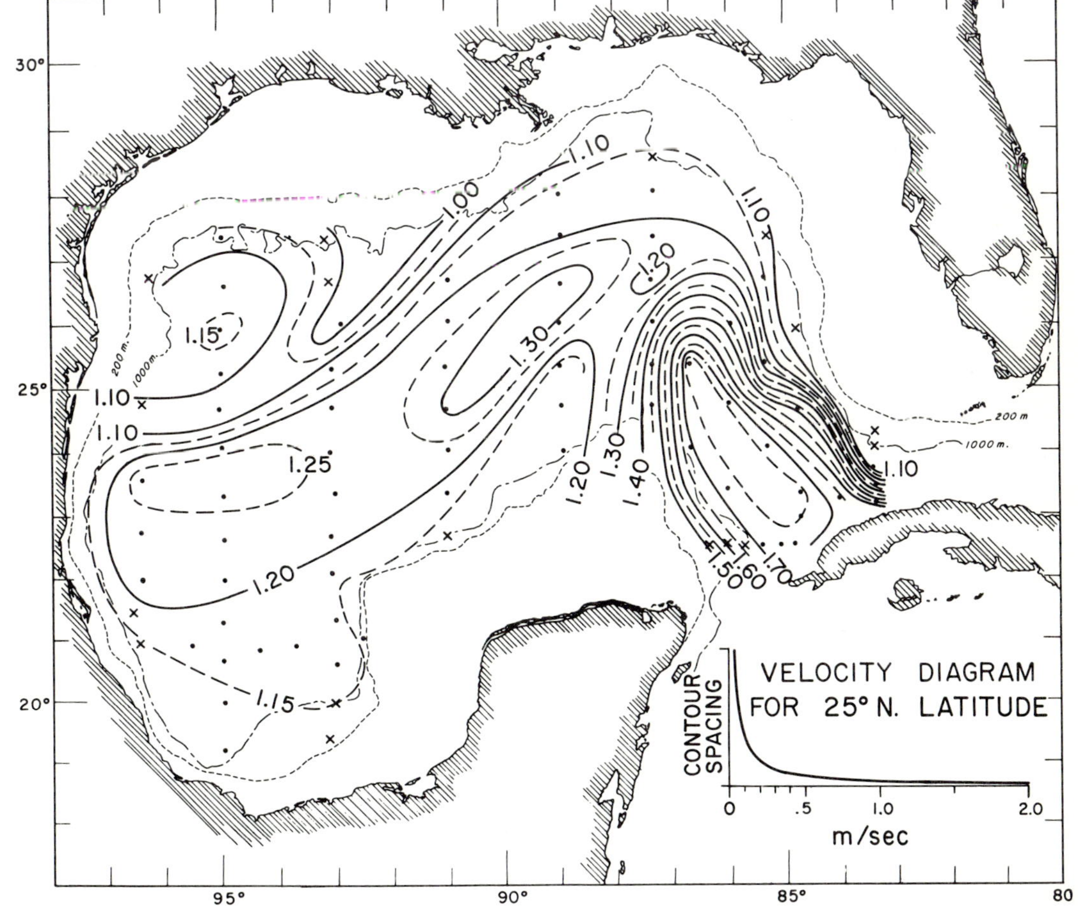

from the main body of the Loop Current. The other principal feature shown in Figure 1-29 is the elongated cell oriented northeast-southwest over the central and western Gulf. That these two anticyclonic circulation systems dominate the circulation in the deep-water regions is also reflected in the mass and property distributions as previously discussed.

In constructing Figures 1-29 and 1-33, some extrapolation was employed for stations at which the maximum sampling depth was between 800 m and 1000 m. In these cases the geopotential anomalies at the maximum pressure sampled, relative to 1000 db, were computed for the two closest deep stations and were then averaged and added to the anomaly for the station in question. The locations of such stations are indicated by *x*'s rather than dots. In constructing Figure 1-32, an analogous procedure was used in extrapolating a few values to 1500 db.

The circulation pattern presented here for the deep-water regions is much simpler than those pre-

Figure 1-30. Dynamic topography (in dynamic meters) of the sea surface relative to the 1000-db surface, Mabel Taylor, *24 January-27 April 1932.*

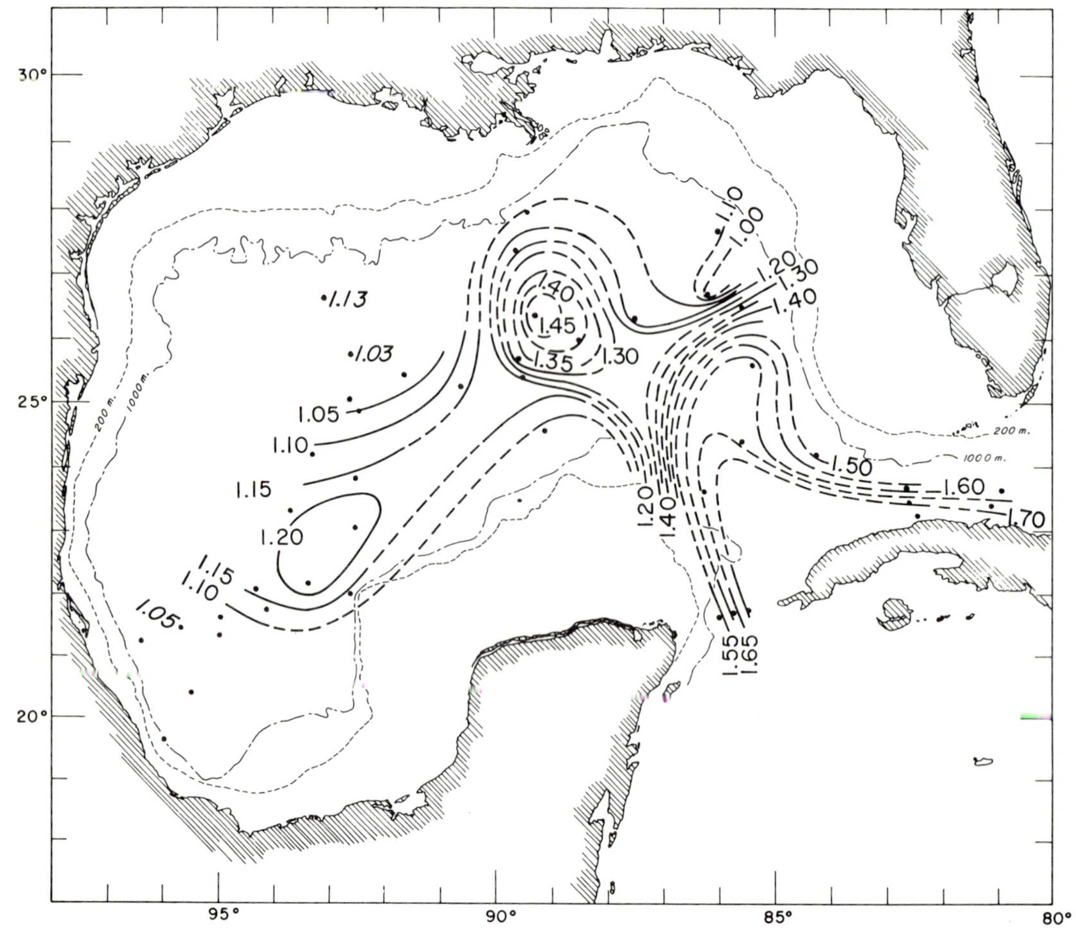

sented by some previous investigators, e.g., Austin (1955) or Duxbury (1962). For comparison with the 1962 data, the 0-db/1000-db geopotential anomalies were computed from data taken during the 25 January-27 April 1932 survey of the Gulf by the *Mabel Taylor* (Parr, 1935) and during the 15 February-13 April 1935 survey by the *Atlantis.* It was found that in both cases the geopotential anomalies could be contoured, with no strain of the imagination, to form patterns (Figures 1-30 and 1-31) consisting of only the same two general features that are evident in Figure 1-29, although the elongated cell in the western Gulf was slightly displaced. The *Mabel Taylor* data have been contoured to emphasize continuity of the high in the central Gulf with that within the Loop Current rather than with the elongated cell over the western Gulf.

Consideration was given to the use of a reference level arrived at by the method of Defant (1961), but I was not able to construct a unique surface with a reasonable depth range in this way.

Figure 1-31. Dynamic topography (in dynamic meters) of the sea surface relative to the 1000-db surface, Atlantis, *15 February-13 April 1935.*

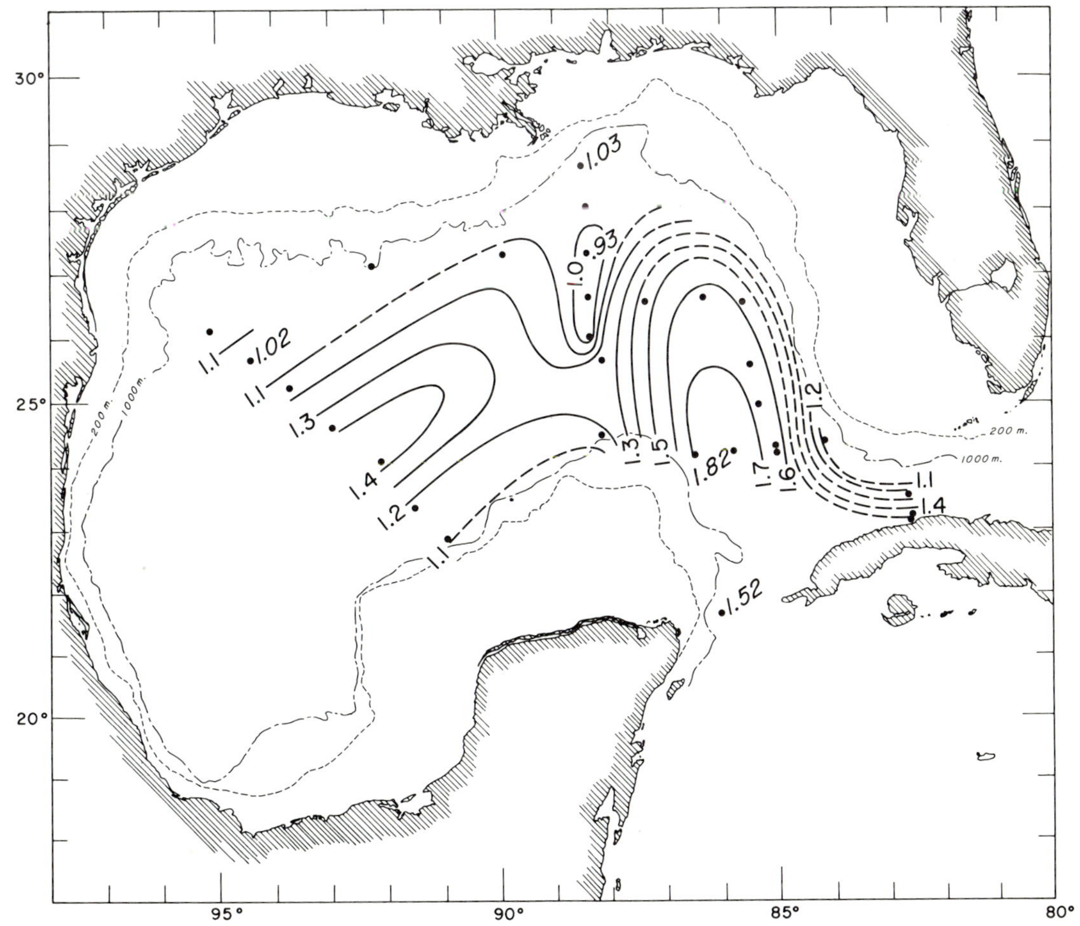

Since the 1962 data showed very little baroclinicity below sill depth, a reference surface below this depth was also considered, but the additional stations lost in the computation made this unattractive.

Geopotential anomalies of the 1500-db surface relative to 2000 db (not shown) indicated a very weak clockwise circulation cell in the western central Gulf and no distinct pattern in the eastern and central Gulf. Gradients on this surface were small, an isolated extreme value being approximately 0.0002 dyn m per nautical mile, which corresponds to a relative current speed of only some 2 cm/sec. Even so, it is interesting to note the indication of some relative motion at and near the sill depth of the basin.

The geopotential anomalies of the 1000-db surface relative to 1500 db (Figure 1-32) show motion with a pattern very much like that in Figure 1-29. However, in the eastern Gulf there is

Figure 1-32. Dynamic topography of the 1000-db surface relative to the 1500-db surface, Hidalgo *62-H-3; x's indicate some extrapolation. Contour interval, 0.005 dynamic meters.*

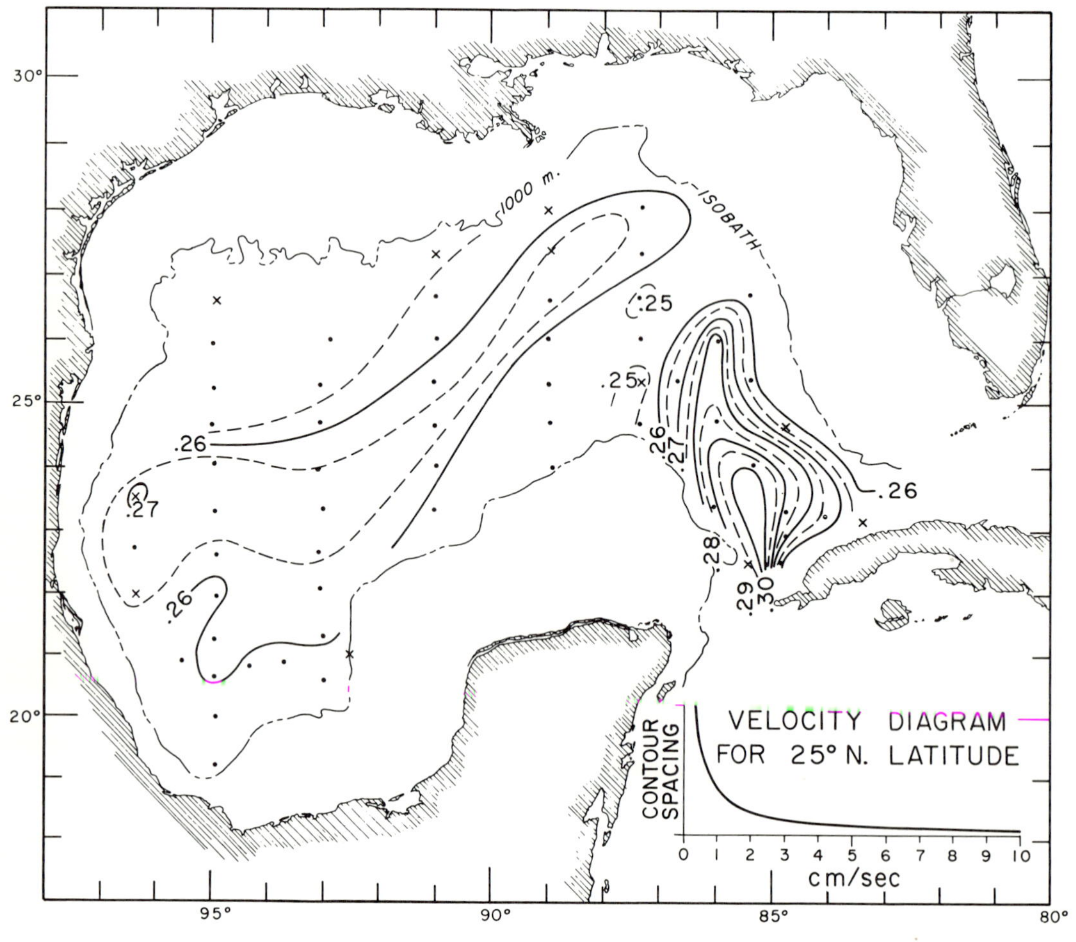

a loop of current that both enters and leaves the Gulf through Section *A* (Figure 1-32). Perhaps this is to be expected, since the controlling depth in the Straits of Florida is thought to be some 800 m (Defant, 1961). This pattern may be partially closed in the Yucatan Strait south of Section *A*. Relative speeds indicated in Figure 1-32 are mostly less than 10 cm/sec. The 1000-db surface may then be taken as a reference for computing geostropic current speeds in the upper layers of the Gulf without introducing errors of much more than 10 cm/sec.

Figure 1-33 presents the contours of geopotential anomaly for the 500-db surface relative to 1000 db. There was considerable motion down to 500 m, and the circulation patterns at the sea surface and at 500 db were quite similar.

Figure 1-34 shows surface currents as inferred from a GEK. No corrections other than those for magnetic-field intensity have been made to these

Figure 1-33. Dynamic topography of the 500-db surface relative to the 1000-db surface, Hidalgo *62-H-3;* x*'s indicate some extrapolation. Contour interval; 0.025 dynamic meters.*

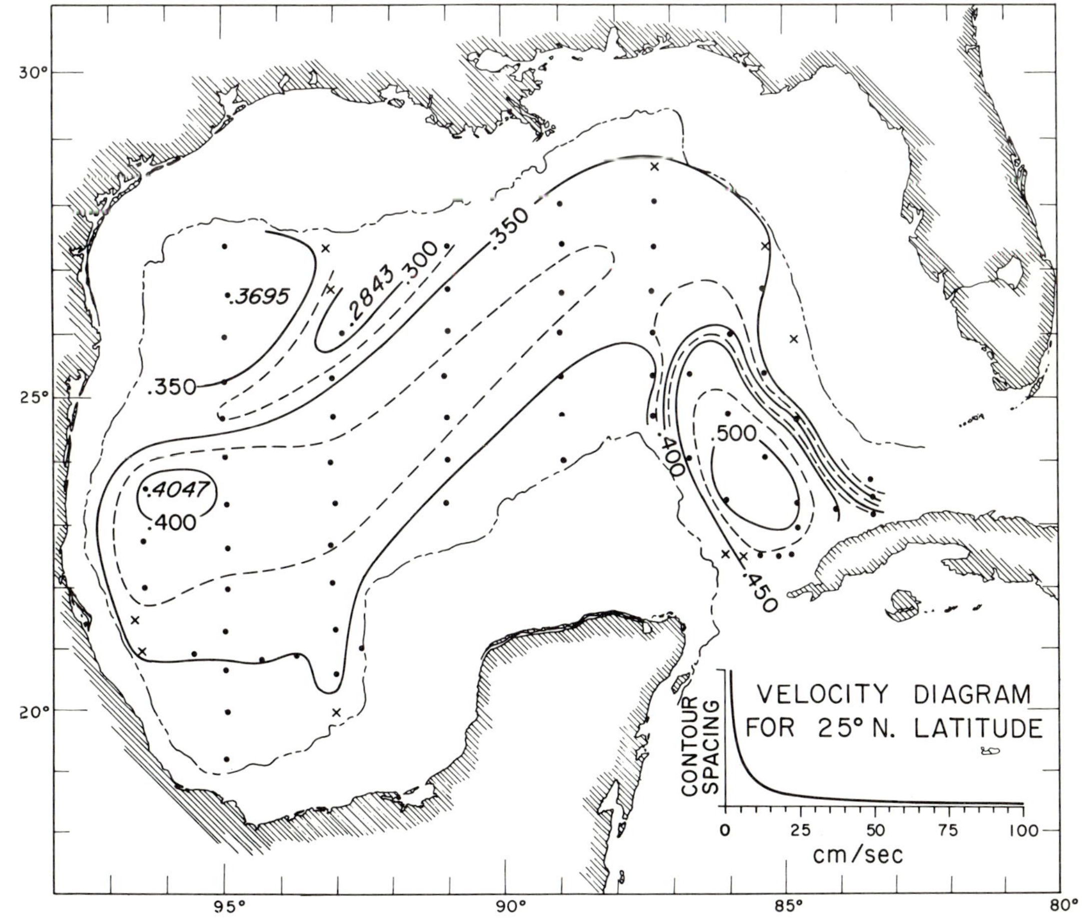

values. Unfortunately, no GEK data were obtained along 89°W.

The GEK measurements are in general agreement with the major features of the computed surface-current pattern. The core of the Loop Current in the eastern Gulf is clearly seen in Figure 1-34. Along the northern boundary of the anticyclonic ridge in the western central Gulf, the core of the northeastward flow is seen, as are the shear zone to its north and the southwestward flow along the outer shelf. The GEK observations show no strong currents along the position of the southern boundary of this ridge, which may be a broad diffuse flow as indicated in Figures 1-29 and 1-33. Many smaller-scale features appear in the GEK data, but these are presumably averaged out or are otherwise not seen in the dynamic topographies.

The geostrophic transport field for the upper 1000 m relative to 1000 db is presented in Figure 1-35. For each station from 1000 m to the surface the goepotential anomaly relative to 1000 db was vertically integrated to obtain the transport po-

Figure 1-34. GEK surface-current observations, Hidalgo *62-H-3.*

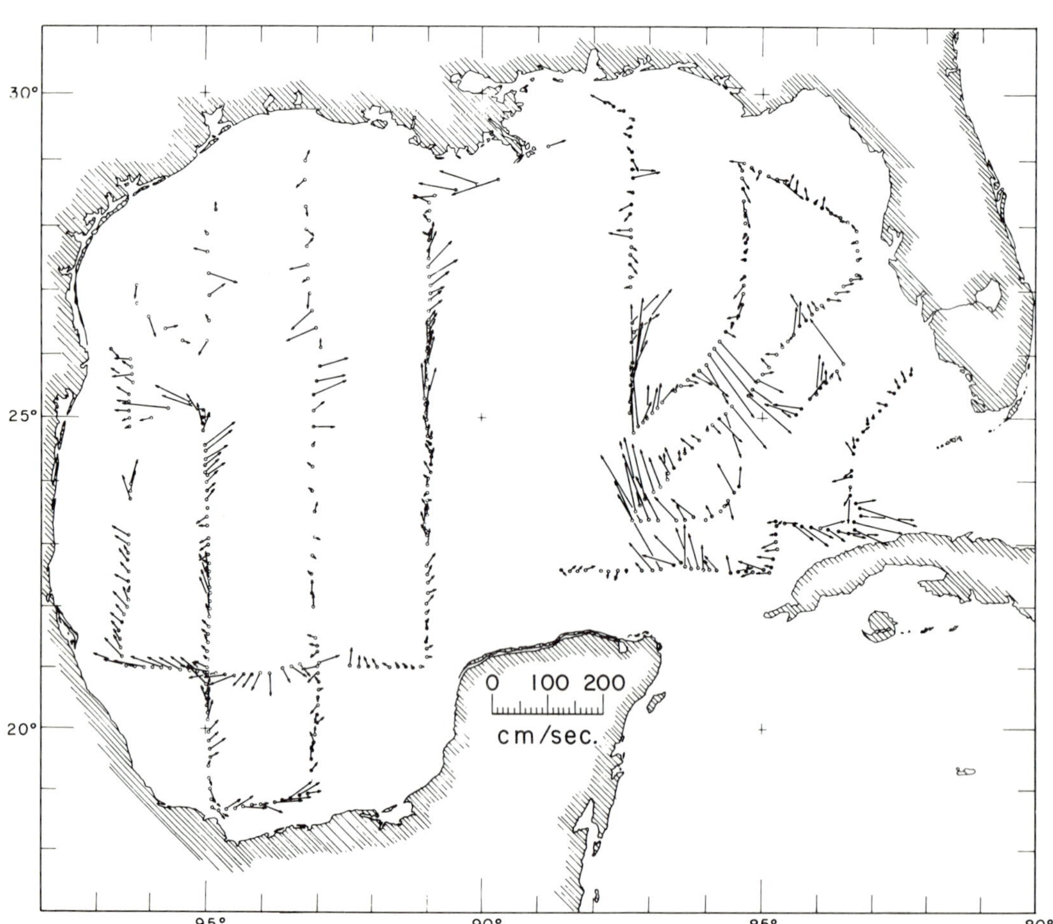

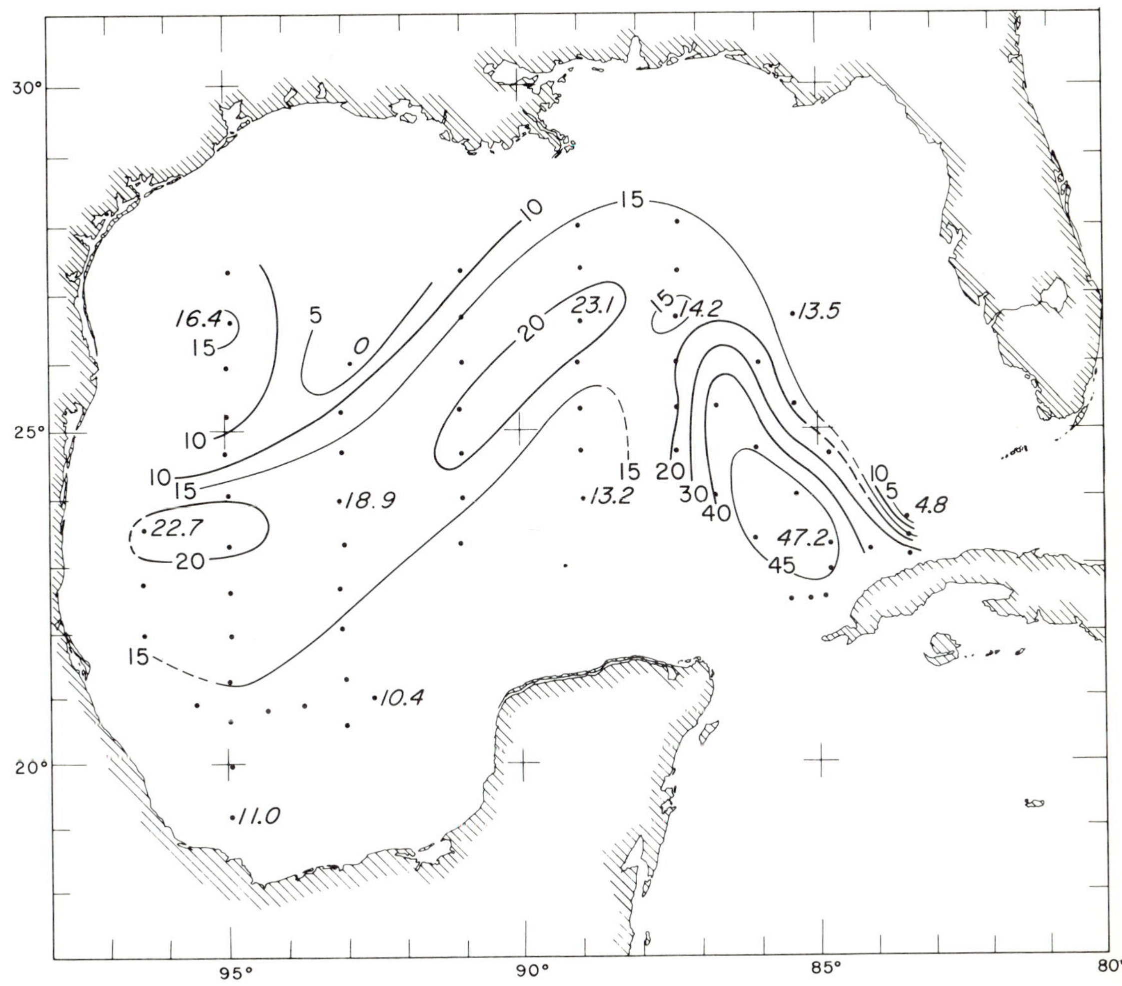

Figure 1-35. Streamlines of geostrophic transport, in 10^6 m^3*/sec, in upper 1000 db relative to the 1000-db surface,* Hidalgo *62-H-3.*

tential. Before contouring this scalar field, the minimum value of this potential function was subtracted from every value and the residual at each station was divided by the value of the Coriolis parameter at 24° latitude. Of course, the field is very similar in appearance to the dynamic topography presented in Figure 1-29.

For further discussion, the Gulf of Mexico is divided into eastern and western subregions.

Eastern Gulf

The Yucatan Current appears to turn clockwise in the eastern central Gulf and then issue into the Straits of Florida, where it is known as the Florida Current. We refer to this current in the turn-around region as the Loop Current. Based on data from any reasonably short time period, the

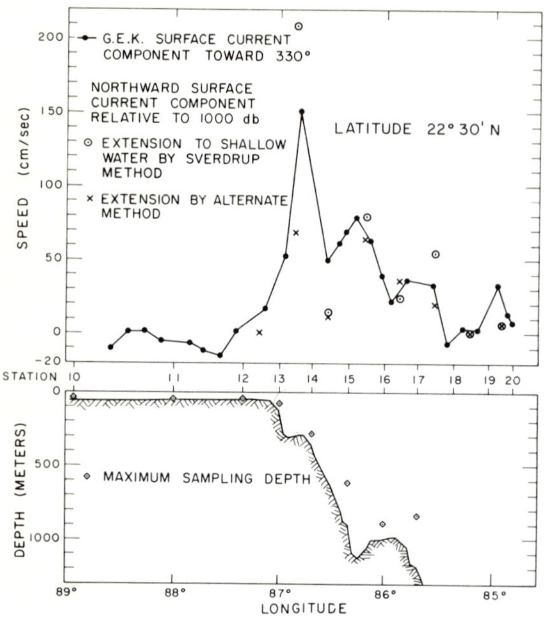

Figure 1-36a. Depth profile and surface currents along Section A *(22°30′N). Shown are GEK surface-current components toward 330° and geostrophic surface current normal to Section*A.

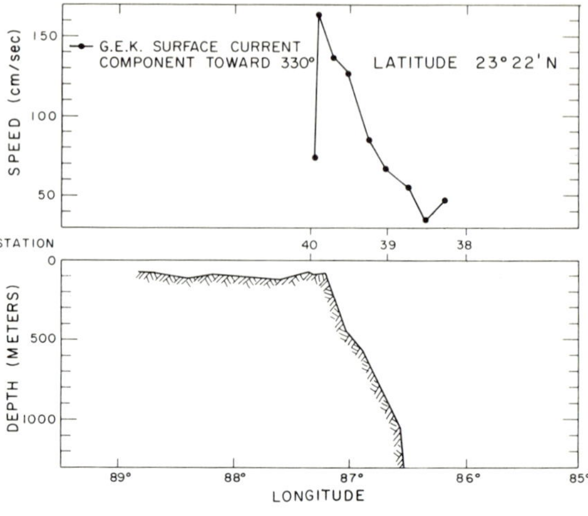

Figure 1-36b. GEK surface current components toward 330° and depth profile along a section at 23°22′N. All data from both sections collected during Hidalgo *62-H-3. For positions of stations, see Figure 1.1.*

configuration of this loop is easily defined by dynamic computations, by salinity values at the salinity maxima, and/or by the characteristic *T-S* relationship of all stations on the Cuban side of the main current.

Cochrane (1963) presented GEK data which indicate that the Yucatan Current in the region of the Yucatan Strait is kinematically banded, showing two or more high-velocity cores. This feature shows in the GEK surface-current profile across Section *A* (Figure 1-36*a*). It is not seen in the profile taken along 23°22′N, which is presented in Figure 1-36*b*. This is not inconsistent with the results of Cochrane, which show a weakening of this intercore shear zone in the downstream direction. Surface velocity components in the direction 330° are shown, since this was judged to be the orientation of the current axis during the period of observation. This banding is also evident in the temperature profile (Figure 1-20) along Section *A*. Referring to the GEK surface velocities for Section *A* in Figure 1-34, it is seen that in the Yucatan Current between the two apparent high-velocity cores there is a considerable component of velocity directed toward the Yucatan Shelf.

The waters at the positions of the two cores in the current are different. At St. 13 and in the high-velocity core just west of St. 14, water of the type that characterizes the eastern Yucatan Shelf (Figure 1-4) was found. At St. 15, within the secondary current core, the waters showed the salinity maximum and smooth *T-S* relationship characteristic of the waters bounded by the Loop Current.

At Section *A*, the main core, with axial velocity components of approximately 150 cm/sec, was situated approximately over the 300-m isobath (Figure 1-36*a*). Cochrane (1965) found a sharp core situated approximately over the 200-m isobath when the surface currents were the strongest (average maximum velocity component of 177 cm/sec), during late spring and early summer. A more diffuse core lies farther offshore in late fall when the current reaches its minimum (average maximum axial-velocity component of 127 cm/sec).

In the downstream profile taken along 23°22′N (Figure 1-36*b*) there was only one high-velocity core (160 cm/sec). This was situated approximately over the 100-m isobath at the edge of the Yucatan Shelf.

Also presented in Figure 1-36 are components of geostrophic surface velocities normal to Section *A*. These have been computed from geopotential anomalies relative to 1000 db for adjacent station pairs. Since the maximum sampling depths for stations west of St. 18 were less than 1000 m, the section was extended into shallow water, using the method of Sverdrup et al. (1942: 451). A simpler method of extending the section was tried and is presented for comparison. In the alternate analysis, the geopotential anomaly of the sea surface relative to the maximum pressure sampled was computed. To this was added the geopotential anomaly (relative to 1000 db) at that same pressure surface for the deeper adjacent station to the east. Perhaps fortuitously, the banded nature of the surface velocity field also appears in these geostrophic computations.

Downstream from the two sections shown in Figure 1-36, the maximum current speeds indicated by the GEK ranged from 100-200 cm/sec. The largest values were observed approximately 160 nautical miles (300 km) north of Section *A*, near St. 56.

Recent results of Schmitz and Richardson (1968) based on the average of three years of direct measurements of transport by free instrument technique give $32 \pm 3 \times 10^6 m^3/sec$ as the mean volume transport of the Florida Current east of Miami. They estimate that the component of this flow derived from the Santaren Channel, and not from the Straits south of the Florida Keys, is only $1\text{-}3 \times 10^6 m^3/sec$. Schmitz and Richardson observed a fluctuation bound of $\pm 12 \times 10^6 m^3/sec$, of which about 75% occurs at frequencies near 1 cycle per day. The principal tidal transport fluctuations were delineated by harmonic analysis; the transport amplitudes obtained were $1.5 \pm 1 \times 10^6 m^3/sec$ for S_2 and $3.5 \pm 1 \times 10^6 m^3/sec$ for each of M_2, 0_1 and K_1 components. They found no evidence for net transport fluctuations larger than $3 \times 10^6 m^3/sec$ at nontidal, e.g., seasonal, periods.

Between Sts. 24 and 26 in Section *C*, the 1962 data give an eastward volume transport relative to 1000 db (Figure 1-35) of $30.0 \times 10^6 m^3/sec$. (The transport values given in the text were computed with the Coriolis parameter corresponding to the average latitude of stations considered.) The agreement between this transport value and the mean transport reported by Schmitz and Richardson is remarkable. This flow is greatly intensified southward, with maximum surface velocities being found near St. 25, as seen in the GEK observations or in the temperature and salinity distributions for Section *C* (Figure 1-23). Much of the outflow must consist of waters that are characteristic of the interior Gulf rather than of the Caribbean, since waters with a salinity maximum typical of the northwest Caribbean were found only at St. 24. In the northern part of Section *C* the data evidence some flow to the west. Relative to 500 db, the westward inflow in the upper 500 m between Sts. 26 and 28 was just over one million m^3/sec—not large when compared with the outflow.

It should not be inferred that the stream is typically found near the Cuban coast. In fact, north-south spatial fluctuations of the edges and axis of the stream are known to be pronounced in the region south of the Florida Keys. Based on GEK transects south of Key West, Hela et al. (1954) have shown the current to sometimes occupy the northernmost part of the strait, sometimes the southernmost part.

No good estimate of inflow can be prepared from the 1962 data across Section *A*. However, based on the outflow at Section *C*, continuity demands that the net inflow in the upper 1000 m across Section *A* must be some $30 \times 10^6 m^3/sec$. The gross inflow in the Yucatan Current must be even larger, since, relative to 1000 m, somewhat over one million m^3/sec is transported southward over Section *A* between Sts. 18 and 20. Most of this southward transport takes place at depths greater than 500 m.

Apparently, about 10 x $10^6 m^3$/sec of this inflow branches westward from the Yucatan Current in the region of the northern Campeche Bank. The transport in the upper 1000 m of the Loop Current is approximately 24 x $10^6 m^3$/sec between Sts. 55 and 57. Along its northeastern portion the transport in the Loop Current increases in a downstream direction as additional waters from outside the Loop join the flow.

In addition to the Loop Current proper, during the winter of 1962 some 5 to 10 x $10^6 m^3$/sec of water having characteristics typical of the northwestern Caribbean were circulating as an anticyclonic gyre within the Loop.

Although it is believed that the description of the eastern Gulf just presented is a fair picture of the circulation as it existed during the winter of 1962, this should/not be accepted as a typical picture. Recent studies have shown without doubt that the current patterns of that region are quite time dependent. The temporal variations seem especially strong from late spring until mid-fall. Perhaps this is because the Yucatan Current begins its intensification in spring, reaching maximum speeds in May (Cochrane, 1966), and according to the pilot charts it is fall before this current reaches its minimum intensification, with a relatively broad core of lesser maximum speeds.

At times the Loop Current extends much further into the northeast Gulf than was observed in the winter of 1962. As an example, in the first thorough study of the Loop Current, which was based on *Alaminos* 66-A-8 data collected during June, 1966, Hubertz (1967) found the current core at the northernmost limit of the Loop to be between 27° and 28°N, just offshore from the 2000-m depth contour along the continential slope (see Chapter 6). At other times the Yucatan Current enters the Gulf of Mexico on a northeastward course and proceeds just north of the coast of Cuba toward the Straits of Florida. In these cases, e.g., the data from *Alaminos* 65-A-12 in September, 1965 (Cochrane, 1966), there may be no extension of the Loop Current into the Gulf, only a continuation of the Yucatan Current directly toward the Straits of Florida.

There have been several occasions on which anticyclonic rings of current were observed distinct from, and outside of, the Loop Current regime. The first of these to be observed in detail and reported in the literature (Nowlin et al., 1968). was studied on *Alaminos* 67-A-4 during June, 1967. Figure 1-37, taken from Nowlin et al. (1968: figure 1), shows the depth of the 22°C isothermal surface as inferred from bathythermograph observations and also shows surface GEK measurements (corrected only for magnetic field intensity). The shear in the zone separating the ring from the Loop Current is seen to be extremely strong. The data show that both within the ring and southeast of the Loop Current the salinity values at the core of the subtropical salinity maximum were in excess of 36.7 per mil, although maximum salinity values along the axis of the shear zone separating the ring and the Loop Current were less than 36.5 per mil. (That this difference is sufficient to delineate waters bounded by the Loop Current from the other waters of the Gulf is readily seen in Figure 1-17*a*.) On *Alaminos* Cruise 69-A-7 during May, 1969, Cochrane (1969, personal communication) first observed conditions just as a ring was in the stage of separating from the Loop Current and then later surveyed much of the separated anticyclonic current ring.

Based on circulation patterns inferred from contours of isotherm depths obtained from BTs on a sequence of cruises within the time period July, 1965, to November, 1966, Leipper (1970) concluded "that there may be a systematic development and breakdown of the pattern of the Loop Current in the Gulf of Mexico," and that "this development may have some seasonal aspects." It is well established through the unpublished work of Cochrane (cited in this chapter) that there are some seasonal aspects to the development of the eastern Gulf circulation. Cochrane has docu-

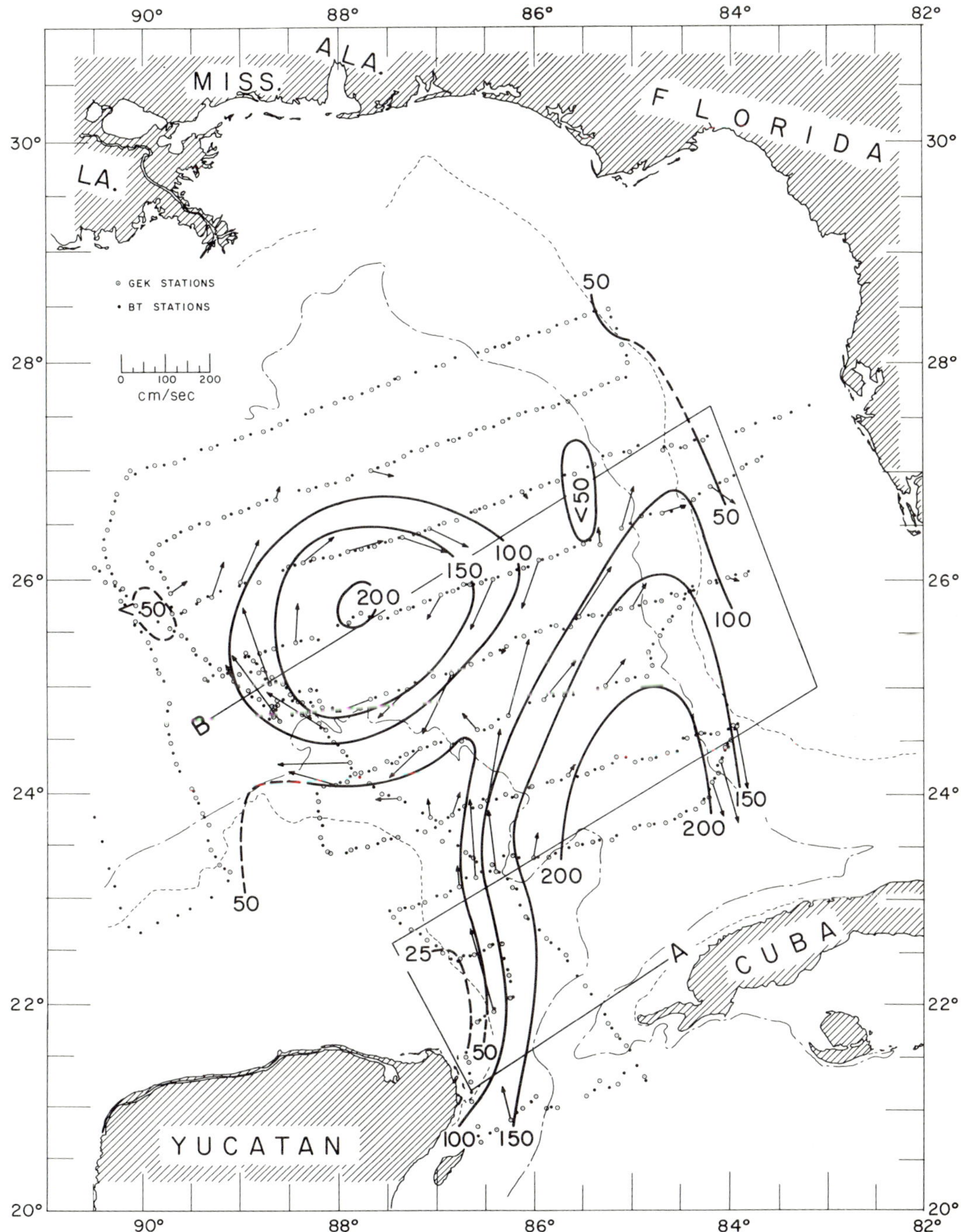

Figure 1-37. Depth (50-m intervals) of the 22°C isothermal surface from BT observations, and selected GEK surface-current measurements; Alaminos *67-A-4, June, 1967.*

mented the characteristics of the Yucatan Current during its period of development to maximum intensity in spring and has likewise established its characteristics in fall during least intensification.

Since the circulation, in the eastern Gulf at least, is driven principally by the Yucatan Current, it is reasonable to expect some seasonal aspects, although a regular seasonal pattern seems unlikely. Very complete pictures in successive seasons, e.g., in June of 1966 and of 1967, have given radically different eastern Gulf circulations: a strong Loop Current extending far into the Gulf in 1966; a large anticyclonic ring detached from the Loop Current in 1967 (see Chapter 6). It seems likely that the specific circulation patterns which develop, while having seasonal aspects derived from the driving forces, will depend strongly on local dynamics and may well not be entirely deterministic.

The faculty and students of Texas A&M University's Department of Oceanography are continuing a series of theoretical investigations of the effects on the eastern Gulf circulation of friction, of topographic control, of stratification and of the westward intensification, total transport and inflow angle of Yucatan Current. Preliminary numerical and analytical studies of the effects of lateral and vertical friction without bottom topography are complete (Wert, 1968 [see Chapter 8] Jacobs and Nowlin, 1968). Molinari (1968) studied the effects of topography of that portion of the Yucatan Current along the east and northeast edge of the Campeche Bank during the time of the year when the current is most intense; data from May, 1962, 1965 and 1966 were compared with calculated results of both graphical and numerical methods for obtaining the stream path based on the conservation of potential vorticity, the actual topography and observed initial vorticity. Some of the other studies just completed or yet in progress were discussed briefly by Reid (1969).

Western Gulf

The connection between the Yucatan Current and the circulation in the western Gulf is not yet clear. From the *Hidalgo* 62-H-3 study, it seems possible that part of the Yucatan Current may leave the main stream and flow across, or along the northern edge of, the Campeche Bank to join the broad slow flow that forms the southern flank of the anticylonic ridge in the western central Gulf. (At 91°W, the computed westward transport relative to 1000 db between Sts. 71 and 73 was 9.5 x $10^6 m^3/sec$.) The sampling pattern did not allow delineation of the western end of this ridge. However, a well-developed current was observed along its northern flank.

For this northern region, geostrophic current speeds at a number of levels relative to 1000 db were computed. From 24°20′N, 95°W downstream to 25°30′N, 93°W, the maximum axial components of velocity were approximately 50 cm/sec. The geostrophic speeds in the core then seemed to decrease downstream to values of some 30 cm/sec at 27°20′N, 91°W. By comparison, the GEK measurements gave a maximum speed of approximately 70 cm/sec at all three of the stream crossings made over this distance. Dynamic computations gave larger speeds at 100 m than at the surface. Along 93°W, the downstream component was almost 70 cm/sec at 100 m. Speeds decreased below 100 m but were still of the order of 20 cm/sec at 500 m. The indications are that the stream widened in a downstream direction from 95° to 91°W; apparently water was being entrained along the northern edge of this flow from a parallel counterflow located over the shelf edge.

In the southwestward counterflow, geostrophic current speeds at the surface and at 100 m were both about 30 cm/sec as measured at 93°W. Surface GEK measurements gave 50 cm/sec at this longitude and showed a current shear between these two flows of more than 100 cm/sec in a distance of 35 nautical miles (65 km).

Other available data from the western Gulf of Mexico were examined for additional north-south sections made during the winter season. In addition to *Hidalgo* 62-H-3, only three cruises appear to have data extending across the entire Gulf with sufficient sampling density that they might be useful in establishing the winter circulation pattern.

These cruises are *Hidalgo* 58-H-1, 23-30 March 1958 (McLellan 1960); *Alaminos* 64-A-2, 17-26 January 1964; and *Alaminos* 64-A-3, 27 Feburary-2 March 1964. Figure 1-38*a* shows locations of stations along north-south sections from these three cruises and from *Hidalgo* 62-H-3. Two other cruises which provided some regional coverage of the western Gulf in winter were *Mabel Taylor* 1932 and *Atlantis* 1935; the resulting sea-surface dynamic topographies relative to 1000 db are shown in Figures 1-30 and 1-31. However, sections comparable with those in Figure 1-38*a* were not made from these two cruises.

For each station in Figure 1-38*a*, the geopotential anomalies, relative to the 1000-db surface, of the sea surface and of the 500-db surface are plotted as functions of latitude in Figure 1-38*b*. Although there is clearly variation from cruise to cruise, the striking new result to be seen in this figure is the remarkable similarity between the baroclinic fields of the western Gulf in three different winters. In each case the dominant feature is a high centered about 23°40′N latitude. Moreover, there is an indication of a westward component to the north of the eastward flowing flank of this generally east-west oriented ridge.

*Figure 1-38*a. *Positions of hydrographic stations occupied on meridional sections in the western Gulf of Mexico during the winter season;* Hidalgo *58-H-1 and 62-H-3,* Alaminos *64-A-2 and 64-A-3.*

*Figure 1-38*b. *Geopotential anomalies (in dynamic m) of sea surface and 500-db surface relative to 1000 db for data from stations with positions shown.*

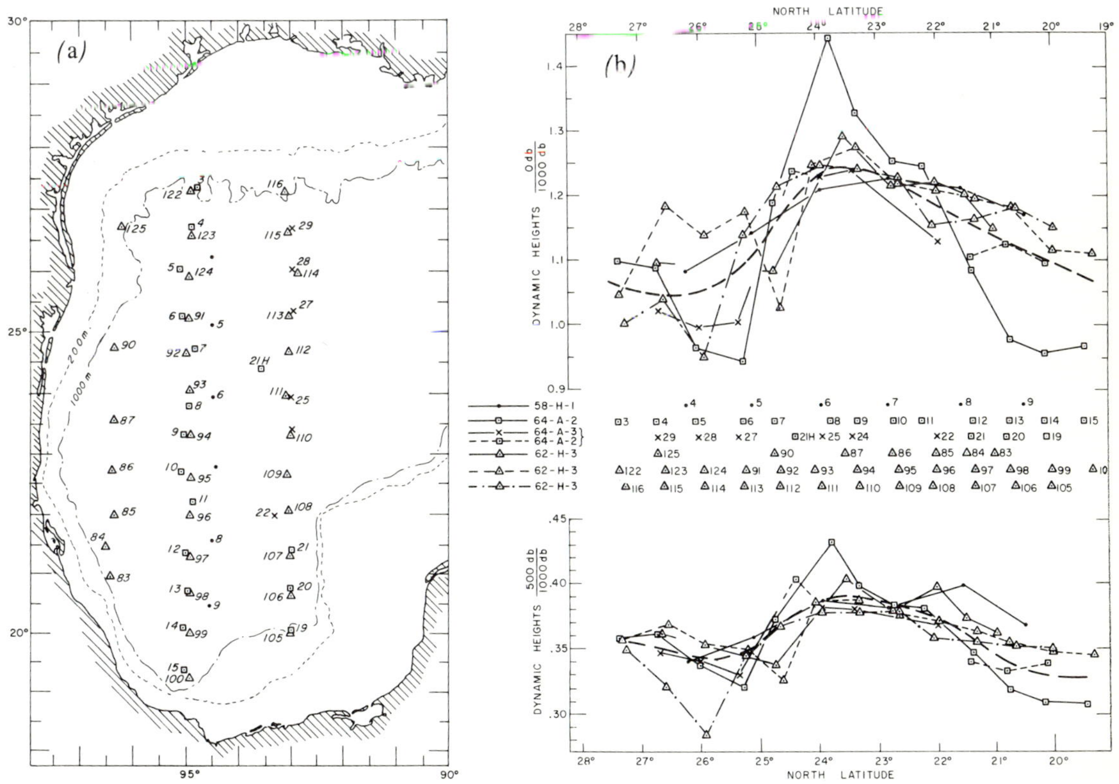

This surface pattern is generally seen also in the relative pressure field at 500 db.

Referring again to Figures 1-30 and 1-31, the results of *Mabel Taylor* and *Atlantis* 1935 are consonant with these four more recent cruises. In fact, in Figure 1-30 there is even a hint of a westward counterflow between the Texas-Louisiana Shelf and the eastward flow along the northern flank of the ridge.

In attempting to obtain a representative winter picture, as well as to quantify the variation therefrom, smooth curves were fit by eye to the geopotential anomaly data of Figure 1-38*b*. Based on data from the 61 stations shown, the standard errors were 0.067 and 0.018 dynamic meters for the relative dynamic topographies of the 0-db surface and the 500-db surface, respectively.

The available evidence points to the existence of the eastward or northeastward current as a semipermanent feature during the winter season, although its position and width are certainly variable. The westward flow over the outer shelf does not appear to be permanent; it may depend on the occurrence of low-salinity water over the outer Texas-Louisiana Shelf.

Acknowledgments

Much of the material presented here was previously presented in identical form in a series of three papers published in the *Journal of Marine Research* and authored by McLellan and Nowlin (1963), by Nowlin and McLellan (1967) and by Nowlin, Paskausky and McLellan (1969). I wish to express my appreciation to my coauthors, Drs. McLellan and Paskausky, and to the *Journal of Marine Research* for allowing me to use text and figures for these publications so freely.

All of the original work, as well as the preparation of these results, was sponsored by the Office of Naval Research through contracts with the Texas A&M Research Foundation.

My special thanks go to Mrs. Ruby Dee Parker for aid in the reduction and analyses of the data, to Mr. Oscar Chancey for preparation of the final figures and to Mesdames Ina Deel and Gwen Stevens for typing of the manuscript.

References

Austin, G. B., Jr. 1955. Some recent oceanographic surveys of the Gulf of Mexico. *Trans. Amer. Geophys. Un., 36*(5): 885-892.

Carpenter, J. H. 1965. The Chesapeake Bay Institute Technique for the Winkler dissolved oxygen method. *Limnol. Oceanogr., 10*(1):141-143.

Caruthers, J. W. 1969. T-S Characteristics of the Gulf of Mexico at intermediate depths. Unpubl. Rept. of Dept. of Oceangr., Texas A&M University, Ref. 69-7-T.

Cochrane, J. D. 1961. Investigations of the Yucatan Current. In Unpubl. Rept. of Dept. of Oceanogr. and Meteorol., The A.&M. College of Texas, Ref. 61-15F:4-7.

—— 1963. Yucatan Current. In Unpubl. Rept. of Dept. of Oceanogr. & Meteorol., The A.&M. College of Texas. Ref. 63-18A:6-11.

——.1965. The Yucatan Current and Equatorial Currents of the western Atlantic. In Unpubl. Rept. of Dept. of Oceanogr. & Meteorol., Texas A&M University, Ref. 65-17T:6-27.

——.1966. The Yucatan Current. In Unpubl. Rept. of Dept. of Oceanogr., Texas A&M University, Ref. 66-23T:14-25.

——.1967. Upwelling off northeast Yucatan. In Unpubl. Rept. of Dept. of Oceanogr., Texas A&M University, Ref. 67-11T:16-17.

——.1969. The currents and waters of the eastern Gulf of Mexico and western Caribbean. In Unpubl. Rept. of Dept. of Oceanogr., Texas A&M University, Ref. 69-9-T:29-31.

Collier, Albert, Drummond, K. H. and Austin, G. B., Jr. 1958. Gulf of Mexico physical and chemical data from *Alaska* cruises, with a note on some aspects of the physical oceanography of the Gulf of Mexico. *Spec. Sci. Rep. Fish and Wildl. Serv., Fish. 249.*

Cox, R. A. and Smith, N. D. 1959. The specific heat of sea water. *Proc. Roy. Soc., A, (London) 252*:51-62.

Defant, Albert 1961. *Physical Oceanography*, Vol. 1. London: Pergamon Press.

Duxbury, A. C. 1962. Averaged dynamic topographies of the Gulf of Mexico. *Limnol. Oceanogr.*, *7*(3): 428-431.

Franceschini, G. A. 1961. Hydrologic balance of the Gulf of Mexico. Unpubl. doctoral dissertation. The A.&M. College of Texas.

Hela, I., Chew, F. and Wagner, L. 1954. Some results of the Florida Current Survey. Unpubl. Rept., Ref. 54-7, University Miami Mar. Lab.

Helland-Hansen, Bjorn. 1930. Physical Oceanography and Meteorology. *Rep. "Sars" N. Atlant. Deep-Sea Exped., 1.*

Hesselberg, Th., and Sverdrup, H. U. 1915. Die Stabilitätsverhältnisse des Seewassers bei vertikalen Verschiebungen. *Bergens Mus Årb. (1914-1915), 15*:1-16.

Hubertz, Jon M. 1967. A study of the Loop Current in the eastern Gulf of Mexico. Unpubl. Rept. of Dept. of Oceanogr., Texas A&M University, Ref. 67-4T.

Jacobs, C. A. and Nowlin, W. D., Jr. 1968. A numerical treatment of steady, frictional boundary Currents in a homogeneous ocean applied to a two-port basin. Unpubl. Rept. of Dept. of Oceangr., Texas A&M University, Ref. 68-19T.

Leipper, D. F. 1970. A sequence of current patterns in the Gulf of Mexico. *Jour. Geophys. Res.*, 75(3):637-658.

McLellan, H. J. 1960. The waters of the Gulf of Mexico as observed in 1958 and 1959. The A.&M. College of Texas, Dept. of Oceanogr. & Meteorol., Unpubl. Rept., Ref. 60-14T.

____.and Nowlin, W.D., Jr. 1963. Some features of the deep water in the Gulf of Mexico. *J. Mar. Res., 21*(3): 233-245.

Molinari, R. L. 1968. The effects of topography on the Yucatan Current. Unpubl. Rept. of Dept. of Oceanogr., Texas A&M University, Ref. 67-24T.

Nowlin, W. D., Jr. and McLellan, H. J. 1967. A characterization of the Gulf of Mexico waters in winter. *J. Mar. Res., 25*(1): 29-59.

____. Hubertz, J. M. and Reid, R. O. 1968. A detached eddy in the Gulf of Mexico. *J. Mar. Res., 26*(2):185-6.

.Paskausky, D. F. and McLellan, H. J. 1969. Recent dissolved-oxygen measurements in the Gulf of Mexico deep waters. *J. Mar. Res., 27*(1):39-44.

Parr, A. E. 1935. Report on hydrographic observations in the Gulf of Mexico and the adjacent straits made during the Yale oceanographic expedition on the *Mabel Taylor* in 1932. *Bull. Bingham Oceanogr. Coll., 5*(1).

Reid, R. O. 1969. Theoretical studies of ocean dynamics. In Unpubl. Rept. of Dept. of Oceanogr., Texas A&M University, Ref. 69-9T:52-58.

Schmitz, W. J., Jr., and Richardson, W. S. 1968. On the transport of the Florida Current. *Deep-Sea Res., 15*(6):679-93.

Sverdrup, H. U., Johnson, M. W. and Fleming, R. H. 1942. *The oceans; their physics, chemistry and general biology*. Englewood Cliffs, New Jersey: Prentice-Hall.

Wennekens, M. P. 1959. Water mass properties of the Straits of Florida and related waters. *Bull. Mar. Sci. Gulf Carib., 9*(1):1-52.

Wert, R. T. 1968. A frictional model of a two-port unbounded ocean basin. Unpubl. Rept. of Dept. of Oceanogr., Texas A&M University, Ref. 68-4T.

Wüst, Georg. 1936. Schichtung and Zirkulation des Atlantischen Ozeans. Die Stratosphäre. *Wiss. Ergebn. Dtsch. Atlant. Exped. Meteor, 6*(1) with Atlas, Berlin.

____. 1964. *Stratification and circulation in the Antillean-Caribbean basins, Pt. 1.* New York: Columbia Univ. Press.

Wilson, R. J. 1967. Amount and distribution of water masses in February and March 1962 in the Gulf of Mexico. Unpubl. M.S. Thesis, Texas A&M University.

2
Water Masses at Intermediate Depths

J.W. Caruthers

Abstract

Potential temperature-salinity characteristics of the Gulf of Mexico are analyzed in considerable detail. The analysis involves a quantitative comparison of θ-S data throughout the Gulf with a "standard" θ-S relation established in the Yucatan Channel. The standard is represented by a least squares polynomial fit to 33 θ-S data pairs. The comparison involves computing the salinity deviations from the standard for various stations throughout the Gulf. The analysis reveals subtle, but distinct, variations in the intermediate water mass of the Gulf in the winter of 1962 and suggests flow patterns and mixing. Analyses of other θ-S data for the Gulf are also discussed.

Introduction

Since first introduced by Helland-Hansen (1916), the study of temperature-salinity (*T-S*) diagrams has developed into a very useful technique for analyzing and identifying a complex system of water masses. On such a diagram a "water type" is represented by the coordinates of a point (*T,S*), and a "water mass" is the distribution of such points, i.e., the *T-S* curve. The *T-S* characteristics of a complex oceanic system have been used in analyzing currents (e.g., Wüst, 1935) and mixing (e.g., Jacobsen, 1927).

The first attempt to establish *T-S* characteristics for the entire Gulf of Mexico was made by Parr (1935) using data of the winter

1932 *Mabel Taylor* cruise. Parr recognized "the presence of two separate water complexes within the Gulf of Mexico proper"; one was characteristic of the Caribbean (Caribbean Complex), the other of the Gulf itself (Gulf Complex). But these distinctions were confined to the upper layers (200-300 m). Apparently, in the intermediate layers the data have sufficient scatter to mask any subtle deviations that may have been present.

Recently, a number of papers concerning the *T-S* relation of the Gulf have appeared. Wenneken's (1959) study included most data available at that time. In this work he also presents an extensive bibliography on the hydrography of the Gulf. However, the data still seemed to be insufficiently accurate to describe more subtle changes in the deeper water. Wennekens writes: "The differentiation between edge water [Gulf Complex] and the Yucatan water [Caribbean Complex] is found mainly in the upper 300 meters. Below about 300 m, both *T-S* characteristics merge within a single narrow envelope."

In an unpublished technical report, McLellan (1960) analyzed the Gulf's *T-S* characteristics. In this analysis he established a reference *T-S* curve by visually averaging data for the western Gulf taken by R/V *Hidalgo* Cruise 58-H-1, and he compared other data to it. There was considerable scatter in the data that established the reference (width of envelope about 0.04 per mil), and there were no definitive variations observed in other data for intermediate depths. But, he did point out the possibility of some variability in the Subantarctic Intermediate minimum. The present analysis shows his reference curve to be characteristic of the western Gulf, but more recent data for that region can be confined to an envelope about 0.02 per mil wide.

The first complete coverage of the Gulf in a short period of time was provided by a hydrographic survey of the *Hidalgo* during the winter (February-March) of 1962. This cruise (62-H-3) provided what was probably the most accurate salinity data for the Gulf up to that time. (Salinities were determined using a conductivity bridge salinometer and have a reported error of ±0.005 per mil.) McLellan and Nowlin analyzed these data in two papers (1963, 1967). The first of these papers was concerned primarily with deep water (below 1500 m) and the effect of the sill in the Yucatan Channel. All the *T-S* data below 1500 m fell inside an envelope about 0.008 per mil wide, indicating very little water mass variation over the deep Gulf and providing some feeling for the accuracy of the data.

The second paper was concerned with the general features of the intermediate and shallow water *T-S* characteristics (among other things). They observed that "for the realm below 17°C the plot [see Figure 2-1 of this chapter] shows a remarkable uniformity. . .if the data are examined on a regional basis the interagreement is even more remarkable."

This chapter reports the results of the first detailed analysis of the *T-S* relation for intermediate waters of the Gulf. Instead of *in situ* temperature (T), potential temperature (θ) as obtained from tables given by Helland-Hansen (1930) will be used. A possible weakness of prior *T-S* investigations is the use of *in situ* temperature rather than potential temperature. It is commonly thought that this is not an important consideration, but a simple example for the Gulf (see Appendix) shows that this is not entirely correct.

The analysis employed for this paper involves:

1. Choosing a suitable "standard" θ-S relation for the Gulf.
2. Describing the standard by a polynomial least square fit.
3. Describing θ-S variations over the Gulf with appropriate number fields of averaged salinity anomaly.
4. Quantitatively comparing various station groups in the Gulf with the standard and plotting the potential temperature-salinity anomaly (θ-δS) curve.

The standard was established and most of the analyses were made using the 62-H-3 data, but data for other winters were also compared. Details of an analysis of individual stations are contained in an unpublished technical report (Caruthers, 1969).

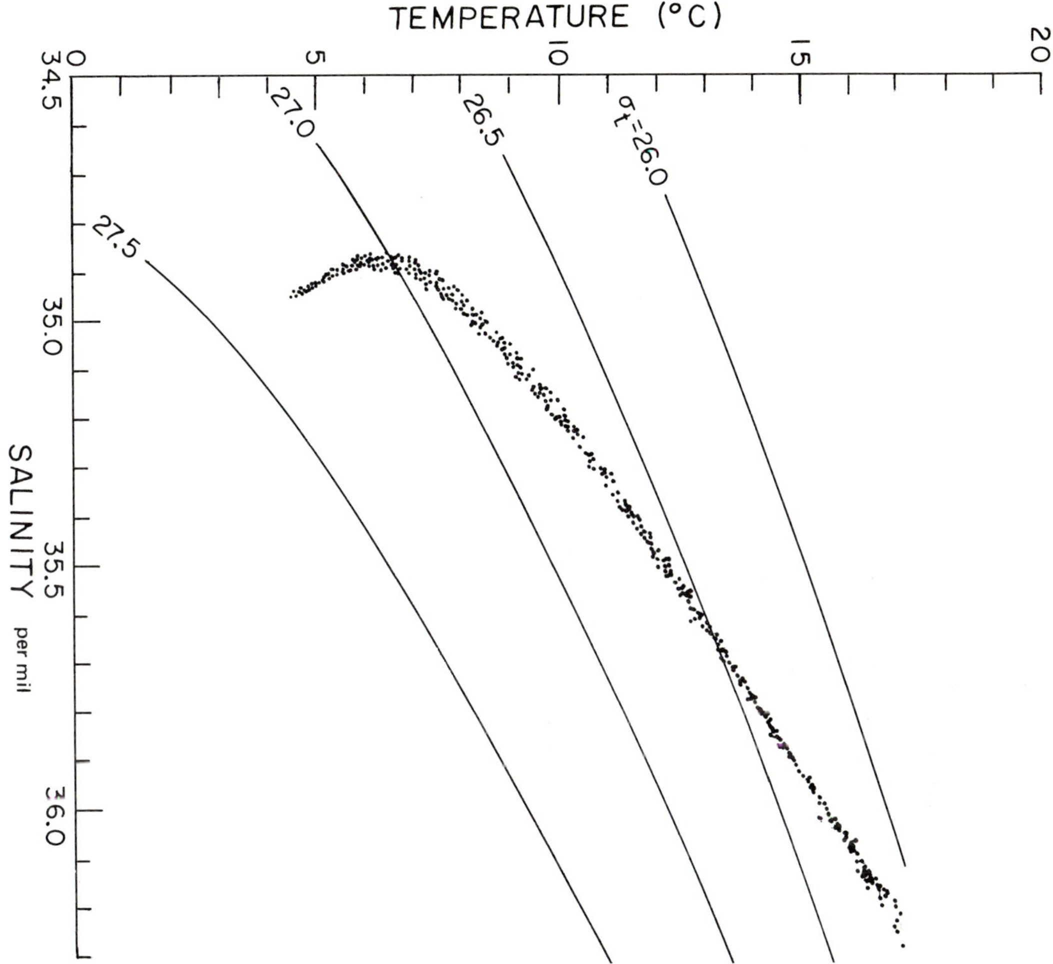

Figure 2-1. T-S *characteristics of the Gulf of Mexico at intermediate depths. (Data from* Hidalgo *Cruise 62-H-3 and graph adapted from Nowlin and McLellan, 1967.)*

Procedure and Analysis

Choosing the Standards

Six stations (subgroup *S* in Figure 2-2) in the Yucatan Current and just inside the Gulf of Mexico were chosen to describe the standard θ-S relation for the Gulf. Although the choice of a standard is somewhat arbitrary, one can attempt to be a little judicious. All deep and intermediate water in the Gulf entered through the Yucatan Channel, and, since the purpose of this analysis is to study the spreading and mixing of the water masses in the Gulf, it is the obvious choice for the standard.

Polynomial Descripton of the Standards

The six stations that make up the standard contained 33 temperature-salinity data pairs in the range of interest—temperatures between 4-20°C and depths between 100-1200 m. These data were used to provide a polynomial least squares fit with potential temperature as the independent variable and salinity as the dependent.

According to the F-test the best fit occurs with the polynomial

$$S = 32.58314 + 23.11518\,(\theta/10) - 82.50041\,(\theta/10)^2 + 143.1953\,(\theta/10)^3 - 135.2648\,(\theta/10)^4 + 71.99103\,(\theta/10)^5 - 20.27993\,(\theta/10)^6 + 2.34921\,(\theta/10)^7,$$

with a standard error of 0.0045 per mil. This polynomial has the general form of Figure 2-1, and the deviations about the polynomial are plotted in Figure 2-3.

Averaged Salinity Anomalies

Individual potential temperature-salinity anomaly points have enough variations due to experimental error to mask the more subtle spatial variations. To reduce the influence of the experimental error the data were, in effect vertically averaged at each station. This was accomplished by fitting a polynomial to the θ-δS points (a smoothing operation) and averaging the polynomial over potential temperature intervals. Although this procedure reduces vertical resolution, it provides better horizontal resolution.

Figure 2-2. Hidalgo *Cruise 62-H-3 stations as grouped for this analysis.* S *represents the subgroup used to establish the standard.*

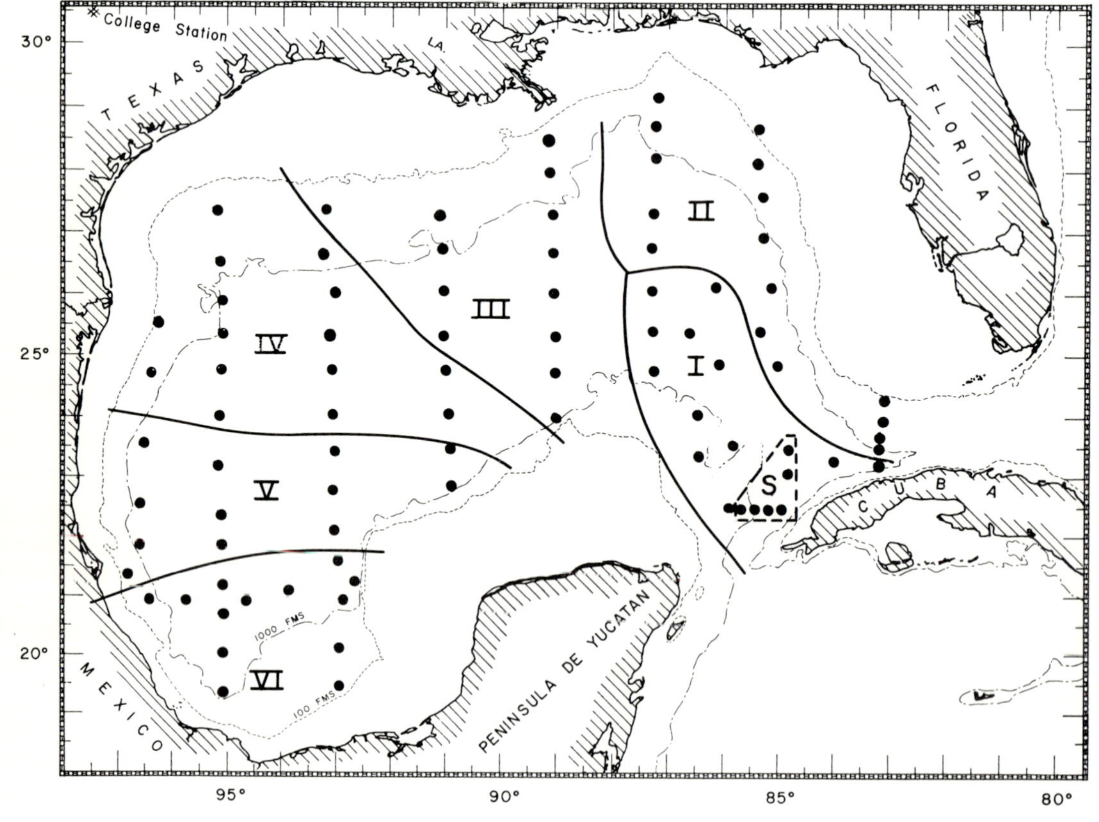

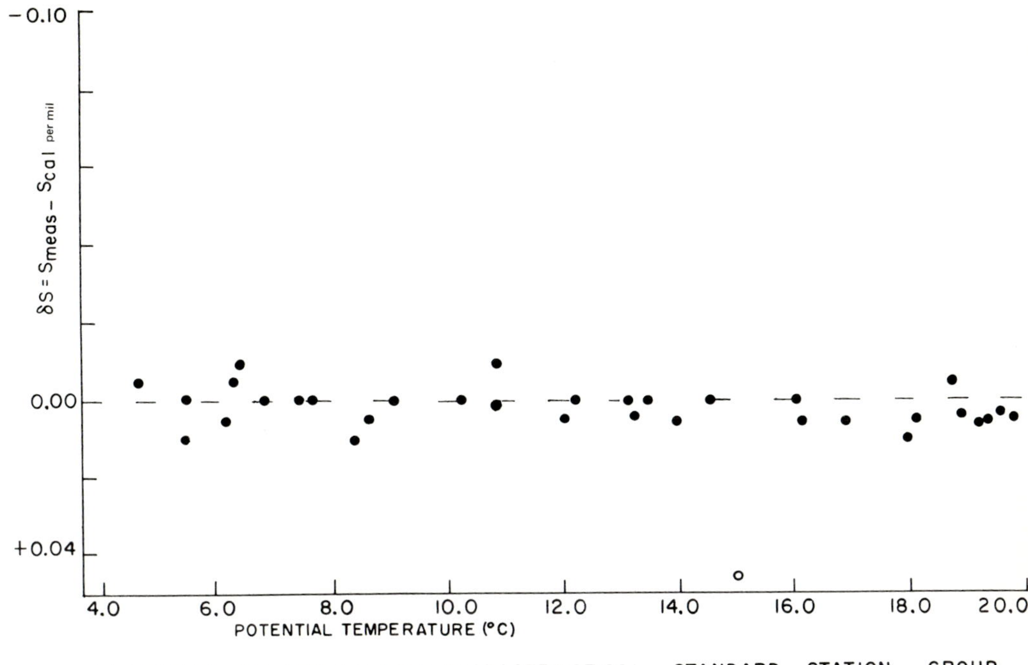

Figure 2-3. Plot of residues from the best polynomial least square fit for standard θ-S relation. S_{cal} *are polynomial values. (The vertical scale is inverted to maintain relative positions when compared to conventional* T-S *diagrams such as Figure 2-1.)*

Figures 2-4 and 2-5 are number fields of the salinity anomalies averaged between 6-12°C and 12-18°C, respectively. The success in contouring the averaged salinity anomalies with intervals of 0.005 per mil testifies to the validity of the procedure. To reiterate this point, neither the actual salinities (or salinity anomalies) at a specific temperature nor the smoothed values taken from the polynomial at a specific temperature could be successfully contoured with this resolution.

These figures indicate a probable deep and intermediate mean flow into the western and southwestern Gulf. In particular, Figure 2-4 suggests a general clockwise flow for depths below about 200-300 m, and Figure 2-5 suggests, possibly, a westward flow for water between 100-300 m.

Regions of water mass formation are shown in Figure 2-5. The possibility of regions of water mass formation along the northern shelf has been considered in the past. In addition to these regions, Figure 2-5 also shows an interesting possibility over the southern part of the western shelf. Additional study is needed in this respect.

Comparison of Station Groups to the Standard Group

To summarize the results of the θ-δS curve analysis for individual stations, the stations were combined into the groups shown in Figure 2-2. The grouping is not fundamentally significant except that it does follow what seems to be the general flow pattern. Some justification of this grouping is found in the results (see Figure 2-7).

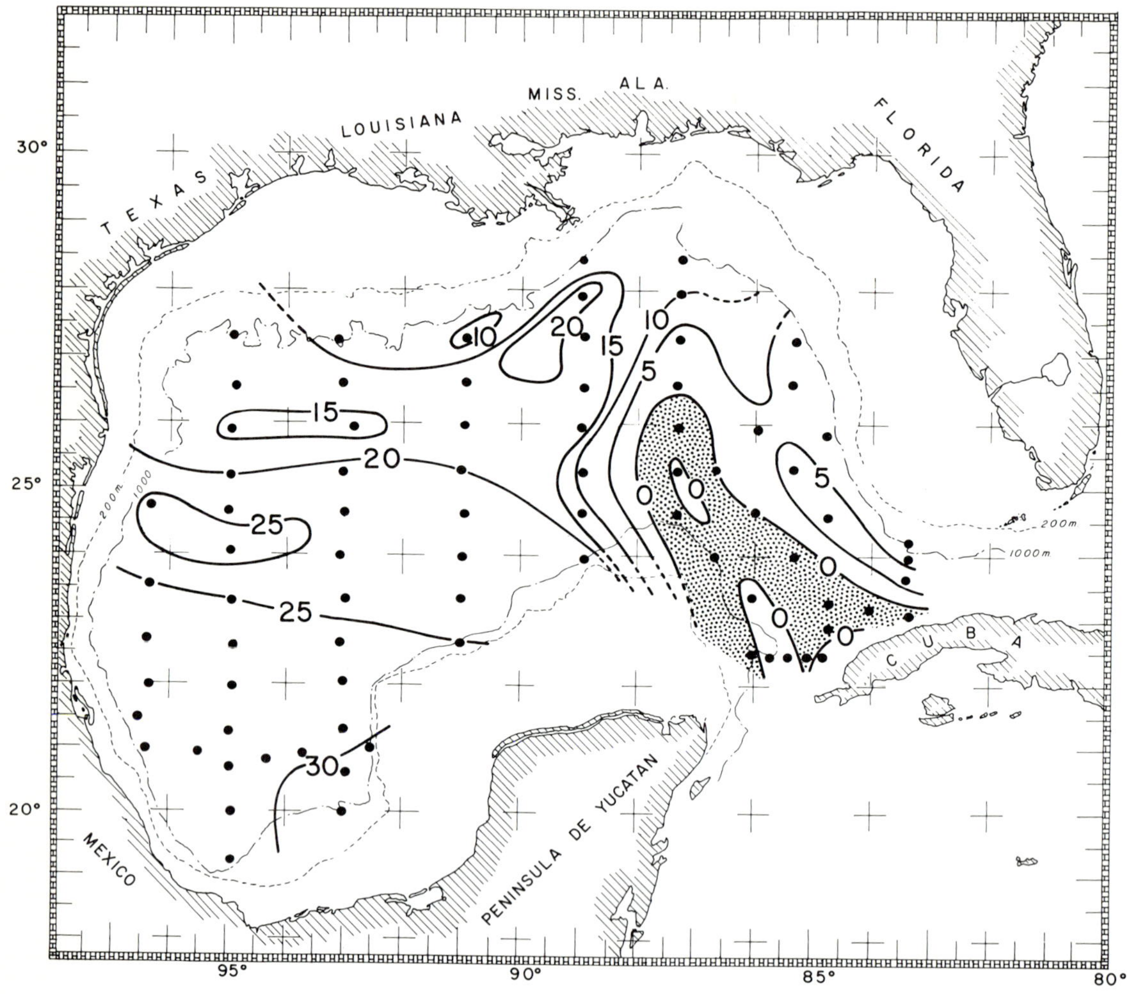

Figure 2-4. Contour plot of salinity anomalies, averaged for each station in the interval 6-12°C.

Graphs of θ-δS points for station groups (Figure 2-6) illustrate deviations from the standard in the data. Polynomials were fitted to these data for further analysis (solid lines on graphs); the degree of these polynomials was rather arbitrarily set at the eighth. Although the data warranted a higher degree, this choice was made because the standard (having fewer data) was of degree seven only. It is obvious that the eighth degree provides a poor fit for station group II.

Figure 2-7—a combined graph of the six polynomials—shows the tendency for salinity of the minimum, which is characteristic of Subantarctic Intermediate Water, to increase uniformly and the tendency for salinity to decrease as the water flows into the northern and western Gulf for temperatures above 14.5°C (water in the upper layers). Although these are not new observations, they are set forth in a more quantitative form than before and may be analyzed and com-

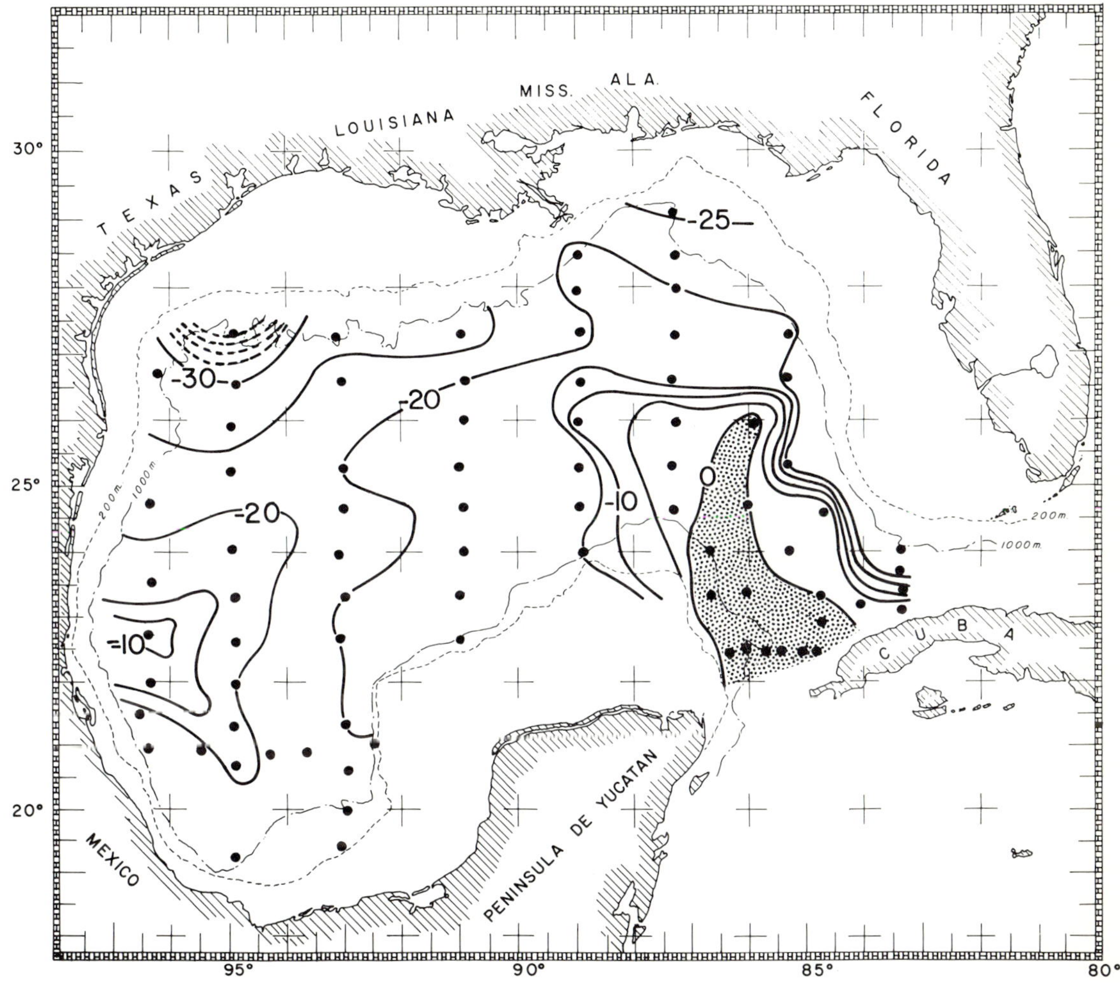

Figure 2-5. Contour plot of salinity anomalies, averaged for each station in the interval 12-18°C.

pared in greater detail. As McLellan and Nowlin (1963) pointed out, there are no significant deviations in water type below 5°C.

Certain specific features not previously mentioned in the literature are worth noting. The maximum salinity change in the Subantarctic Intermediate Water minimum region occurs around 9°C rather than at the minimum, and the water type found at 14.5°C is very constant and has a salinity and sigma t of about 35.87 per mil and 26.76, respectively. This constant water type is found at depths between 150-450 m with the greater depths associated with water of the Caribbean complex in the eastern Gulf.

Analyses of Other Data

A study of the temporal variations in the water masses of the Gulf was also made. The data analyzed in this study suggest that in some cases

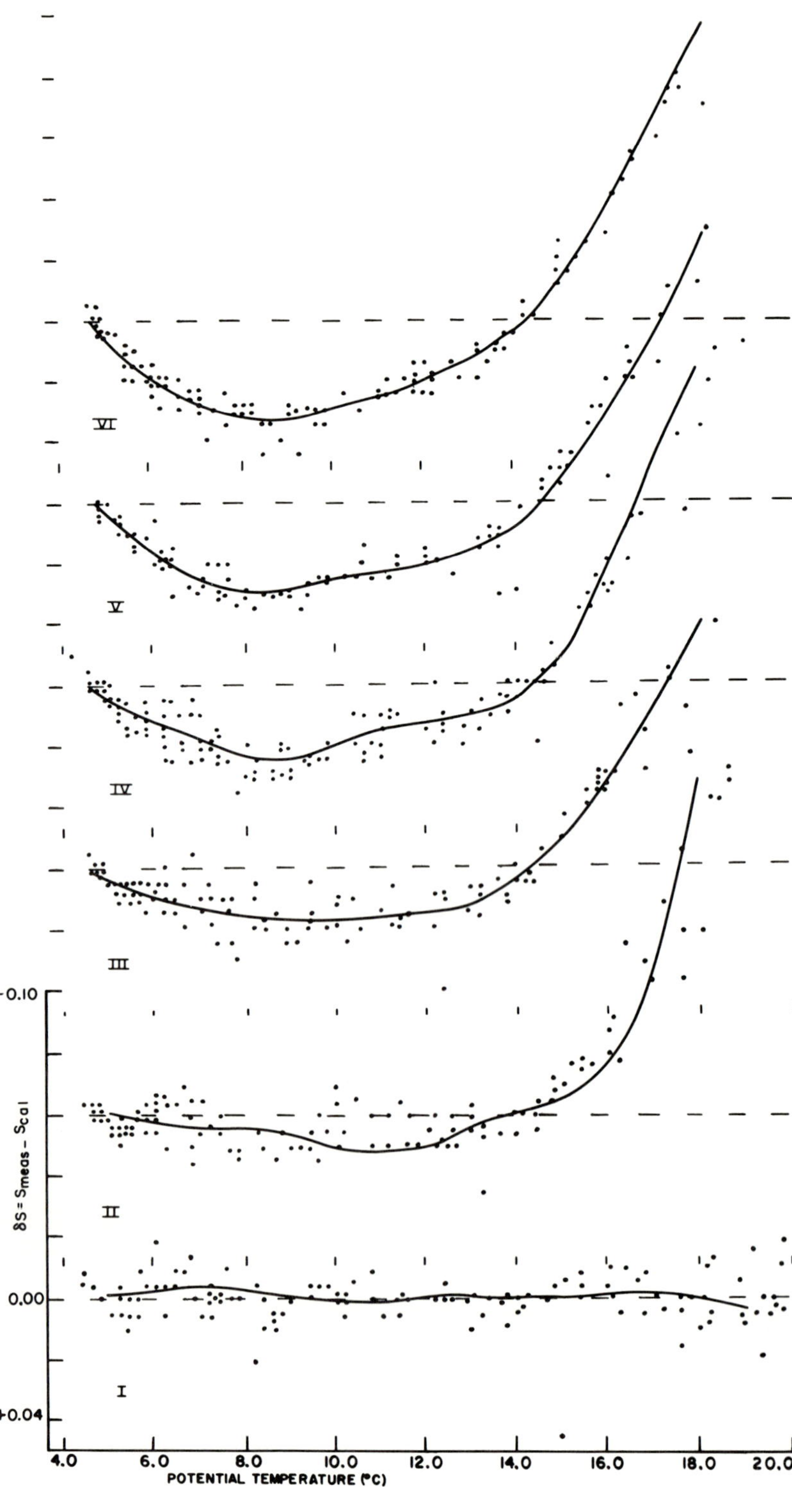

Figure 2-6. θ-δS points and fitted eighth degree polynomials for station groups as labeled.

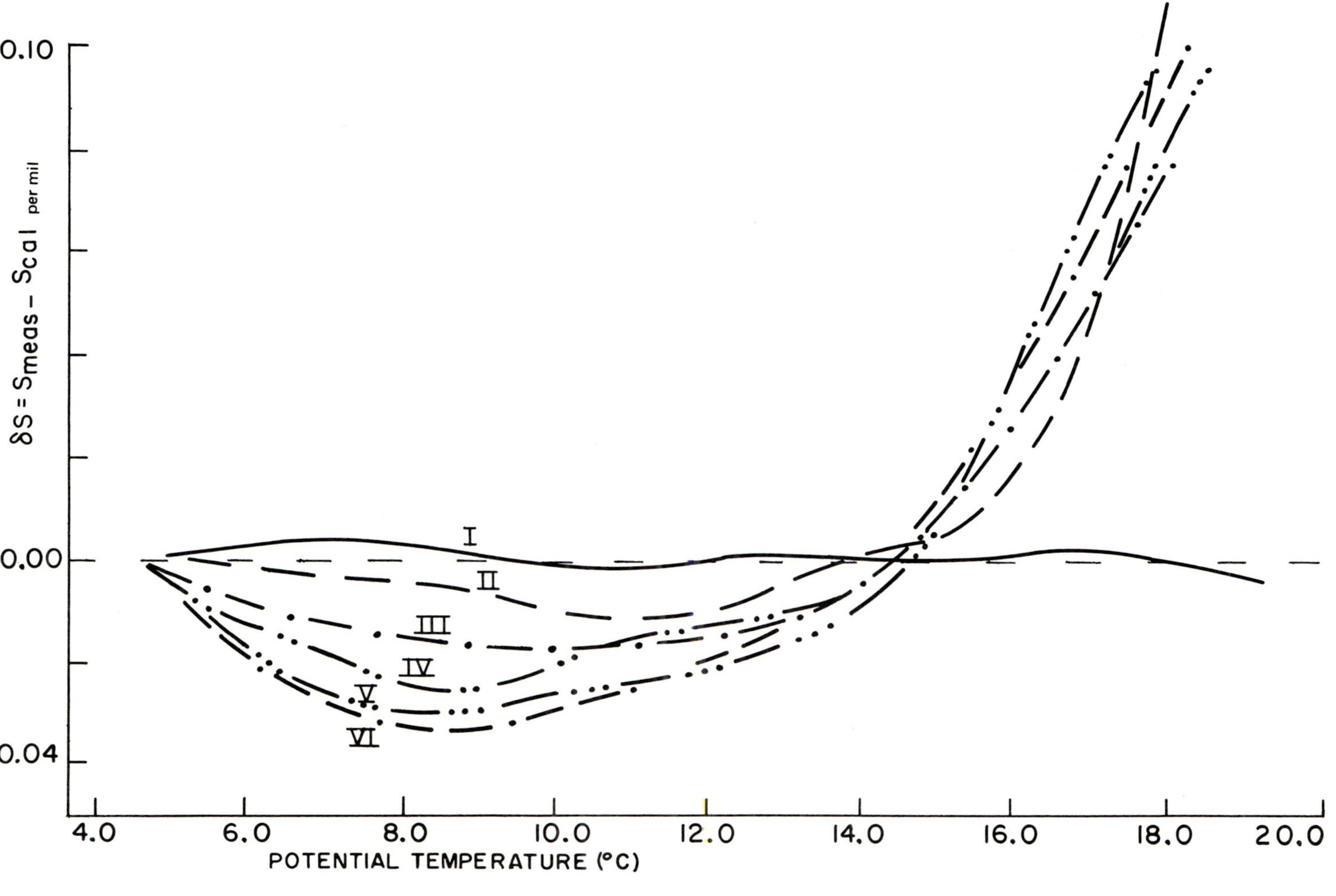

Figure 2-7. Comparison of θ-δS polynomials between various groups. Note the slow evolution through consecutive groups for temperatures below 14.5°C but the abrupt change between group I and all the others for temperatures above 14.5°C.

the intermediate water masses present during the winter are remarkably similar to those of the winter of 1962 (see Figure 2-8, *Alaminos* Cruise 68-A-2), while in other cases either the water masses were significantly different or there were systematic errors, e.g., miscalibration and drift in the salinometer. An analysis of winter data for the western Caribbean—along the meridian at 84.5°W—indicates a major alteration (compared to the 62-H-3 standard) in the Subantarctic Intermediate core (see Figure 2-8, *Crawford* Cruise 17). It is believed that the alteration is due mainly to

Figure 2-8. θ-δS points for Alaminos *Cruise 68-A-2 (above) and* Crawford *Cruise 17 (below). Eleven stations in region of Group IV of 62-H-3 (polynomial for that group plotted as dashed line for comparison) analyzed for the* Alaminos *cruise and nine stations in the western Caribbean for the* Crawford *cruise.*

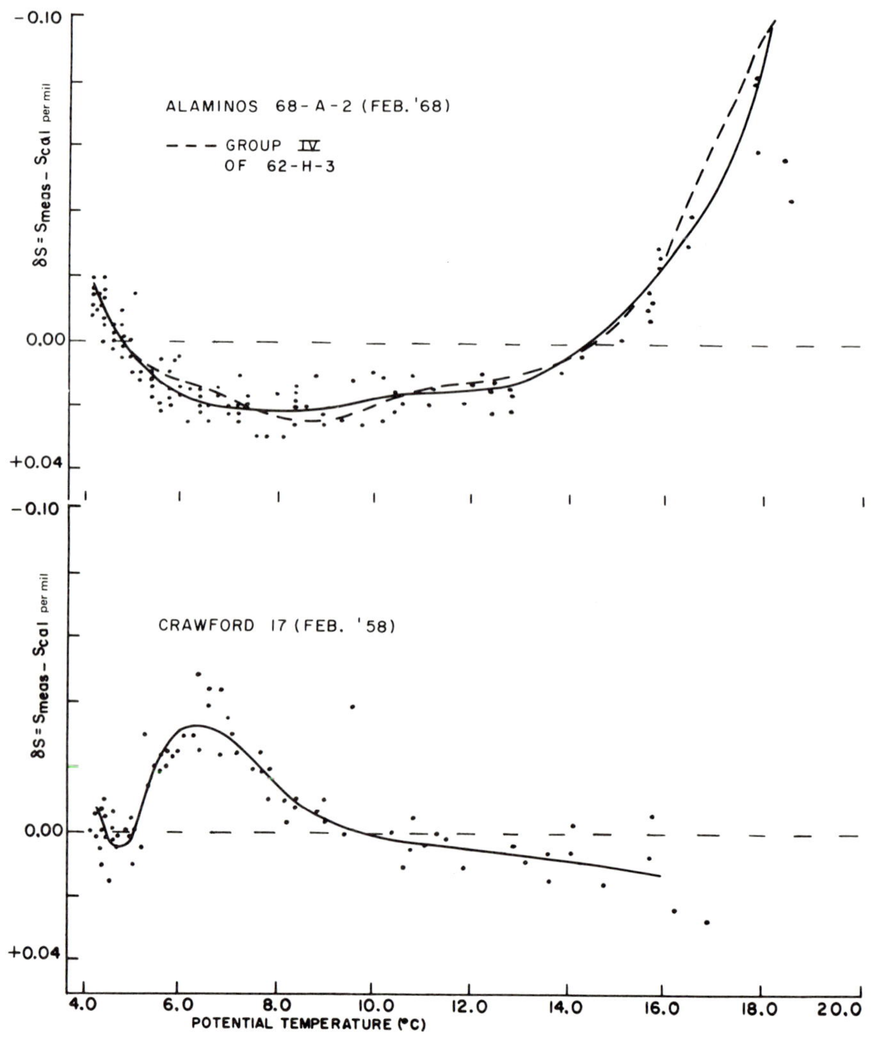

mixing in the Yucatan Channel and not to temporal variations, but the latter cannot be discounted as a possibility.

In an attempt to compare modern with classical data and to compare this θ-S treatment with prior treatments, the data from the *Mabel Taylor* cruise (February, 1932) were analyzed. As Figure 2-9 indicates, the results are, on the average, in excellent agreement with what may be expected on a regional basis, but the scatter is somewhat larger than for more recent data.

Figure 2-9. θ-δS points for Mabel Taylor *cruise in the Gulf of Mexico. Above are for stations 913-916 and 1001-1006 in the southwestern Gulf; below are for stations 601-605 in the Yucatan Channel.*

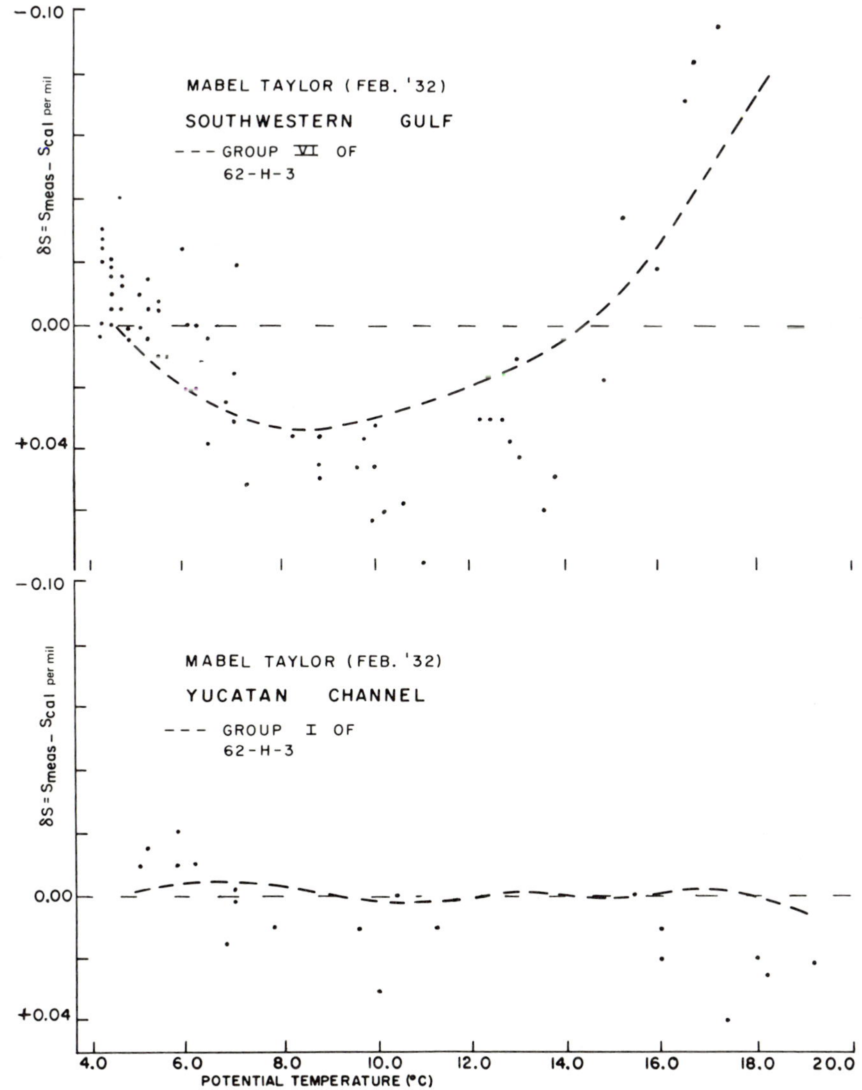

Conclusion

It is believed that standard *T-S* diagrams and interpretations made from them are not utilizing the *T-S* data as accurately as is actually possible. When a scheme employing a greatly expanded salinity scale is followed, and if potential temperature is used, very subtle features of the diagram become plainly visible. Such a scheme also provides a means by which questionable data can readily be recognized.

θ-S diagrams of station data indicate systematic trends in the *θ-S* curves of the Gulf. It is concluded from the present study of recent data that differences do exist in the *θ-S* characteristics of the Gulf below 300 m and that these differences are, in fact, characteristic of motion and mixing in the intermediate layers. Though not as distinct, these regional differences are also present in older data.

Appendix

The following example illustrates the fact that greater regional variations in the *T-S* characteristics may arise from the use of *in situ* temperature than from experimental error.

Suppose that a *T-S* relation for the region near the Yucatan Channel were established from a very large set of data. Suppose further, as is quite nearly true, that the *T-S* characteristics near 14.5°C are defined by a linear relation between *T* and *S* with a slope of 0.17 per mil per °C and a water type that is (14.5°C, 35.870 per mil). Such a water type is found in the Yucatan Channel at about 450 m.

Now, as this water type moves into the western Gulf it ascends to 150 m and the temperature changes adiabatically to 14.45°C. Since its salinity is unchanged the water type is found to deviate from the standard curve by 0.009 per mil (=0.05 x 0.17).

Therefore, if a standard *T-S* relation were established in the Yucatan Channel an attempt to compare the *T-S* characteristics of the western Gulf to it may result in an apparent error as large as 0.009 per mil, while the experimental error is ±0.005 per mil.

Note that for regions of the *T-S* curve where the slope is small or where a given water type is not found with a large depth variation, *in situ* temperature would serve as well as potential temperature, but in general it will not.

References

Caruthers, J.W. 1969 *T-S characteristics of the Gulf of Mexico at the intermediate depths.* Technical Report No. 69-7T. Dept. of Oceanog., Texas A&M University.

Helland-Hansen, Bjorn. 1916. Nogen hydrografiske metoder. *Scand. Naturforsker Möte. Kristania*

________. *1930. Physical oceanography and meteorology.* Rep. "Sars" N. Atlant. Deep-Sea Exped., I.

Jacobsen, J.P. 1927. Eine graphische methode zur bestimmung des vermischungskoeffizienten im Meer. *Gerl. Beitr. Geophys.*, 16.

McLellan, H.J. 1960. *The waters of the Gulf of Mexico as observed in 1958 and 1959.* Technical Report No. 60-14-T. Dept. of Oceanog., Texas A&M University.

McLellan, H.J. and Nowlin, W.D. 1963, Some features of deep water in the Gulf of Mexico. *J. Mar. Res.*, 21:233-245

Nowlin, W.D. and McLellan, H.J. 1967. A characterization of the Gulf of Mexico waters in winter. *J. Mar. Res.*, 25:29-59.

Parr, A.E. 1935. Report on hydrographic observations in the Gulf of Mexico and adjacent straits made during the Yale Oceanographic Expedition of *Mabel Taylor* in 1932. *Bull. Bingham Oceanogr. Coll.*, 5(1):1-92.

Wennekens, M.P. 1959. Water mass properties of the Straits of Florida and related water. *Bull. of Mar. Sci. Gulf Carib.*,9:1-52.

Wüst, Georg. 1935. Die strato sphare. *Deut. Atl. Exped. "Meteor" 1925-27, Vol. 6, Parti, Lief. 2 (Berlin).*

3
A Deep Bottom Current on the Mississippi Cone

Willis E. Pequegnat

Abstract

A preliminary study of a swift bottom current discovered from biological evidence in the eastern Gulf of Mexico has been carried out by means of cameras and current meters. Studies thus far have been confined to the Mississippi Cone at depths between 3000-3300 m. Short time series measurements made from an anchored vessel and an independently mounted current meter yield current speeds up to 19 cm/sec. These values are compared with ripple marks, lineations, scour and other manifestations of bottom currents in photographs. Observations were made in close geographical proximity from 1967 to 1969. It is proposed that this current be named the East Gulf Deep Bottom Current.

Introduction

Evidence continues to mount that presently active bottom currents of substantial velocities and sustained continuity are normal attributes of the environment of at least parts of many deep-ocean basins. This concept of the dynamic role of deep-ocean basin waters in the scheme of general oceanic circulation seems to be acceptable today, but even a cursory review of the literature reveals that it was not generally accepted even 15 years ago. Moreover, the flow of water along the deep bottom of a semiclosed basin such as the Gulf of Mexico may defy solution for some time. Nevertheless, the purpose here is to present diverse evidences that such currents do exist over parts of the bottom of the eastern Gulf at depths in excess of 3200 meters.

Supportive data from which this conclusion was drawn were derived from three methods of investigation: a critical study of types of animals obtained by dredging, analysis of selected bottom photographs, and employment of current meters with *in situ* and acoustically monitored recording systems. Data from the latter were taken from short time series, primarily because of urgent demands on ship time for other more pressing purposes. The author concurs with Webster (1969) that short time series cannot be expected to characterize ideally this or any other current on the deep-ocean floor—a fact that arises from the presumed high degree of variability of the speed and direction of these currents. But economic factors alone may preclude carrying out deep-ocean studies under ideal conditions. Accordingly, here we can do no more than present a modicum of proof that the East Gulf Deep Bottom Current exists today, that it has occurred in about the same place for three consecutive years, that it seems to vary on a very short time scale, and allude to the fact that photographic evidence indicates that such currents appear to be absent from other sectors of the Gulf.

Fortunately, in the case of the East Gulf Deep Bottom Current, direct measurements were made in the same general area over a span of several years—each time with positive results. This fact serves to heighten interest in formulating a theoretical model that will explain the interaction of forces causing the observed flow and the effects described here.

The fact that bottom currents do or do not exist in specifiable parts of the Gulf can be useful to, among others, both biological and geological oceanographers. The causes of observed patterns of distribution of both sediments and certain species of bottom life are far from understood, but little stretching of the intellect is required to recognize the importance of bottom currents in both processes. Furthermore, the distribution of land-derived plant materials appears to this author to be the key to the distribution of some deep-sea animals. Here again, although surface currents have the first influence on rafting such vegetation, it is in the last analysis the bottom current, when present, that will determine where it will come to rest.

First Evidence of a Deep Bottom Current

The author's first indirect evidence of a deep bottom current in the Gulf took the form of a rather scanty epifauna living on indurated, iron-rich rock dredged in July, 1965 from the Mississippi Cone at a depth of 3255 meters. The kind and size of these sessile animals suggested that a current might well be available to bring suspended food and to sweep the rock relatively clean of at least certain components of the anticipated sediment cover.

During ensuing years, the author has devoted as much time as possible to studying this current. Reasonable attempts have been made to improve upon study methods; yet considerable room for improvement remains. In time, however, the combined data have yielded a good picture of certain aspects of this current. But, as is often the case when dealing with deep-sea phenomena, as many questions arose as were answered by each probe into the environment. This can come as no surprise when one becomes aware of the recent date of most studies of deep-sea bottom currents.

Brief Historical Review

As late as the 1950s oceanographers often assumed that very weak or no currents at all existed in the near-bottom environment of the deep ocean (Skoda, 1970). In the mid-1950s Wüst (1955, 1957) predicted on theoretical grounds that currents having velocities up to 10 cm/sec existed along the bottom of parts of the South Atlantic. For some time, however, these calculations were viewed with only mild interest or ignored altogether. But near the close of that decade they were viewed with greater interest as photographic evidences began to reveal that bottom flow had occurred at one time or another in the deep sea (Hunkins et al., 1960). Pursuant to these findings, interest in determining the common characteristics of these deep currents has increased, particularly among geological and physical oceanographers (Menard, 1964; Morgenstern, 1967; Reid, 1967, and others). Biologists are no less interested in the existence of bottom currents, especially those density currents referred to as turbidity currents, inasmuch as they modify the physical and chemical natures of the bottom water and seafloor, control the transport of quantities of organic matter of several types from shallow to deep water, and may even influence the dissemination mode of several animal species (Menzies et al., 1967; Rowe and Menzies, 1968).

Until well into the 1960s much of the evidence for the existence of deep-water currents was obtained from neutrally buoyant floats (Swallow, 1955) as well as from photographs. The effects of currents on bottom sediments can be observed in photographs as ripple marks, scour around stones or other fixed objects, current lineations and somewhat less positive evidence as bare rock and coarse residual debris (Hollister and Heezen, 1966). Most pre 1960 current measuring instruments were designed for operations in shallow waters (e.g., the Ekman meter) and thus proved very unsatisfactory for deep-water investigations. In recent years the Savonius rotor (in one configuration or other) has been the most widely used instrument for deep-bottom current measurements (Knauss, 1965; Isaacs et al., 1966; Maloney, 1967; Nowroozi et al., 1968; Reid, 1969, and others). Television has been coupled with various sensors (Sternberg and Creager, 1965) with varying success. And some good measurements of bottom currents have even been made from deep submersible vessels (LaFond, 1962). On balance, however, photographic techniques are still the most widely employed method of detecting and determining the direction of deep bottom currents (Bruce and Thorndike, 1967; Rowe and Menzies, 1968). Unfortunately, still photography will not yield more than approximate maximum velocities of flow, and at best considerable interpretation may be required to determine how recent a current established by photography may actually be.

As yet no studies of bottom currents in the deep aspects of the Gulf have been published. Therefore, this chapter is intended to partially fill a significant gap in our knowledge about an area where the surface currents (the East Gulf Loop Current) have been studied so extensively (see Leipper, 1970). It is my hope, also, that this rather modest pioneer effort will stimulate appropriate investigators to look further into what must be judged a promising area of research.

Instrumentation and Methods

Current Meters

Two different speed and direction sensors were used to measure the bottom current reported on here. The first of these was fabricated under the direction of Dr. Worth D. Nowlin and proved to be the prototype of a commercial model (Hydro Products In Situ Current Speed and Direction Sensor). This instrument was used on a large tripod in the summer of 1967. The other meter is Hydro Products Model 501; it was employed in the field during 1968 and 1969. We found, however, that the instrument, as delivered, was unfit for critical work in the field. As a consequence, it was completely disassembled and brought up to our performance standard prior to being used in the deep-water program. Nevertheless, since the essential current-measuring elements of the prototype and Model 501 are very similar, only the latter will be described here for those who are unfamiliar with their operation.

The speed measuring sensor is a Savonius rotor containing a ring of 10 magnets imbedded in a PVC disk mounted on the rotor's top. A direct-current source activates a magnetic (reed-type) switch that is closed as each magnet nears the switch during a rotor rotation. The sensor's output is a pulse train whose frequency is necessarily proportional to the rotor's rotation rate and thus to the driving current's speed. A pulse-rate counter converts the pulse train output to a direct-current mode that is displayed on a strip-chart recorder calibrated in tenths of knots. Two recorder scales

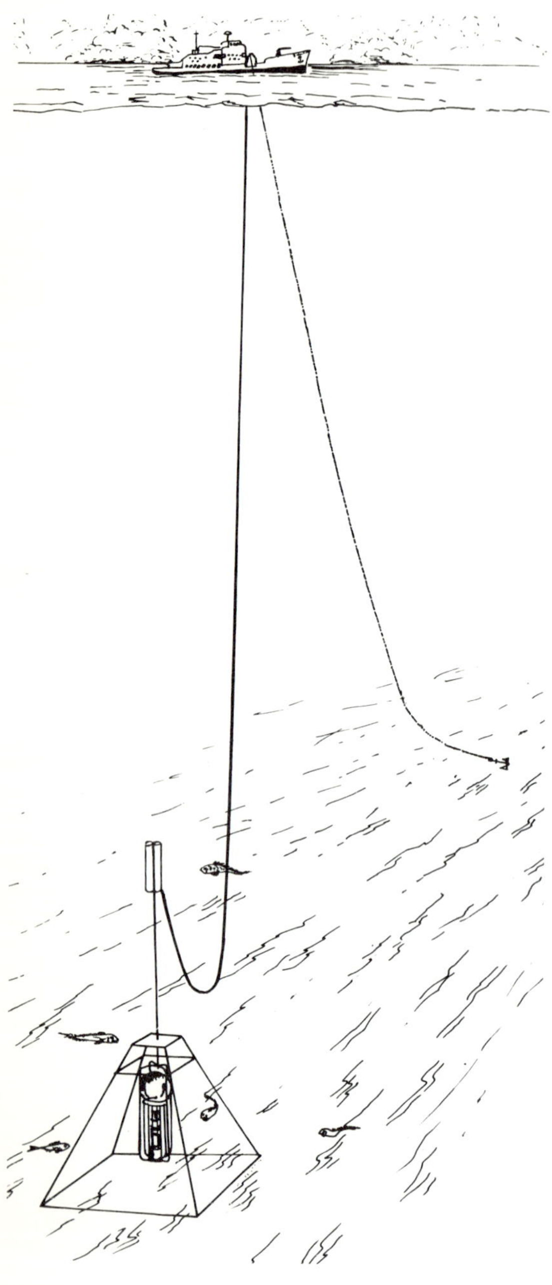

Figure 3-1. Current meter suspended in aluminum tetrapod. Note gasoline-filled floats and slack wire that prevent any movements of anchored ship from disturbing meter. Tetrapod is 2-1/2-m high; the Savonius rotor is 1 m above bottom.

are available for low-speed and swifter currents, but in the present study only the lower scale (0-1 kt) was used.

The meter described here was modified in the Texas A&M laboratory to transmit current speed information from the bottom to the ship's electronics laboratory via acoustic telemetry (Gutierrez and Pequegnat, 1968). To accomplish this, a second magnetic switch was mounted adjacent to the first so that it too was energized by passage of the magnets. Its output was sent to an electronic counter designed to give good pulse separation, dividing the train by 10 or 100 in accordance with estimated current speed. The counter's output was used to trigger an acoustical transmitter (in this case, a bottom pinger modified for the purpose), the signal of which was then picked up by the ship's PDR (precision depth recorder) transducer and relayed either to an oscilloscope or numerical counter. At present direction data is sent by acoustical means also, but the system had not been fabricated in our laboratories in time for the present study.

Data from the two sensors are printed on pressure-sensitive paper rolled by a single-channel Rustrak strip-chart recorder.

Mounting of the Current Meter

To obtain current speed measurements free of movements of the anchored ship, the meter was mounted on an aluminum tetrapod that stands 8 ft. tall and has a leg spread of 8 ft., 4 in. To minimize the influence of tilt on current direction measurements, the meter is hooked to a gimbal suspension that permits it to hang vertically with a tetrapod tilt of 30°. The system is lowered to the bottom by means of 3/16-in. wire rope spooled on a hydrographic winch. When the unit has been lowered 30 m, two tubular plastic floats carrying 10 gals. of gasoline each are clamped to the wire. When the tetrapod touches bottom, these floats keep the 30-m run of wire taut, thereby permitting the winch operator to pay out an additional 30-60 m of wire without danger of fouling the tetrapod or suspended instrument (Figure 3-1).

Prior to lowering the meter, the ship is anchored securely by means of a 1½-ton anchor cast on 7/16-in. wire rope spooled on the dredging winch. The latter is run through a 12-in. sheave hung from the A-frame on the forward quarter of the starboard side. Ordinarily, the ship comes to "rest" with the surface current approaching from the starboard beam and the wind from the stern quarter of the port side. When both surface current (1 kt or more) and wind (10 kt or more) are strong, the ship swings less than 5° after the initial 30 minutes at anchor. But even when yawing occurs it is a simple matter to adjust slack in the tetrapod wire to prevent overturning the meter. Experience has dictated a policy that the tetrapod is not permitted to touch bottom until the captain advises that the ship has ceased drifting, insofar as this is ascertainable by "no change in Loran readings" made at 15-minute intervals during instrument lowering.

The current sensor is attached very securely to the gimballed part of the tetrapod and adjusted so that the Savonius rotor will be held 1 m above the bottom. It was found advisable to tape 12-kilo lead bars to the four basal members of the system when on the bottom to stabilize the tetrapod and to minimize vibratory movements.

Photographic System

Two still cameras were employed for bottom studies. Both are standard models offered on the open market; hence, little need be said about them except where special modifications of the system or of the configuration of use were made in our laboratories.

One camera is a 35 mm multi-exposure model (Alpine Geophysical Associates, Model 314) that is triggered by bottom contact. It is equipped with a Wollensak F 11 water corrected lens with a fixed focus of 10 ft. and with a single 200 watt-second strobe light. The focus was modified to options of 3 or 5 ft. by introducing appropriate brass shims behind the lens. An additional 200 watt-second strobe was added to double the available light and permit use of fine-grain black and white or color film.

The second camera is a 70 mm deep-sea camera (Hydro Products Model PC-700) that is equipped with a single 200 watt-second strobe light. The focus of this camera was ordinarily set at 5 ft., 8 in.

These cameras are used in a horizontally oriented aluminum frame of our design. This construction permits camera and lights to be tilted toward one another with consequent better definition of animal subjects.

Dredging

Most biological samples germane to this study were taken during *Alaminos* cruises in 1967 and subsequent years by means of the Benthic Skimmer, which the author designed for high-speed dredging operations. This device has been described elsewhere (Pequegnat et al., 1970). Suffice it to say, therefore, that it is a rugged, all-metal dredge with a 3-m gape capable of collecting large rock samples without sustaining damage. The epifaunal species living on the rock were preserved in 70% alcohol or buffered 7% formaldehyde solutions and returned to the laboratory for identification, enumeration and photographic depiction. Samples of the ironstone were subjected to an exhaustive chemical and mineralogical study that will be reported on later (Pequegnat, Fredericks and McKee, in preparation).

The Study Site

The observations reported in this study were made in that area of the Gulf lying roughly between the 25th and 27th parallels and the 85th and 87th meridians (Figure 3-2). This places it directly over the Mississippi Cone, which is a deep-sea fan that has its apex at the delta of the Mississippi River. From here the cone extends gulfward in a large-radius arc that sweeps westward from near the Florida Straits to its terminus in the Sigsbee Abyssal Plain (Figure 3-2). Beyond question, the cone is the predominant physiographic feature of the Gulf basin, covering about 3.5×10^5 km^2 (Ewing et al., 1958) and extending through a vertical range of over 3700 m.

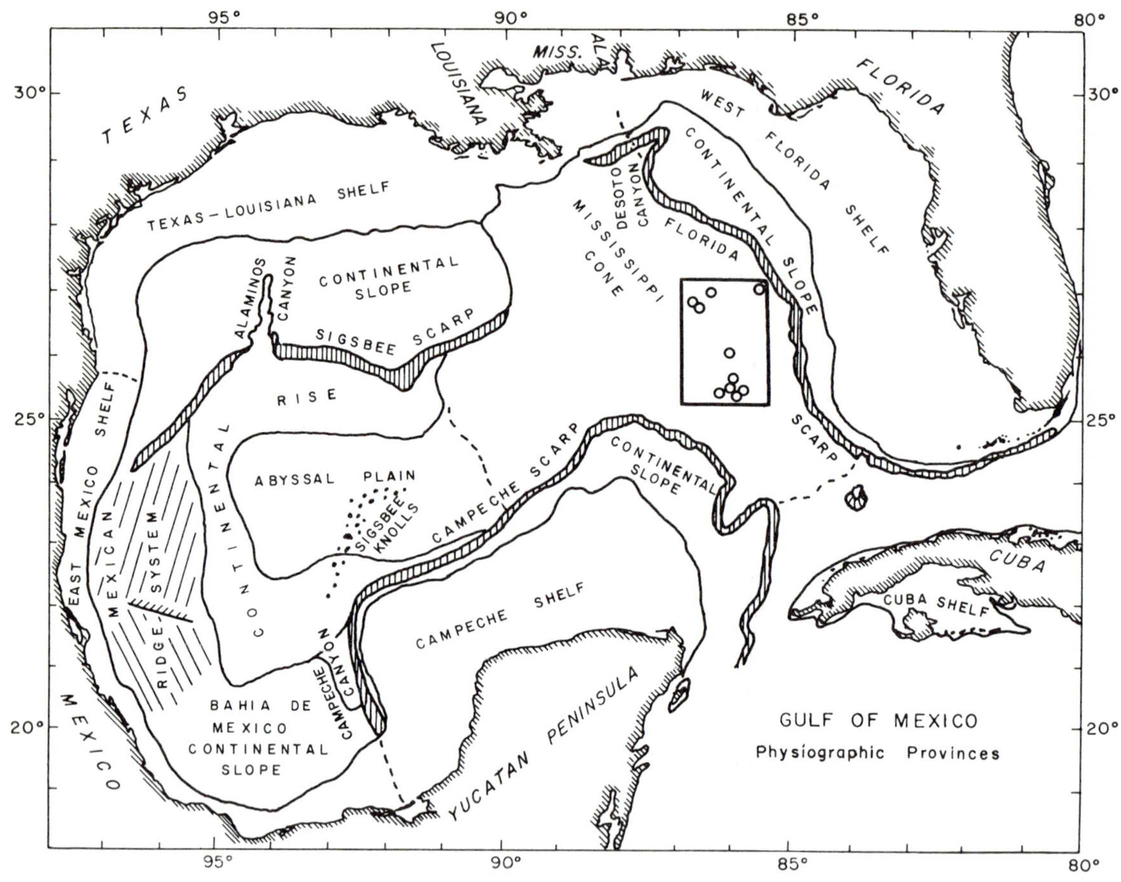

Figure 3-2. General location of the study site (in rectangle) on the Mississippi Cone in the eastern Gulf of Mexico.

At the present time the eastern Gulf receives sediments principally from the Mississippi River, but lesser amounts do enter via other sources (e.g., the Mobile River system). Moody (1967) estimates that each year these two systems deliver about a thousand million tons of land-derived materials into the Gulf. Perhaps as much of this as 75% is detrital. Near the delta the latter is comprised of 28% fine sand, 29% silt and 43% clay (Scruton, 1960; Shepard, 1960; Moody, 1967). Seaward of the delta the surface sediments are reported to be primarily foraminiferal clay (Huang, 1969).

In those areas on the cone where current measurements were made, shallow cores yielded large quantities of well-graded quartz sand, photographs demonstrated the presence of large aggregations of loose shell material, and dredging yielded substantial amounts of ironstone.

The primary shallow current in the Gulf of Mexico is the East Gulf Loop, which enters through the Yucatan Channel and leaves through the Florida Straits. Although this loop penetrates northward into the Gulf for varying distances, depending on season and year (Leipper, 1970), the

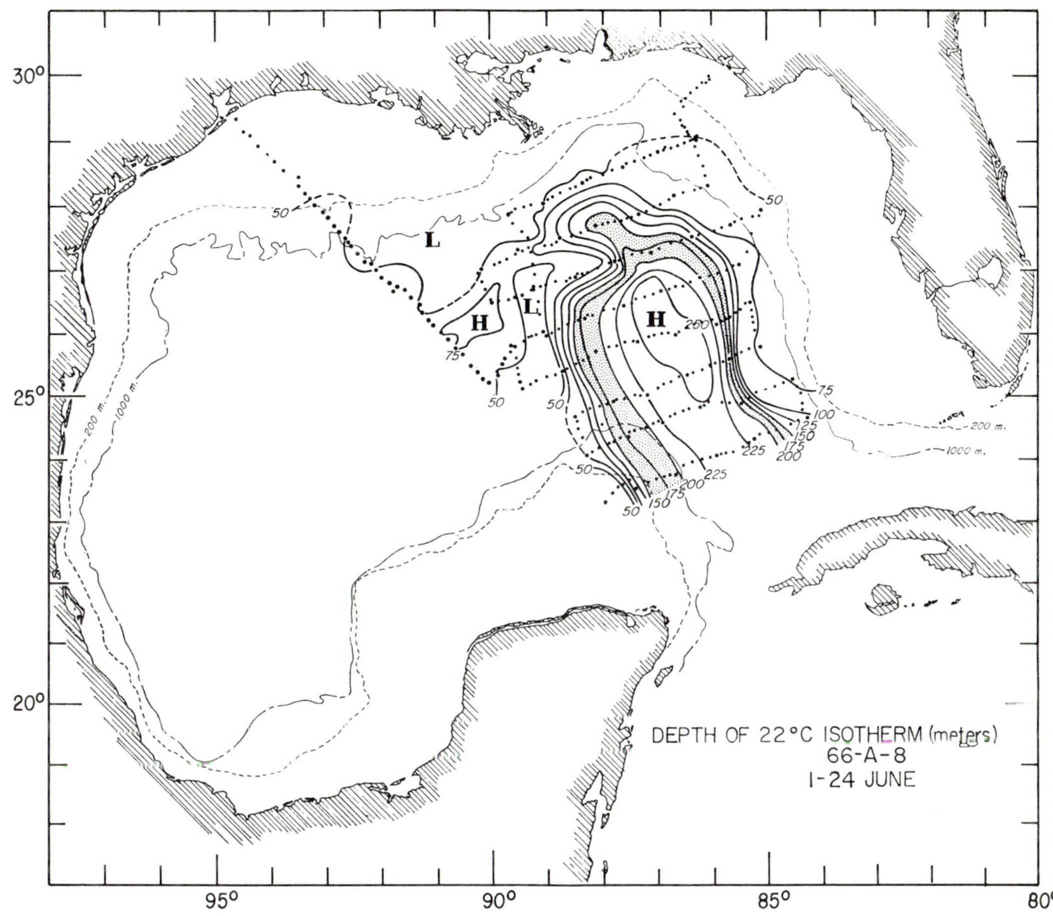

Figure 3-3. Topography of the 22°C isothermal surface around June 24, 1966, defining the position of the East Gulf Loop Current. Comparison with Figure 3-2 shows that the present study site is generally correlated with the Loop's south flowing limb (after Leipper, 1970).

study site lies principally in the eastern arm of the loop where the flow is southward (Figure 3-3). The flow is well represented throughout the year by the topographies of the 22° isothermal surfaces (Table 3-1).

A rather complete hydrographic profile of the water column in the region of the Loop Current and the Deep Bottom Current is given in Table 3-1. In this cast, however, the lowermost Nansen bottle was tripped about 36 m above the bottom. Near-bottom conditions are given in Table 3-2, where values from the three bottommost bottles of each cast are given. The three parameters of temperature, salinity and dissolved oxygen were sampled between 2 and 6 m above the sediments. Some differences among stations are observed for the stations given, but no consistent pattern appears to emerge from the limited number of samples.

Table 3-1
Surface-to-Bottom Hydrographic Profile in Region of Bottom Current
Cruise 66-A-9, St. 9, 25°28′N, 86°17′W
July 10, 1966

Depth (m)	Temperature (°C)	Salinity (per mil)	Dissolve O_2 (ml/1)	Remarks
1	29.21	36.186	4.55	
10	28.81	36.183	4.78	
25	28.50	36.180	4.85	
74	27.27	36.169	4.91	
148	25.06	36.395	4.26	
198	21.74	36.741	3.44	Core of Loop Current
297	17.64	36.388	3.75	
395	16.48	36.192	3.40	
493	14.06	35.802	3.09	
738	7.92	35.001	2.72	0_2 minimum
981	5.80	34.886	3.54	Salinity minimum
1224	4.73	34.943	4.40	
1471	4.29	34.992	4.77	
1719	4.23	34.976	4.94	
1969	4.20	34.982	5.05	Temperature minimum
1997	4.27	34.981	5.08	
2197	4.26	34.977	5.02	
2396	4.24	34.978	5.11	
2596	4.26	34.983	5.08	
2797	4.26	34.980	5.10	
2997	4.28	34.991	5.15	
3097	–	34.986	5.15	
3147	4.26	34.980	5.11	
3172	4.28	34.980	5.18	
3192	4.28	34.984	5.15	
3197	4.29	34.982	5.11	
3233	Bottom depth			

Results

Biological Dredging

As indicated above, first evidence of the eastern Gulf deep current came from the nature of the fauna living on ironstone dredged in July, 1965, from a portion of the Mississippi Cone at 25°31′N and 86°08′W and at a depth of 3200 m. These rock chunks and fragments supported a wide variety of attached epifaunal organisms whose maximum dimensions were little more than 5 mm (Figure 3-4). The presence of these small epifaunal species, some of which were sessile or sedentary, revealed that the rock when dredged had to have been devoid of a sediment cover. This condition must result from passage of a bottom current strong enough to sweep the rock surface

Table 3-2
Hydrographic Data from Four Stations with Measurable Bottom Current
Bottom Three Nansen Bottles of Hydrocast

Alaminos Cruise No. and St.	Date	Position	Bottom Depth (m)	Depth of Nansen Bottles (m)	Temp. (°C)	Salinity (per mil)	Dissolved 0_2 (ml/1)
66-A-5-3	31 March	25°25′,86°13′		3179	4.28	34.979	5.090
				3194	4.30	34.975	5.110
			3210	3204	4.28	34.974	5.090
68-A-7-4C	29 July	25°25.3′,86°05.3′		3240	4.30	34.985	5.331
				3243	4.29	34.969	5.433
			3248	3246	4.32	35.143	5.388
69-A-13-31	12 October	25°26′.86°09′		3171	4.33	34.837	5.065
				3215	4.32	34.879	4.880
			3226	3221	4.33	34.876	4.990
69-A-13-34	13 October	26°50.5′,86°40′		3023	4.32	34.872	5.030
				3073	4.30	34.890	4.969
			3079	3074	4.32	34.780	5.024

clean of the lighter components of the sediment suite of this region.

Analysis of this dredging revealed a mixture of animals living on the rock (epifauna) and those living in the adjacent soft bottom (infauna). This indicated that the ironstone did not form a uniform cover along the 2000-m path taken by the dredge. Laboratory analysis of the catch indicated that the principal infaunal species were distributed as shown in Table 3-3.

The epifaunal species on the rock are closely correlated with water movement. Thus, considerable attention is focused on them (Table 3-4).

It is likely that most of the organisms in Table 3-4 are suspension-feeders, i.e., types that are capable of and do remove particulate organic matter (living and/or nonliving) directly from the water. Their shallow water relatives are for the most part abundant only where there is considerable water movement. We have no evidence to preclude application of this proclivity in deep water as well.

Interestingly enough some of these species shared another attribute with their shallow water counterparts. In particular the sagartid anemone, sponges, and maldanid and spionid polychaetes covered their integuments with a dense incrustation of quartz grains. The grains ranged in diameter from 0.04 to 0.79 mm. Each kind of animal is apparently somewhat selective as to grain size. It seems that in this region water movement next to the bottom is sufficient to bring sand grains across the rock within easy reach of some of these sedentary and/or sessile species of animals.

Photographic Evidence of Bottom Currents

Bruce and Thorndike (1967) describe several ingenious methods for measuring both the speed and direction of bottom currents by means of still photography. So far as the author is aware, however, their methods for measuring speed by photographing the deflection of suspended balls or the movement of buoyant drops of liquids immiscible

A

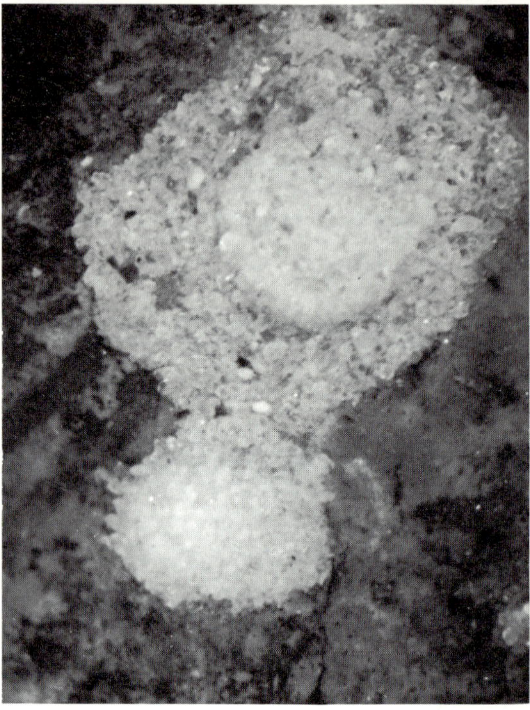

B

Table 3-3
Distribution of Infaunal Species

	Mollusca
Protobranch Pelecypods:	*Nucula verrilli* (Dall)
	Nucula obliterata (Dall)
	Neilonella sp. (new)
	Pristigloma nitens (Jeffreys)
	Malletia aeolata (Dall)
Scaphopods	*Dentalium ensiculus* (Jeffreys)
	Dentalium meridionale verrilli (Henderson)
	Annelida
Polychaeta	
Onuphidae	*Nothria* sp.
	Paradiopatra sp.
Oweniidae	*Myriochele* sp.
Ampharetidae	*Amphicteis* sp.
	Nicomache sp.
	Echinodermata
Echinoidea (sea-urchin)	*Aceste bellidifera* (Wyville Thomson)

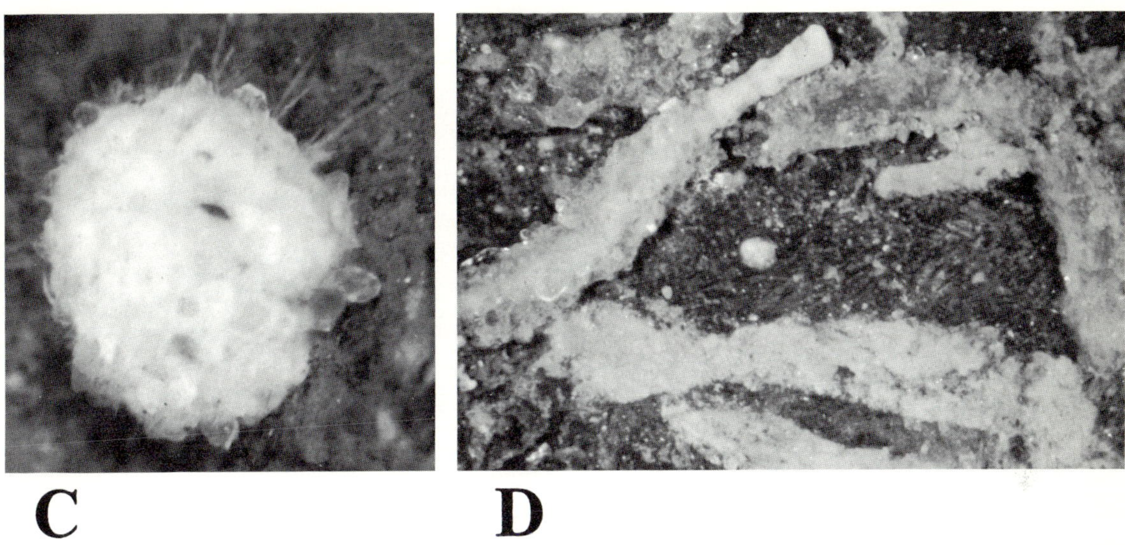

Figure 3-4. (A&B on page 72) Epifaunal animals on ironstone rock; A. *Ironstone with small contracted sponges and anemones attached (3x);* B. *Enlargement of anemones from upper portion of* A *showing incrustation of quartz sand grains and undergoing reproduction in the form of pedal laceration (10x);* C. *Sponge from upper portion of* A *incrusted with sand grains (20x);* D. *Tubiculous polychaete worm tubes incrusted with sand grains. Note worm protruding from tube near upper margin (15x).*

Table 3-4

Living Epifaunal Species on the Ironstone

Species	Measurements (mm)	Remarks
Sponges (two species)	4, diam; 2, ht.	Common, with quartz grains
Sagartiogeton sp. (sea anemone)	4, diam; contracted	Common, covered with quartz grains
Stephanoscyphus simplex Kirkpatrick (scyphozoan polyps)	1, diam; 4-5, ht.	Part of medusa life cycle
Levinsenella brasiliensis (Busk) (ectoproct)	35, ht.	Colonies branched
Pelagodiscus atlanticus (King) (brachipod)	4, 5, diam; 1.5, ht.	Attached by broad pedicle
Bentharca asperula (Dall) (pelecypod mollusk)	4.5, max. length	Attached by strong byssus
Spionid polychaete (annelid worm)	0.9, diam; 12, long	Tube covered with quartz grains
Maldanid polychaete	1.0, diam; 10, long	Tube covered with quartz grains
Hyalopomatus langerhansi Ehlers (serpulid polychaete)	0.5, diam; 15, long	Calcareous tube cemented to rock
Verruca hoeki Pilsbry (barnacle)	3, ht.	Attached to ectoproct

with water have not been employed very widely. On the other hand, the practice of placing a liquid-filled compass in the photographic field is in common use today.

In the present study a nonphotographic current meter was employed in place of the above methods. But bottom photographs were taken at or near both dredging and current meter stations. In this instance we were searching for various types of physical manifestations of past or present currents in the sediments. Furthermore, we hoped that it would be possible to obtain the speed of the current recorded by the current meter. So far, this has been only partially successful.

The criteria used here to detect the presence of bottom currents are:

1. Lineations. These show up as fine streaks that are more or less parallel and are obviously parallel with the axis of current flow. The lineations may result from alignment of small objects (shells or stones) and the development of intervening sediment streamers. If the edges of the lineations are sharp, it is assumed that the current was flowing when the photograph was taken.
2. Scour marks. These occur around rocks or some other solid object. Frequently the object, especially rocks, will come to lie beneath the general surface of the sediments due to winnowing of the sediment from the incipient trough.
3. Sediment streamers. These are essentially large, smooth and streamline ridges that cover over and trail behind buried objects; they parallel the axis of flow.
4. Ripple marks. These may be short- or long-crested and symmetrical or asymmetrical.
5. Non-random orientation of bottom animals or linear pieces of vegetation that have fallen to the bottom.
6. The presence of turbid water adjacent to the bottom.

Bottom photographs are shown in this report from only five stations, as follows:

67-A-5-16B 25°29.2′N, 86°07.0′W (Figure 3-4)
68-A-5-16F 25°25.3′N, 86°05.3′W (Figures 3-5 and 6)
69-A-13-35 26°50.0′N, 86°42.9′W (Figures 3-7, 8, 9)
69-A-13-22 25°23.6′N, 86°03.5′W (Figure 3-10)
70-A-10-57 25°22′N, 86°06′W (Figure 3-11)

It is fortunate, however, that Huang (1969) presents three photographs that show ripples similar to those in Figures 3-7, 3-8 and 3-9 above. These were taken at the following locations from a cruise of R/V *Eastward* into the eastern Gulf in 1967:

Station E2 27°00′N, 86°29.6′W 3129 m
Station E3 26°58′N, 85°29.3′W 3254 m
Station E4 26°00′N, 86°15′W 3173 m

All of the photographs in Figures 3-4 to 3-11 show one evidence or another of a past or extant current having operated at the site. Some evidences are much stronger than others. For instance, the ripple marks in Figure 3-7 leave little room for doubt as compared, say, with Figure 3-10, where various types of shells have been caused to aggregate. Slightly more difficult to pin down from the photographs alone is the problem of whether or not a current was flowing at the time of picture taking. The likelihood that a current was actively photographed in Figure 3-6 seems greater than in Figure 3-8. But what about Figure 3-9? Here it seems that those ripples with round ridges on the left half of the picture have formed on top of and at right angles to very faint ripples in the center of the photograph. Needless to say, it seems imperative that one bring the photographic and current meter results together for a concomitant view. This we can do after examining some of the current meter results.

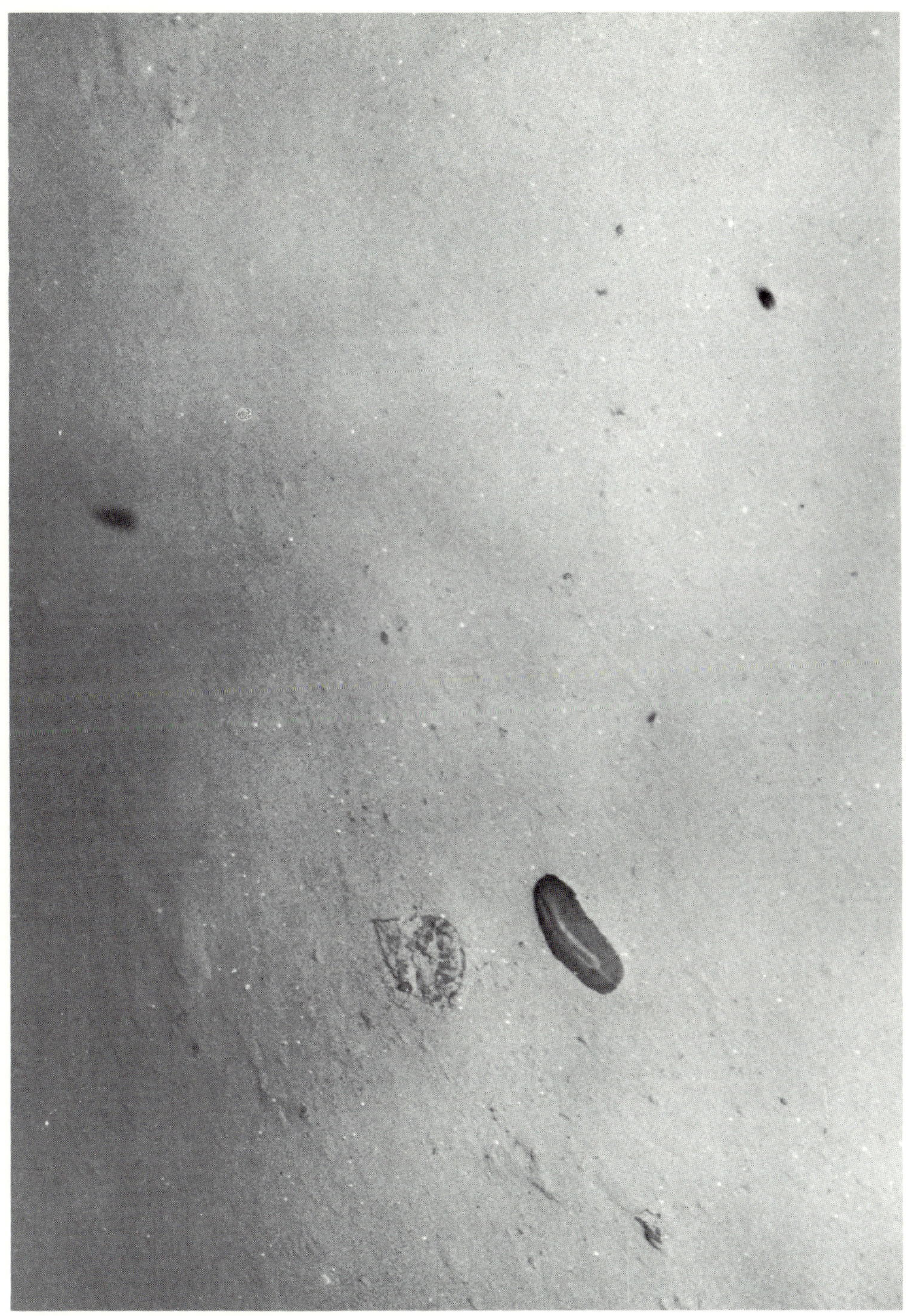

Figure 3-5. Alaminos *Cruise 67-A-5 (St. 16-F) 25° 25.3'N, 86°05.3'W. Depth: 3046 m. Note faint lineations, some scour and a few sediment streamers. The sea cucumber (*Benthodytes typica*) just below piece of ironstone is headed upstream more or less parallel to axis of current.*

Figure 3-6. Alaminos *Cruise 67-A-5 (St. 16-F) 25°25.3'N, 86°05.3'W. Depth: 3046 m. Lineations and lenticular streamers over and beyond submerged objects indicate a substantial current, which is moving from upper right to lower left.*

Current Meter Results

On six occasions from July, 1967, to October, 1969, the current meter was lowered successfully from *Alaminos* at anchor, as described previously (Figure 3-1). The results of these lowerings are presented in Table 3-5. Four of the lowerings are presented more graphically in diagrams showing the ship's position at anchor relative to the wind and surface current as well as the speed and direction of the bottom current (Figures 3-12, 3-13, 3-14 and 3-15).

Throughout the measurements the current speed ranged from 6-19 cm/sec and took direc-

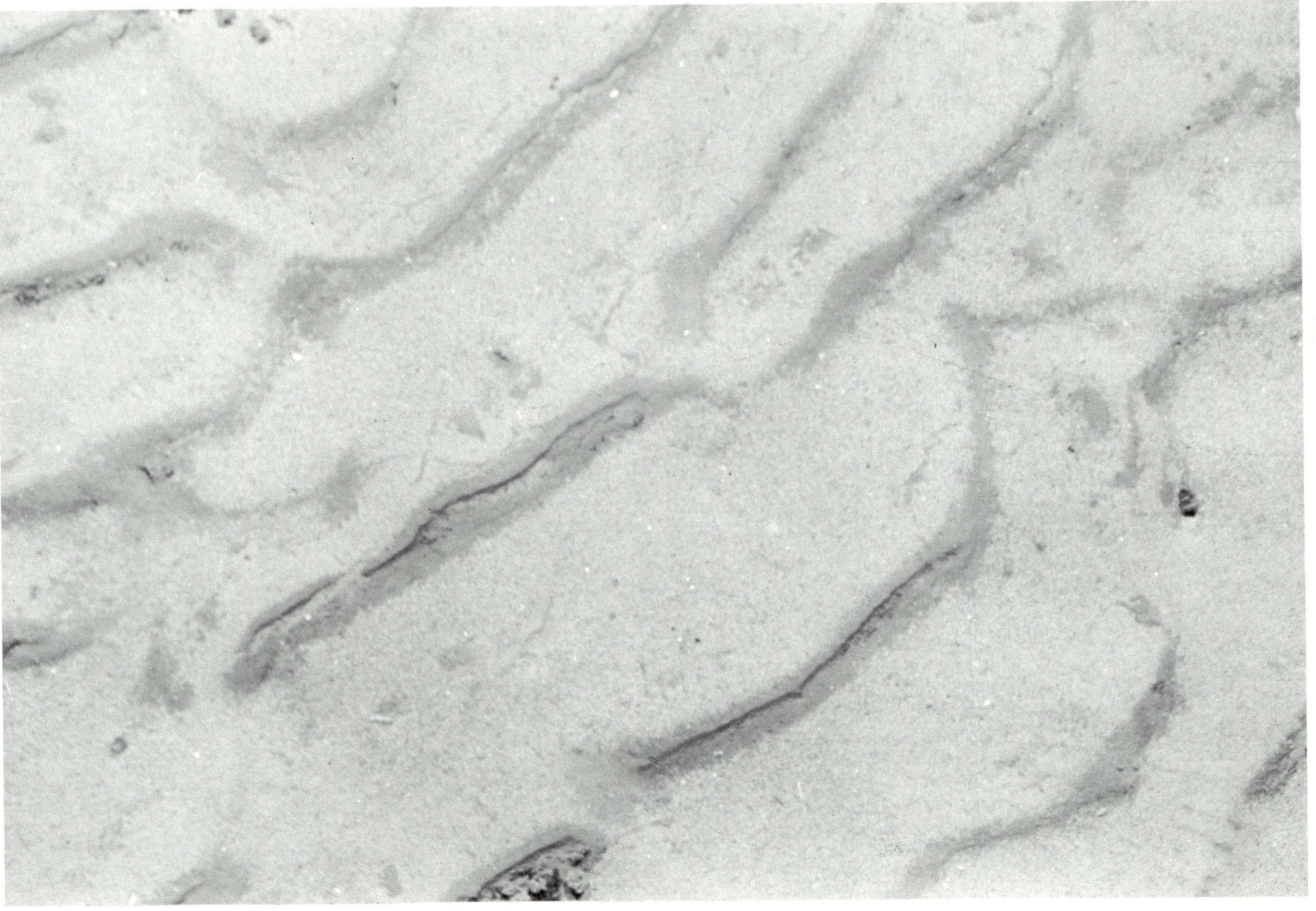

Figure 3-7. Alaminos *Cruise 69-A-13 (St. 35) 26°50.0'N, 86° 42.0'W. Depth: 3091 m. Ripple marks vary in width from 6-12 in. They appear to be formed as loose sand material overlying a hard substrate. A current measuring as much as 0.35 kt toward 191° recorded on day picture was taken.*

Figure 3-8. Alaminos *Cruise 69-A-13 (St. 35) 26°50.0'N, 86°42.9'W. Depth: 3091 m. Faint ripple marks 4-10 in wide crossed by numerous worm tracks and burrows of various sizes.*

tions varying from 078-248°. On only one occasion was the current speed wholly steady, and this was at the lowest speed recorded (6 cm/sec). Occasionally, there was a temptation to read into the record (as at 68-A-7-5A) a periodicity of range from high to low speed. In fact, in this case the period, from a high reading (19 cm/sec) to a low (16 cm/sec) and back to a high, ran about 6-7 minutes for an hour or more. This can hardly be attributed to an intrinsic function of the equipment employed, because it was noted by acoustic transmission through a separate reed switch to pinger to PDR. Throughout the study, the concept of channelization of the current crept in as a strong biasing factor. It arose as a result of three things: (1) finding small scale channel-like structures on the PDR records in this study region, (2) the fact that only some and seldom all of the photographs of a lowering revealed the presence of a current, and (3) pictures such as Figure 3-10 that reveal the great accumulation of shell material. Two things about the current itself are suggested, viz., (1) that it has wave characteristics and/or (2) that it meanders, perhaps on a fairly regular basis.

It is difficult to rid oneself of the bias that the bottom current is a meandering one in view of the author's observations of the Loop Current itself. On several occasions, anchored in 3200 m or more of water with the Loop Current coursing hard against the ship in the early morning, the author has been surprised to see it shift its position a mile or more leaving the ship by afternoon in relatively quiet water—only to have it back at the ship by early morning.

The author has no good conception of the vertical dimensions of the bottom current. So far as could be ascertained there was only a slight increase in the speed of the current several meters

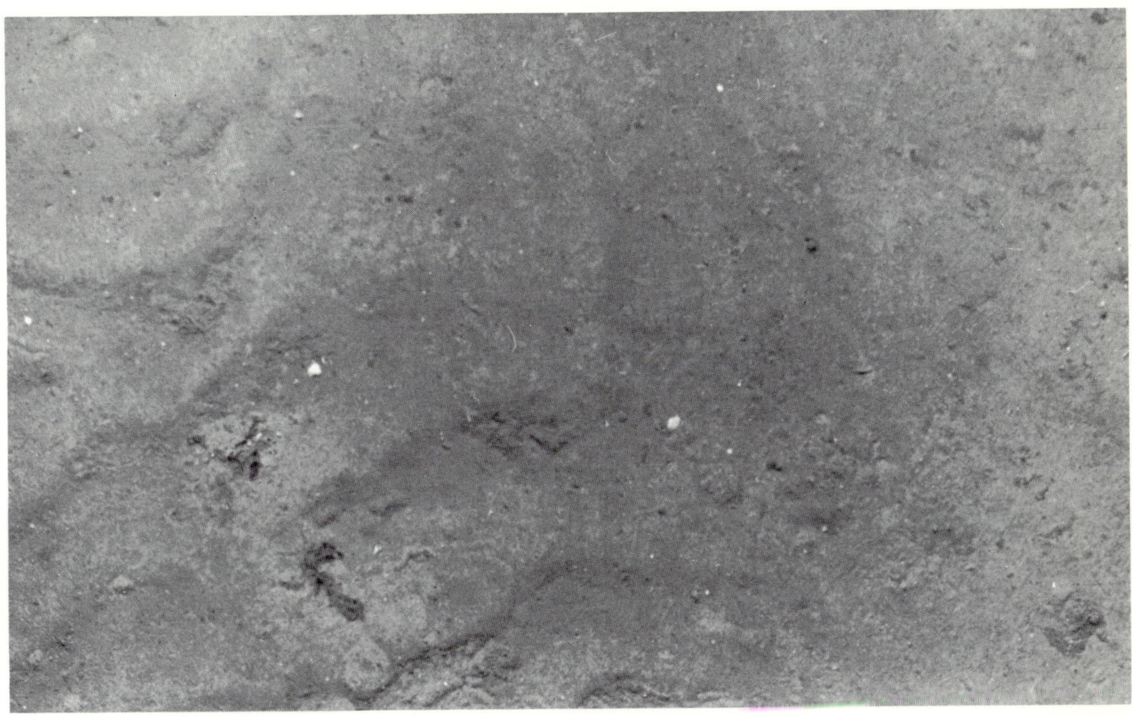

Figure 3-9. Alaminos *Cruise 69-A-13 (St. 35) 26°50.0'N, 86°42.9'W. Depth: 3091 m. Note ripple marks on left part of frame (5-1/2 to 7 in wide). Their rounded crests and animal workings in troughs suggest that they are not in active formation.*

above the bottom. For instance, when the current recorded at 26°50'N, 86°44'W was flowing rather steadily at 16 cm/sec 1 m above the bottom, the tetrapod was pulled up to a position 10 m above the bottom and held for a few minutes. The readings here increased to a steady 18 cm/sec. Subsequently during this series the meter was pulled up and held in 20 m increments with the same speed and direction results. This test was abandoned at 110 m off the bottom when the current proved to be the same. During this time there was very little ship yawing.

Perhaps the most useful correlations between photographs and current meter readings are those between Figures 3-5 and 3-6, which were taken on July 27, 1967, where current speeds of 17-19 cm/sec were recorded. The other is between Figure 3-7 and the current meter readings of October 13, 1969, near the site where current speeds of 14-17 cm/sec were recorded. The interesting point here is that the rather obsolescent ripple marks appearing in Figures 3-8 and 3-9 were taken in the same lowering. It should be pointed out, however, that the ship did drift at least 1 nautical mile during the taking of pictures in Figures 3-8 and 3-9 and the well-formed ripples in Figure 3-7. This observation is consistent with the concept of channelization of the currents on the Mississippi Cone.

Discussion

The facts presented above clearly establish that bottom currents of substantial velocities up to 19 cm/sec exist in a defined area of the eastern Gulf of Mexico. The author proposes herewith to call this current the East Gulf Deep Bottom Current.

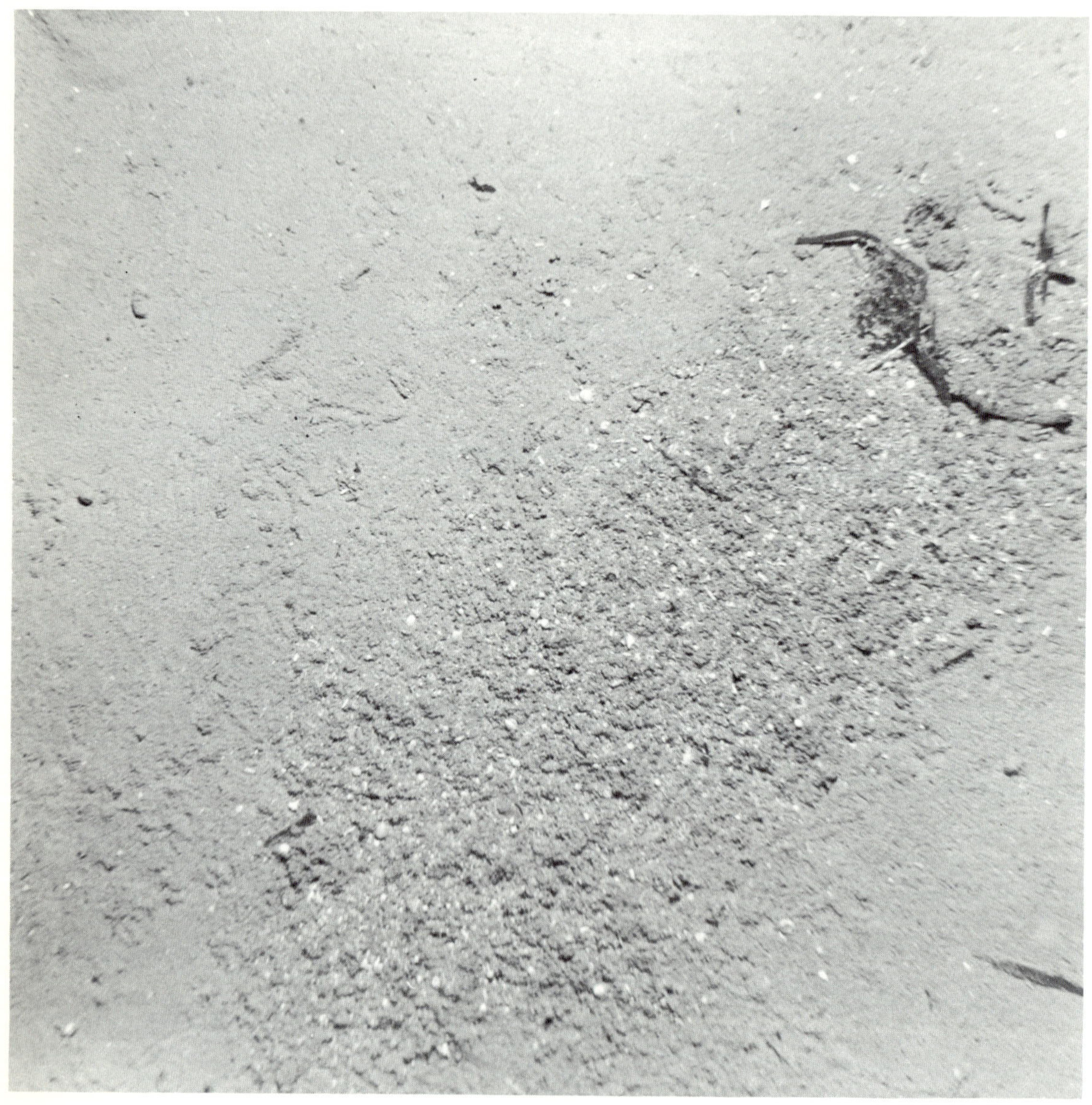

Figure 3-10. Alaminos *Cruise 69-A-13 (St. 22) 25°23.6'N, 86°03.5'W. Depth: 3246 m. Shells and plant debris deposited in a large ripple trough (picture covers about 9 sq ft). A bottom current of 0.15 kt was measured about 3 miles from this site on the day picture was taken.*

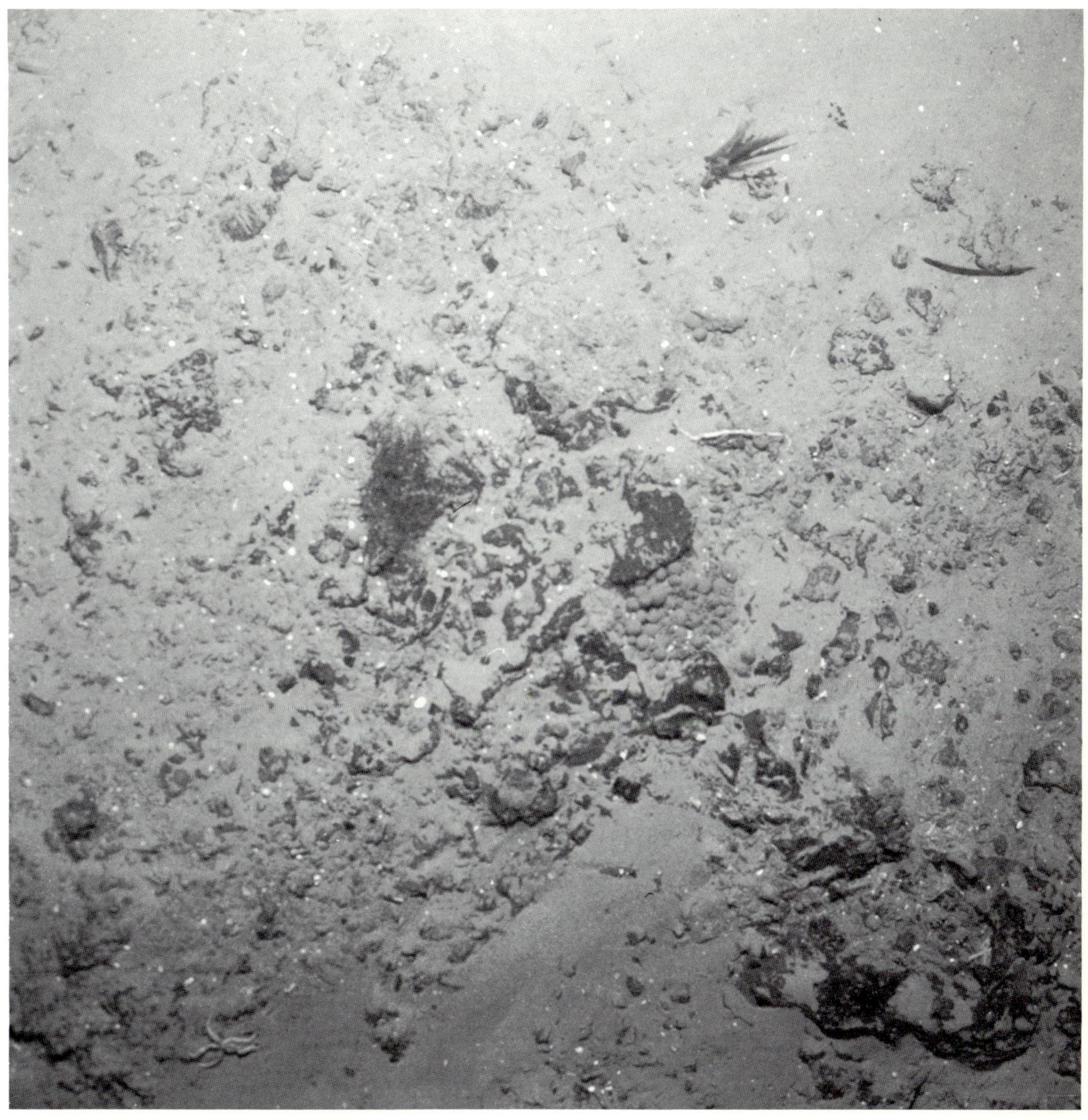

Figure 3-11. Alaminos *Cruise 70-A-10 (St. 57) 25°22′N, 86°06′W. Depth: 2346 m. Ironstone on side wall of a small canyon (picture covers about 16 sq ft) swept free of sediments on the more exposed parts. Current approaching from the right. Current measured near this site had a speed of 0.32 kt and direction of 078°.*

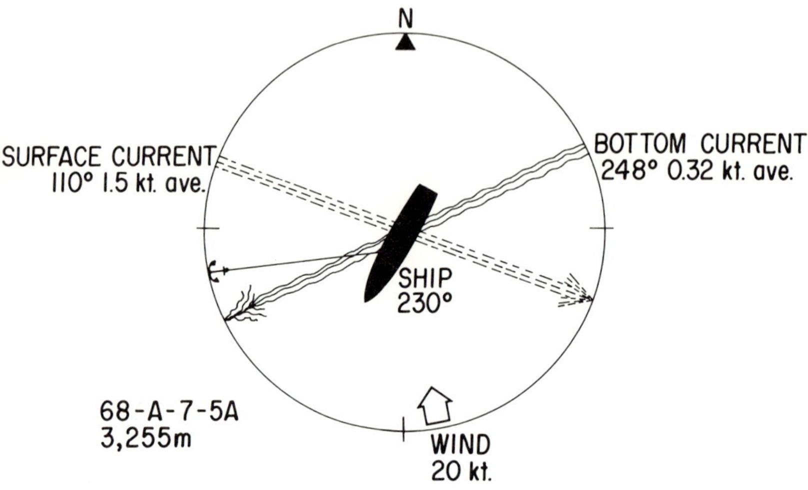

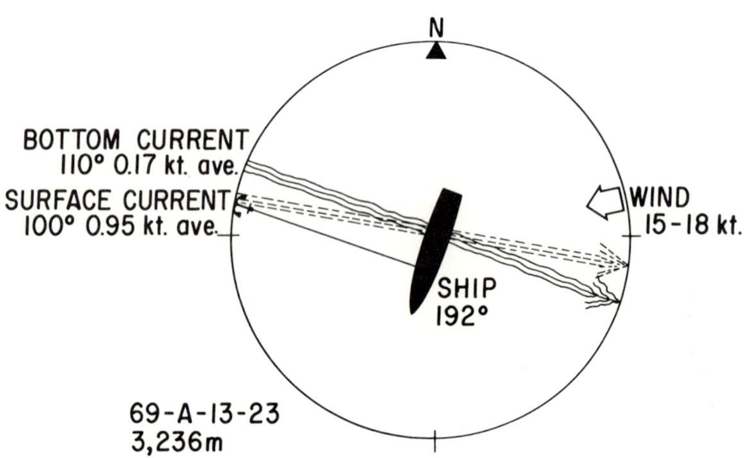

Figure 3-12. Relationships between surface current, wind and anchored position of Alaminos *while measuring bottom current (wavy arrow). For position and other data see Table 3-5.*

Figure 3-13. Relationships between surface current, wind and anchored position of Alaminos *while measuring bottom current (wavy arrow). For position and other data see Table 3-5.*

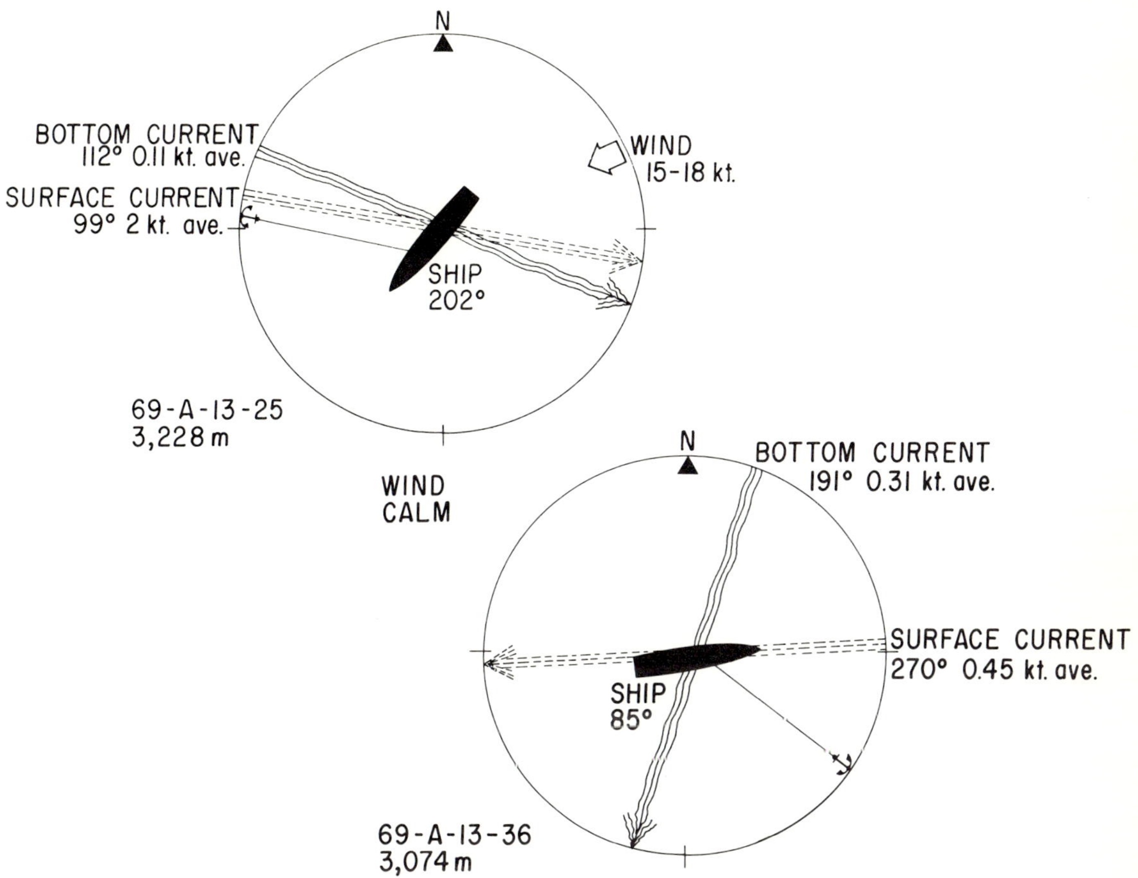

Figure 3-14. Relationships between surface current, wind and anchored position of Alaminos *while measuring bottom current (wavy arrow). See Table 3-5 for additional data.*

Figure 3-15. Relationships between surface current, wind and anchored position of Alaminos *while measuring bottom current (wavy arrow). See Table 3-5 for additional data.*

Table 3-5.

Current Speed and Direction Data from Eastern Gulf of Mexico

Cruise	Date	Time of Measurement (CDT)	Position	Depth (m)	Speed Ave.	Range	Direction
67-A5-16A	July 25	0640 to 0955	25°29′N, 86°06′W	3255	15 cm/sec	14-17 cm/sec	080°
67-A-5-16G	July 27	0815 to 1115	25°25′N, 86°08′W	3286	18 cm/sec	17-19 cm/sec	078°
68-A-7-5A	July 30	1750 to 1920	25°24′N, 86°07′W	3255	17 cm/sec	16-19 cm/sec	248°
69-A-13-23	Oct. 9	1620 to 1930	25°23′N, 85°59′W	3236	9 cm/sec	7-11 cm/sec	110°
69-A-13-25	Oct. 10	1125 to 1405	25°23′N, 86°07′W	3228	6 cm/sec	steady	112°
69-A-13-36	Oct. 13	1525 to 1909	26°50′N, 86°44′W	3074	16 cm/sec	14-17 cm/sec	191°

Less positive evidence indicates that this current is channelized. Commensurate with this view is the fact that only rarely do all bottom photographs in one camera lowering show evidence of a current.

Although there are some evidences that the current may be transitory in some places, based on pictures of what appears to be old ripple marks, it is possible that this results from meandering. On the other hand, the fact that a bottom current was found in about the same place in the years 1967, 1968 and 1969 argues for its constancy.

As yet, very little can be said about its direction. In four of the six current meter measurements it has an easterly trend, whereas in two it was moving either southerly or southwesterly. Point measurements in a meandering stream would be expected to yield varied directional indications. It is interesting to note, however, that the two measurements during Cruise 67-A-5 and the one of 68-A-7 are perfect reversals—a fact that is of increased interest because the stations were apparently (ship position determinations being what they are) quite close together. It is left to the reader to speculate as to the cause of the observed bottom flow.

Acknowledgments

I thank A.D. Fredericks, R.V. Pittman and A. Gutierrez for technical aid with the acoustical monitoring devices; A.D. Fredericks and B.M. James for photographic assistance; Jack Grant for original art; and the crew of *Alaminos* for excellent cooperation during deep-sea anchoring and other nonroutine procedures.

Acknowledgment is also made of the support of the Biology Program of the Ocean Science and Technology Group of the Office of Naval Research through contracts Nonr 2119(04) and N00014-68-A-0308(0001). Some ship support is gratefully acknowledged from the National Science Foundation under grants GA-1296 and GA-4544.

References

Bruce, J.G., Jr., and Thorndike, E.M. 1967. Photographic measurement of bottom currents. In: J.B. Hersey, ed., *Deep-sea photography.* Baltimore: The Johns Hopkins Press, 107-110.

Ewing, M., Ericson, D.B. and Heezen, B.C. 1958. Sediments and topography of the Gulf of Mexico. In: *Habitat of oil.* Tulsa, Okla.: Amer. Assoc. Petroleum Geol.

Gutierrez, A. and Pequegnat, W.E. 1968. *An acoustical system for remote bottom contact and current speed determinations.* Dept. of Ocean., Texas A&M Univ., Ref. 68-21T. (Unpublished technical report.)

Hollister, C.D. and Heezen, B.C. 1966. Ocean bottom currents. In: Rhodes W. Fairbridge, ed., *Encyclopedia of oceanography.* New York: Rheinhold Publishing Corp., 576-583.

Hunkins, K.I., Ewing, M., Heezen, B.C. and Menzies, R.J. 1960. Biological and ecological observations on the first photographs of the Arctic Ocean deep-sea floor. *Limnol. and Oceanogr.,* 5:157-161.

Huang, Ter-Chien. 1969. *The sediments and sedimentary processes of the eastern Mississippi Cone, Gulf of Mexico.* Contr. No. 31, Dept. of Geol., Florida State Univ.

Isaacs, J.D., Reid, J.L., Jr., Schick, G.B. and Schwartzlose, R.A. 1966. Near-bottom currents measured in 4 kilometers depth off Baja California Coast. *Jour. Geophys. Res.,* 71:4297-4303.

Knauss, J.A. 1965. A technique for measuring deep ocean currents close to the bottom with an unattached current meter and some preliminary results. *J. Mar. Res.,* 23:237-245.

La Fond, E.C. 1962. Deep current measurements with the bathyscaphe "Trieste." *Deep-Sea Res.,* 9:115-116.

Leipper, D.F. 1970. A sequence of current patterns in the Gulf of Mexico. *Jour. Geophys. Res.,* 75:637-657.

Maloney, W.E. 1967. *A study of the Antilles current using moored current meter arrays.* Naval Oceanographic Office, TR-199.

Menard, H.W. 1964. *The marine geology of the Pacific.* New York: McGraw-Hill.

Menzies, R.J., Zaneveld, J.S. and Pratt, R.M. 1967. Transported turtle grass as a source of organic enrichment of abyssal sediments off North Carolina. *Deep-Sea Res.,* 14:111-112.

Morgenstern, N.R. 1967. Submarine slumping and the initiation of turbidity currents. In: A.F. Richards, ed., *Marine Geotechnique,* 189-220.

Moody, C.L. 1967. Gulf of Mexico distributive province. *Amer. Assoc. Petroleum Geol. Bull.,* 51:179-199.

Nowroozi, A.A., Ewing, M., Nafe, J.E. and Fleigel, M. 1968. Deep ocean current and its correlation with the ocean tide off the coast of northern California. *Jour. Geophys. Res.,* 73:1921-1932.

Pequegnat, W.E., Bright, T.J. and James, B.M. 1970. The benthic skimmer, a new biological sampler for deep-sea studies. In: W.E. Pequegnat and F.A. Chace, eds., *Contributions on the Biology of the Gulf of Mexico,* Texas A&M University Oceanographic Studies, voi. 1. Houston: Gulf Publishing Co., 17-20.

Reid, J.L. 1967. Intermediate and deep circulation. *Trans. Am. Geophys. Un.,* 48:572-575.

________. 1969. Preliminary results of measurements of deep currents in the Pacific Ocean. *Nature*, 22:848.

Rowe, G.T. and Menzies, R.J. 1968. Deep bottom currents off the coast of North Carolina. *Deep-Sea Res.,* 15:711-719.

Scruton, P.C. 1960. Delta building and the deltaic sequence. In: F.P. Shepard, F.B. Phleger and Tj.H. van Andel, eds., *Recent sediments, northwest Gulf of Mexico.* Tulsa, Okla.: Amer. Assoc. Petroleum Geologists.

Shepard, F.P. 1960. Mississippi delta: marginal environments, sediments, and growth. In: F.P. Shepard, F.B. Phleger, and Tj.H. van Andel eds., *Recent sediments, northwest Gulf of Mexico.* Tulsa, Okla.: Amer. Assoc. Petroleum Geologists, 56-81.

Skoda, J.D. 1970. A survey of literature pertaining to sediment transport on the deep-sea floor. In: H.A. Einstein and R.L. Wiegel, eds., *A literature review on erosion and deposition of sediment near structures in the ocean floor (U). Sec. 3, 1-39*. Berkeley, Calif.: Hydraulic Engineering Laboratory, College of Engineering, Univ. of Calif.

Sternberg, R.W. and Creager, J.S. 1965. An instrument system to measure boundary-layer conditions at the sea floor. *Mar. Geol.,* 3:475-482.

Swallow, J.C. 1955. A neutral-buoyancy float for measuring deep currents. *Deep-Sea Res.,* 3:74-81.

Webster. F. 1969. On the representativeness of direct deep-sea current measurements. In: R.W. Steward, ed., *Progress in oceanography,* vol. 5, pt. 1. New York: Pergamon Press, 3-15.

Wüst, G. 1955. Stromgeschwindigkeiten im Tiefenund Bodenwasser des Atlantischen Ozeans. *Deep-Sea Res.,* 3(suppl.):373-397.

________. 1957. Quantitative Untersuchungen zur Statik und Dynamik des Atlantischen Ozeans: Stromgeschiwindigkeiten und Strommungen in der Tiefen des Atlantischen Ozeans. Deut. Atlantische Expedition *Meteor,* 1925-1927. Wiss. Ergeb., 6(2):420 pp.

Section 2
The Loop Current and Its Variability

4

Separation of an Anticyclone and Subsequent Developments in the Loop Current (1969)

John D. Cochrane

Abstract

In May, 1969, the process of detachment of an anticyclonic current ring from the Gulf of Mexico Loop Current was observed for the first time. In early May, the loop was strongly constricted by two meanders, one protruding into the loop from Campeche Bank, the other from Florida Shelf. Roughly 10 days later, the meander off Campeche Bank had grown more than 80 nautical miles and joined the meander on the east to form a cyclonic shear zone which cut an anticyclone off from the loop. The anticyclone center moved little between mid-May and mid-June; but by mid-July, it had reached 90 nautical miles west of its May position. By mid-September, the center had migrated still farther west, while the loop had pushed northward.

In three well-documented cases of detached anticyclones found in late spring or summer, the separating shear zone was located near the smallest distance from Campeche Bank to Florida Shelf and was a deep-reaching feature characterized by low-salinity Shelf Water in the thermocline.

During spring or summer of each year, a meander appears to form off southeastern Florida Shelf. In some but not all years, a meander has formed in late spring or summer off eastern Campeche Bank. In such years, the two meanders have grown together to cut an anticyclone off from the loop. Such anticyclones have usually moved west and, although interannual differences are great, some of these anticyclones may migrate to the western Gulf.

Introduction

On a number of occasions prior to 1969, an anticyclonic current ring has been found close to, but separated from, the Loop Current of the Gulf of Mexico. On these occasions, the anticyclone had apparently broken off from the loop, although the separation process was not observed. Anticyclones of moderate strength have been encountered sometimes in the central and often in the western Gulf, where they constituted the strongest circulation features. Some anticyclones in the central and western Gulf may originate by detachment from the loop. However, neither the detachment process nor the subsequent movement of a detached current ring appears to have been observed closely before 1969.

In 1969, the *Alaminos* made four cruises in the eastern Gulf of Mexico. The cruises form a series in which the circulation changes can be followed from May to mid-September. Three surveys during the May cruise provide a picture of the process by which an anticyclonic ring becomes detached from the loop. Subsequent cruises provide indications of the westward movement of the ring and the concurrent changes in the loop.

Figure 4-1 shows the geographical features important to the discussion and, in schematic form, the major aspects of the Loop Current during May and June. The circulation is overwhelmingly dominated by the Gulf Loop Current, the segment of the Gulf Stream system connecting the inflow into the Gulf at Yucatan Strait with outflow at Florida Strait. (See, for example, Nowlin and McLellan, 1967.) The gross features of the current pattern clearly are controlled by the topography of the eastern Gulf basin. The loop is normally confined between Campeche Bank and Florida Shelf, with its core pressed close to the bank on the west but less close to the shelf on the east. The anticyclonic arc forming the north part of the loop extends farther north than Campeche Bank. During May, the arc has been found only as far north as 27°N, roughly, so that the penetration of the loop into the Gulf is about 300 nautical miles north from the Yucatan Strait at 22°N. Thus, while the east and west sides of the loop are adjacent to shelf edges, the northern limb of the loop lies considerably to the south of the shelf edge.

The northward flow which constitutes the western side of the loop is continuous with the northward flow which originates near 17° or 18°N along the western side of the Cayman Sea. It is often convenient to call the entire segment of northward flow the Yucatan Current, as Defant (1957) and others have done. In some contexts, it is convenient to refer to the east side of the loop as the West Florida Current.

The periphery of the loop may extend far to the north and west in the Gulf, but usually appears to involve little transport. An appreciable portion of the loop's interior flow seems to return to the Yucatan Strait after reaching the offing of Cuba and leaves the Gulf through that strait rather than through Florida Strait. Normally, a closed anticyclonic circulation is found in the interior of the loop. The interior anticyclone is elongated, as is the loop, along an axis parallel to the bounding shelf regions on either side of the loop. The axis normally lies over the deep region between Campeche Bank and Florida Shelf somewhat west of the deepest area, as Figure 4-1 indicates.

Over the shallow shelf regions, the thermocline is often sharp and shallow so that, for its depth, the bottom water on the shelves is exceptionally cold (and, therefore, dense). Consequently, the sea-surface geopotential is normally rather small over shelves. While regions of high geopotential have never been encountered over shelves, regions of low geopotential (cyclonic regions) are not infrequently found over the deeper parts of the eastern Gulf.

Like the Gulf Stream proper, the Loop Current and its mass field extend to considerable depths, as Nowlin and McLellan (1967) indicate. But since the current speed decreases rapidly with depth, experience has shown that an isothermal surface may usually be found whose topography mirrors fairly well (in expanded form) the sea-surface geopotential and so gives a representation of the geostrophic flow pattern. For the eastern

Gulf, the 20°C or 22°C surfaces behave in the manner described, although both, particularly 22°C, are subject to modification in the north due to the seasonal warming and cooling of the upper layers.

Nowlin and McLellan (1967), as well as others, have described the waters of the Gulf of Mexico. Below about 17°C, the *T-S* and other characteristic curves vary only a little with time and position. Above 17°C, however, the range of salinity at a given temperature may be considerable. Much of the water above 17°C falls into two convenient categories. The larger part of the water imported into the Gulf is characterized by a salinity maximum near 22°C. Since the high salinity appears to originate in the subtropical regions of high salinity at the sea surface in the North Atlantic (Montgomery, 1937), the water is called North Atlantic Subtropical Water. (Wüst, 1964, uses the name Subtropical Underwater.) Differing from the latter by having lower salinity within the thermocline is the water usually found in shelf regions and occasionally in the deeper regions of the eastern Gulf. This less saline water is conveniently called Gulf Shelf Water. Nowlin and McLellan (1967) give examples of the *T-S* relationships of the two types of water just described.

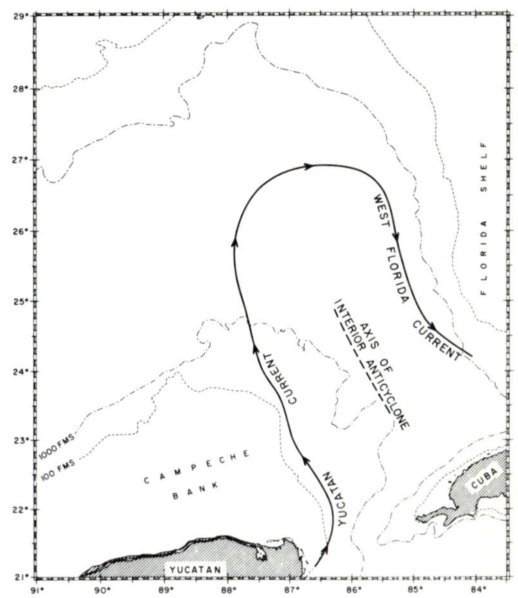

Figure 4-1. Schematic representation of the core and the axis of the Loop Current's interior anticyclone, typical of May or June. Positions are based on observations made from R/V Alaminos *in May, 1968.*

Method

In the four *Alaminos* cruises from May to mid-September, 1969, bathythermograms, very often of the expendable type, were obtained. These bathythermograms form the basis for the representations giving the most complete coverage of the period. Since, as noted above, the topography of the 22°C isothermal surface gives an indication of the geostrophic flow pattern and has been used extensively in earlier papers for that purpose, that isotherm is used in the present study.

Where STDs or bottle casts are available, more complete analysis is, of course, possible. Station curves were drawn in the manner described by Montgomery and Stroup (1962). From these, vertical sections of thermosteric anomaly were constructed. To provide a quasi-horizontal picture, the depth and salinity at various thermosteric anomaly (Montgomery and Wooster, 1954) values were determined and their distributions mapped. The distributions at 400 cl t^{-1} and 250 cl t^{-1} were given special attention because of their nearness to the sea surface and to the salinity maximum, respectively.

The geostrophic flow in the isosteric surfaces, as Montgomery (1937) has shown, may be represented by means of an acceleration potential (anomaly) $\Upsilon - \Upsilon_0$, which is defined by

$$\Upsilon - \Upsilon_0 = \int_{\delta_0}^{\delta} p \, d\delta + p_0 \delta_0$$

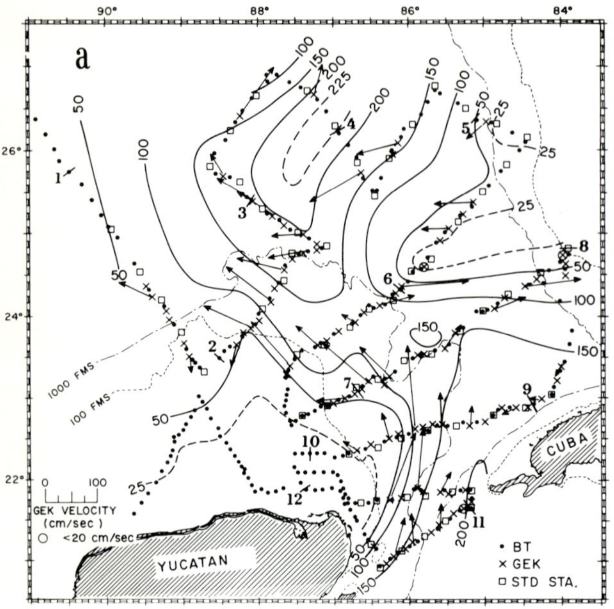

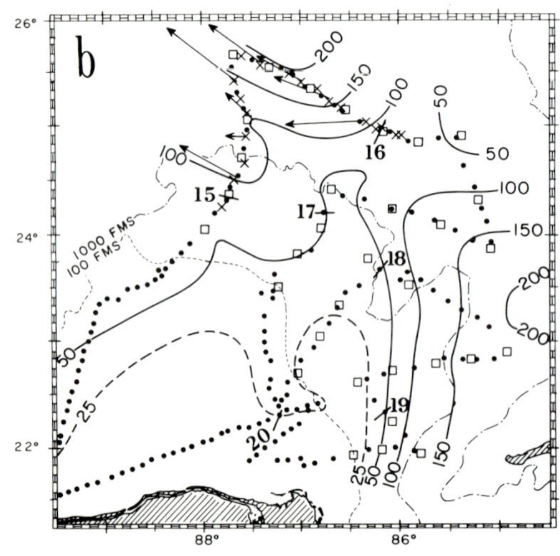

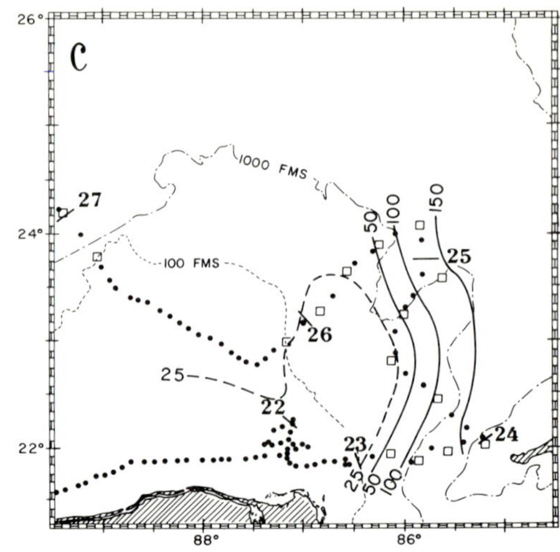

*Figure 4-2. Topography in meters of the 22°C surface observed in May, 1969, during the first (*a*), second (*b*) and third (*c*) surveys. Position and date at the beginning of each day are indicated by a line segment across the track and a large numeral. Representative GEK indications of surface current are also shown.*

where the zero subscript specifies the reference level, δ is the steric anomaly, and the p the pressure. Montgomery and Wooster (1954) have shown that when, as in the geostrophic relationship, differences between such integrals are required, there are certain regions for which thermosteric anomaly may be used in place of steric anomaly in the integral. In the Gulf of Mexico, the variation in the *T-S* relationship is so small below about 17°C that the pressure-dependent terms of steric anomaly lead, when integrated, to a nearly constant value for any part of the Gulf and so may be omitted when differences are required.

In the present study, acceleration potentials are referred to 1000 db. At some stations, the bottom depth is less than 1000 m, particularly along the side of Campeche Bank. To obtain values for these stations, the integration was continued along the sea floor toward the next station on the line reaching to 1000 db. Traces of the isosteres were extrapolated to the sea floor in vertical sections. In a few instances, XBT observations together with *T-S* relations extrapolated or interpolated from nearby stations were used in obtaining an estimate of the acceleration potential in shallow water.

In the first two May surveys, GEK measurements of sea-surface current were obtained. For regions where these measurements were made, a representative sampling was plotted on the maps of the topography of 22°C. Near the edges of the regions where temperature data did not determine the topography completely, the contours were

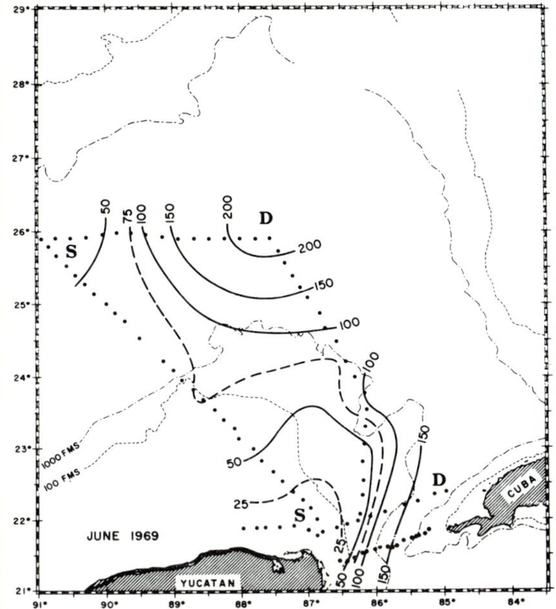

Figure 4-3. Topography in meters of the 22°C surface observed between 4 and 22 June 1969. Data from this cruise were kindly supplied by Robert O. Reid and Donald Durham, who supervised the cruise.

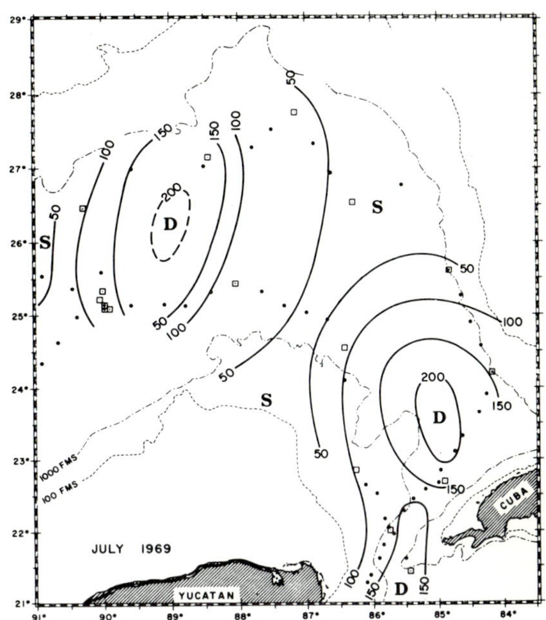

Figure 4-4. Topography in meters of the 22°C surface observed between 12 and 25 July 1969. Data from this cruise were kindly supplied by Leo Berner and William W. Schroeder, who supervised the cruise.

drawn to agree as far as possible with the GEK currents. This procedure was followed also in constructing the maps of acceleration potential at 400 cl t^{-1}.

Results of Observations in 1969

Between 30 April and 28 May 1969, observers aboard the *Alaminos* made three successive detailed surveys of the Yucatan Current and surrounding regions; Figure 4-2 gives the locations of pertinent observations during these surveys. During June in a cruise to the Yucatan Strait region, bathythermograph observations were taken at the locations shown in Figure 4-3. During July, the *Alaminos* covered the eastern Gulf along the track shown in Figure 4-4. Another rather intensive reconnaissance of the eastern Gulf was made in September by the *Alaminos*. At nearly the same time, R/V *Pillsbury* worked in a region farther east. Figure 4-5 shows the locations of stations for both ships. In each of the figures just noted, the 22°C topography provides an approximate representation of the geostrophic flow.

First May Survey

A more exact indication of the near-surface geostrophic flow for the first May cruise is given in tial at 400 cl t^{-1}. Because of the large contrasts in salinity at 250 cl t^{-1}, the acceleration potential and salinity for that surface are shown in Figures 4-7*a* and 4-8*a*, respectively. The depth of the 250 cl t^{-1} surface is roughly the same as that of the 22°C isothermal surface shown in Figure 4-2*a*.

The flow pattern encountered during the first May survey was unusual for that month. The

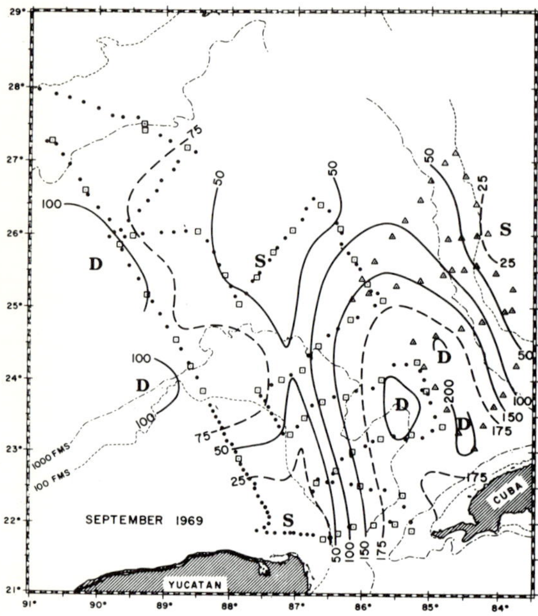

Figure 4-5. Topography in meters of the 22°C surface observed between 10 and 19 September 1969. Triangles indicate bathythermograph observations taken from R/V Pillsbury *of the University of Miami. Data from the latter cruise were kindly supplied by Walter Düing, who supervised the cruise.*

Yucatan Current entered the Gulf with a small eastward component as it passed through the strait. This direction persisted northward to about 23°N, where the current core, as indicated by the maximum gradient and GEK speed value, was displaced nearly 50 nautical miles from the edge of Campeche Bank; in May, the core is usually found near the edge of the Bank. At about 23.5°N, the Yucatan Current branched. The western portion of the current recurved westward toward the northern edge of the bank forming a "meander" out from the bank which passed around a cold ridge or region of low acceleration potential pushing north from the bank. The meander will be called the Yucatan meander. The eastern, or inner, branch of the Yucatan Current at 23.5°N curved eastward toward Florida Strait.

Northwest of the branching, the outer, or western, portion of the Yucatan Current continued to the northwest and then passed around the north side of a strong interior anticyclone centered near 26.0°N, 87.5°W.

On the northeast side of the loop, southward flow along the edge of the Florida Shelf was interrupted by a sharp turn to the west at about 25.5°N. This interruption was part of a meander in the flow around a rather narrow cold ridge protruding westward from the edge of the Florida Shelf. It is convenient to refer to the latter meander as the Florida meander. The 400 cl t^{-1} surface rose to the sea surface around the center of the ridge (Figure 4-6*a*). The sea-surface temperature within the 400 cl t^{-1} isopleth was quite low, ranging down to 22.5°C, and the surface salinity was relatively high, up to 36.4 per mil. The Yucatan meander, on the other hand, did not exhibit such low temperature or high salinity at the sea surface. Between the meanders on either side, the loop was considerably constricted to form a rather strong anticyclonic shear zone.

The eastward flow on the south side of the Florida meander was joined by the eastward extension of the inner, or eastern, branch of the Yucatan Current. Thus, it might be said that there were two loops, an inner weaker loop confined to the south of the constriction and an outer, stronger loop, which passed north of the interior eddy.

The salinity at 250 cl t^{-1} (Figure 4-8*a*) indicates that the two cyclonic meanders along the shelf regions were characterized by the low salinity of Shelf Water. The anticyclonic regions north and south of the constriction and a connecting neck in the constriction were characterized by relatively saline water. This saline water was encountered on all three of the lines crossing the constriction.

The GEK speeds agreed reasonably well with the acceleration potential (Figures 4-2*a* and 4-6*a*). However, north of the constriction, the flow indicated by the GEK crossed the contours toward the west generally, while south of the constriction the GEKs indicated flow with an eastward component across the small anticyclonic eddy implied by the acceleration potential.

Second and Third Surveys in May

In the second survey, a part of the original region was again covered, as Figure 4-2*b* indicates. The track followed in the second survey was similar in shape and direction to that of the first survey. The time interval between successive observations of a given region was consequently roughly constant, actually about 10 or 11 days.

Despite a weakening of the Florida meander, it had apparently joined with the Yucatan meander to form a ridge running from the Campeche Bank to the Florida Shelf. In large degree, this ridge separated an anticyclonic current ring on the northwest from the main portion of the loop on the southeast. The path of the greater part of the flow from the Yucatan to the Florida Strait was thereby shortened. It is possible that a connection between the loop and the anticyclone had formed east of the region of the second survey. But this would require a very contorted and complex current pattern; the occurrence of such complexity is not suggested in even the most detailed cruises.

Confirming evidence of the separation is provided by the distribution of salinity at 250 cl t-1 shown in Figure 4-8*b*. Relatively low-salinity Shelf Water evidently had moved northward with the Yucatan meander and probably also westward from the Florida meander so as to form a band separating the high-salinity water of the loop from

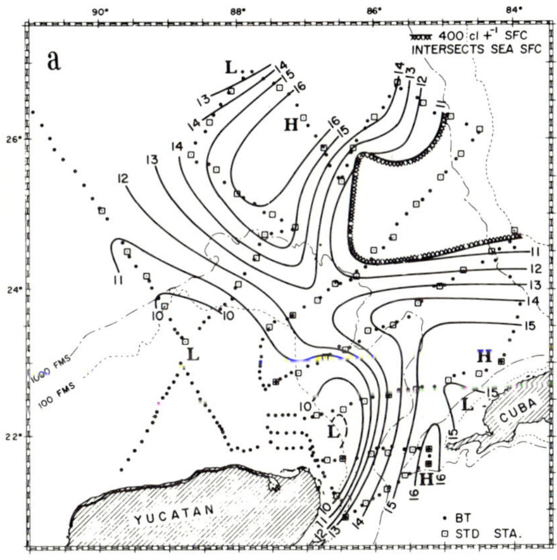

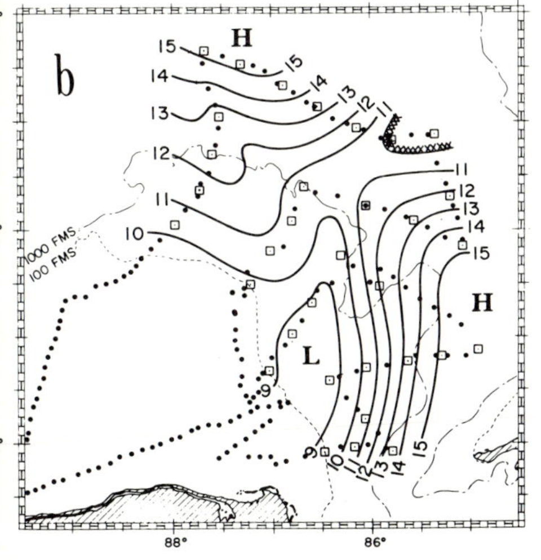

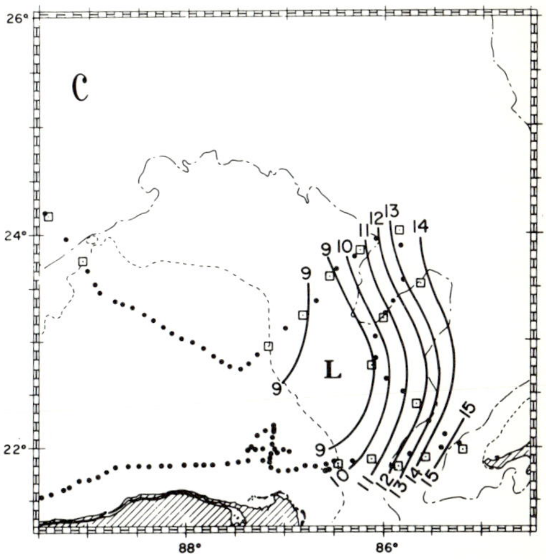

*Figure 4-6. Acceleration potential at 400 cl t-1 relative to 1000 db in J kg-1 (dynamic decimeters) based on observations taken in May, 1969, during the first (*a*), second (*b*) and third (*c*) surveys.*

that of the anticyclone. The low-salinity band agrees in position with the ridge in the 22°C topography and the corresponding trough in the acceleration potential.

Within the short interval of 10 or 11 days, major features of the flow pattern had changed markedly. While the Florida meander had weakened, the Yucatan meander had extended far to the north of its earlier position as its associated ridge lengthened and intensified. The change is illustrated by successive positions of various contours of the 22°C surface shown in Figure 4-9. The successive positions of the northern extreme of the Yucatan meander indicate a movement of more than 80 nautical miles. This value is comparable to the upper limit of lateral movement of the Gulf Stream found by Fuglister and Worthington (1951) during Operation Cabot, 11 nautical miles per day.

The vertical sections of thermosteric anomaly shown in Figures 4-10*a* and *b* indicate that the change in the Yucatan meander was not confined to the upper layers, but extended at least to 1000 m, or the bottom where the latter was shallower than 1000 m. The locations of the vertical sections are indicated in Figure 4-7. The section for the first survey crossed the northern part of the meander. The section for the second survey, although farther north over much of its length, clearly indicates the rise of the isanosteres which accompanies the development of the meander.

Another measure of the change is provided by the difference in acceleration potential from the

*Figure 4-7. Acceleration potential at 250 cl t $^{-1}$ relative to 1000 dm in j kg^{-1} (dynamic decimeters) based on observations taken in May, 1969, during the first (*a*), second (*b*) and third (*c*) surveys. Locations of lines along which vertical sections of Figures 4-10*a *and* b *run are shown in parts of* a *and* b*, respectively.*

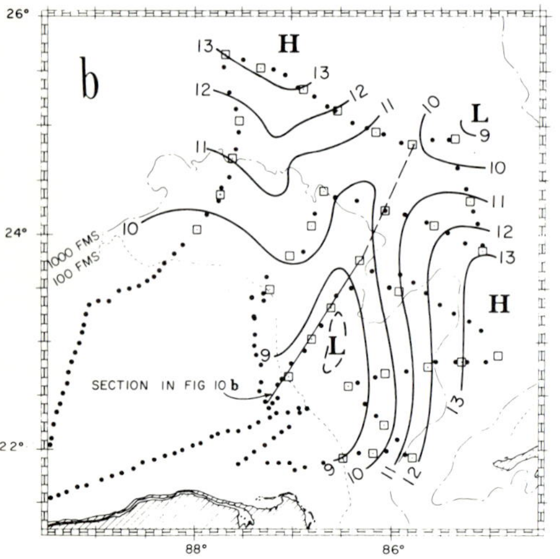

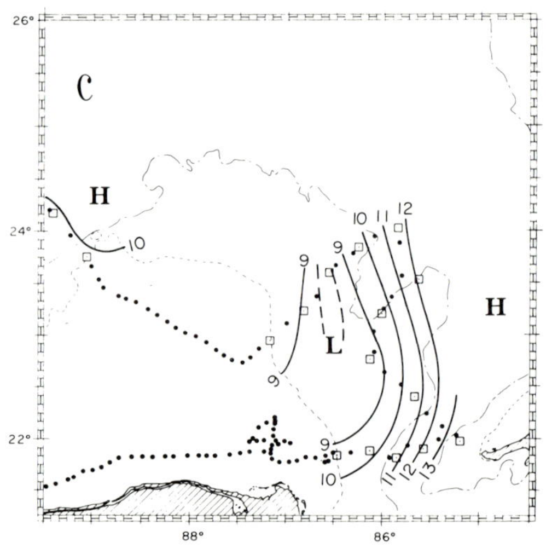

first to the second survey. The difference for 250 cl t-1 is given in Figure 4-11. The pattern of change at 400 cl t-1 is similar in form, although magnitudes at 400 cl t-1 are somewhat larger than those at 250 cl t-1. Large decreases in the potential had occurred in the northern portion of the Yucatan meander and at the southwestern side of the anticylonic ring which became separated. Sizable rises in the potential occurred with the weakening of the Florida meander.

The mean rate of change can be obtained from the difference and the time interval of approximately 11 days between surveys. With such rates, it is clear that the circulation pattern may have altered appreciably even during the first survey. However, the rapid changes during May occurred primarily in the cyclonic meanders. Where high, the acceleration potential changed rather little, particularly near high centers, as Figure 4-11 shows.

The third survey of May (Figures 4-2*c*, 4-6*c*, 4-7*c* and 4-8*c*), which followed the second by about one week, was very brief, covering only the region of the Yucatan meander. The meander had maintained its position or grown slightly. The trough in the acceleration potential associated with the meander extended out from the shelf, with its shallowest portion lying some distance off the shelf, as Figure 4-7*c* illustrates.

Developments Observed in June and July

Observations in June (Figure 4-3) indicate that the separation between the anticyclone and loop persisted into June. The center of the anticyclone had moved only a little to the west. In the region west of the anticyclone center, the 22°C

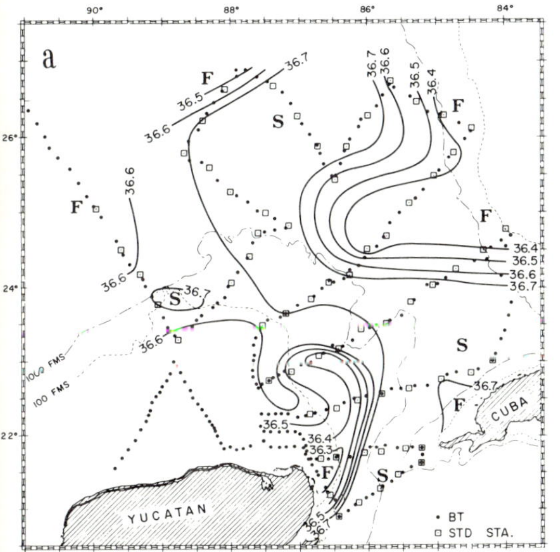

*Figure 4-8. Salinity at 250 cl t-1 based on observations taken in May, 1969, during the first (*a*), second (*b*) and third (*c*) surveys.*

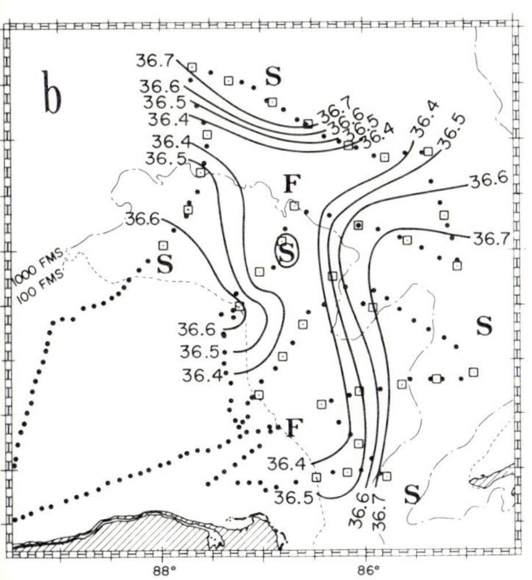

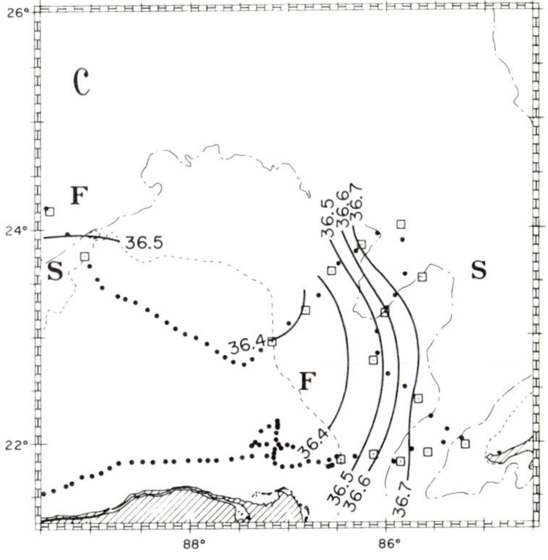

surface had become somewhat shallower than it was in late May, while at the center the surface had remained at about the same depth.

The observations made in July (Figure 4-4) show that the center of the anticyclone had advanced rather quickly to the west since the middle of June. The center position, estimated from the lines to its north and south, had moved approximately 75 nautical miles from its mid-June position. The speed of the center toward the west was thus between 2 and 3 nautical miles per day. Some confirmation of this speed is provided by the deepening of the 22°C isotherm near 25°N, 90°W between the 13th and 25th of July. If the gradient of depth west of the anticyclonic center remained nearly constant between the mid-June and mid-July fixes of the center position, the speed of center to the west was 2 or 3 nautical miles per day.

Figure 4-9. Changes in the Loop Current and the meanders indicated by positions of the 50-m, 100-m and 200-m contours of 22°C during the first (a) and second (b) surveys of May, 1969.

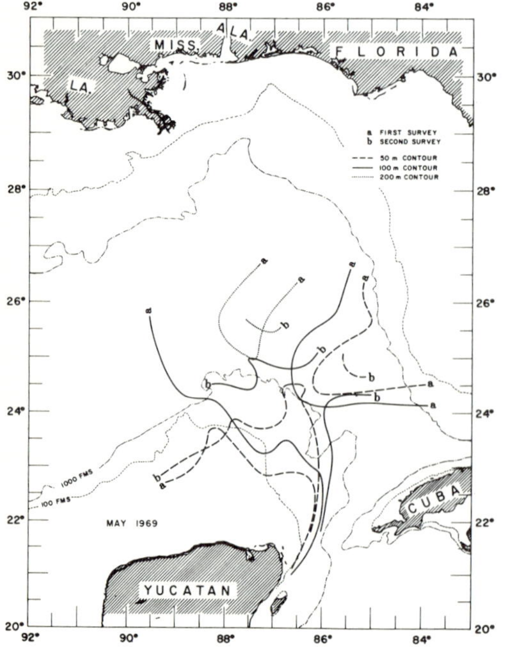

The core of the loop, as indicated by the 100-m contour of the 22°C surface, had moved northward since May, but not so rapidly as the eastern limb of the anticyclone had moved to the west. Consequently, the separation between the anticyclone and the loop had broadened.

September Observations

Figure 4-5 shows the topography of the 22°C surface observed in September, 1969. Because of the similarity in form of the acceleration potential field at 400 cl t^{-1} to that at 250 cl t^{-1}, only the latter is shown (Figure 4-12). By September, the center of the anticyclone had either moved westward out of the region, or the anticyclone had decayed rapidly. An anticyclonic wedge, possibly related to the anticyclone, pushed from the west into the region north of Campeche Bank. The major loop and the separation were still easily identifiable at the sea surface and down to 1000 m. The separation between the loop and anticyclonic region to the west had become narrower than in July, particularly in the south, and was marked in the north by a weak, closed cyclonic circulation. The loop had pushed north of its July position, but the core of the Yucatan Current still remained well to the east of the Campeche Bank edge, except at Yucatan Strait. As in earlier cases, the water at 250 cl t^{-1} in the loop was rather saline, while that in the separation and along the shelves was relatively fresh, as Figure 4-13 shows.

Summary of Changes in the Flow Pattern Observed in 1969

At the beginning of the series of observations in early May, an anticyclonic constriction in the loop was formed by the presence of two cold ridges, one on either side of the loop. The ridges developed so as to join by mid-May and form a cyclonic shear zone (separation) which detached an anticyclonic current ring from the loop. Subsequently, in June, July and September it remained possible to identify the major features: the loop, the detached anticyclone and the separation

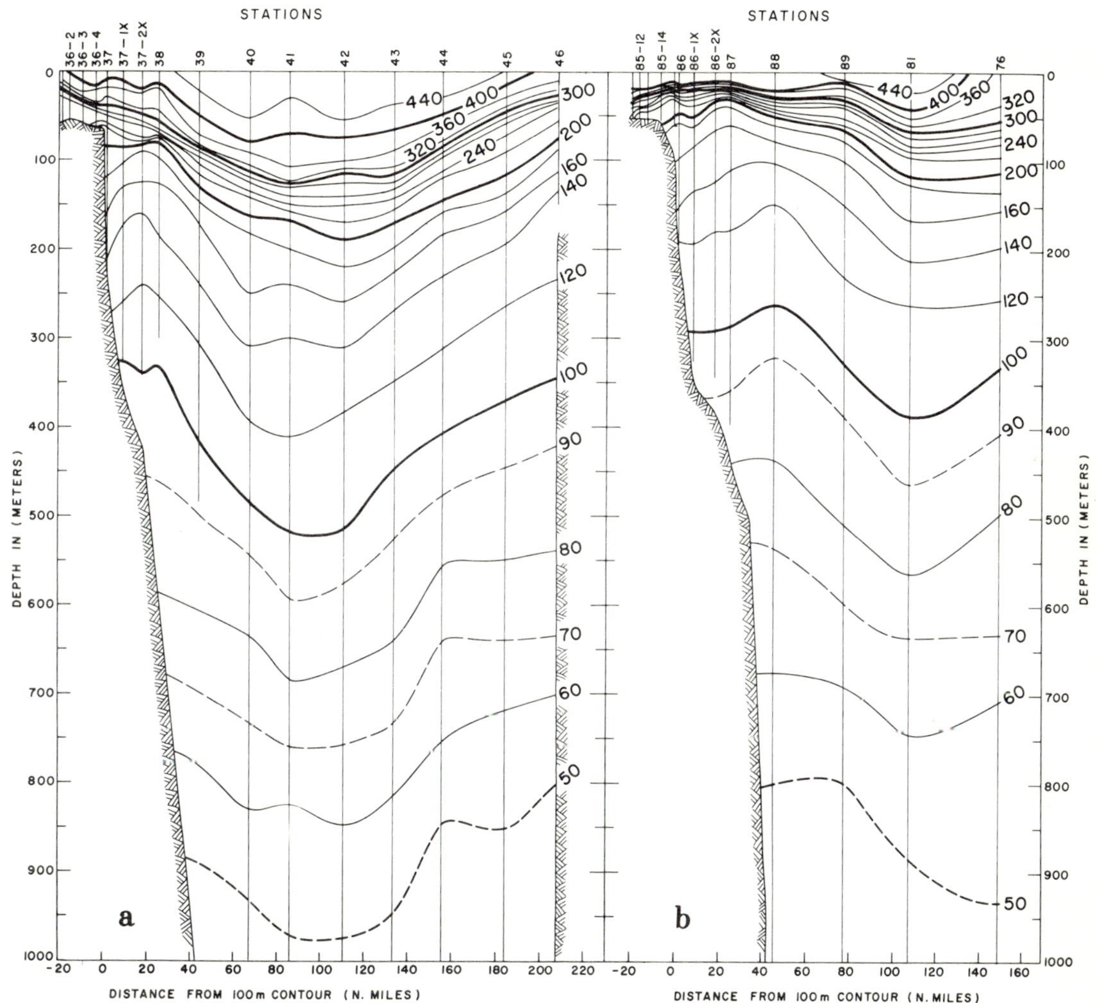

*Figure 4-10. Thermosteric anomaly at the constriction during the first survey (*a*) for the vertical section along the line indicated in Figure 4-6*a*, and during the second survey (*b*) for section along the line indicated in Figure 4-6*b*. Vertical extent of observation is indicated by vertical line. Vertical exaggeration is 370, approximately.*

between them. Apparently, changes like that in the formation of the separation did not occur so frequently or extensively as to obscure or obliterate the major features. The successive positions of the loop, separation and anticyclone after the separation formed in May are shown in Figure 4-14. The anticyclonic center moved little between the May and June surveys but in July was found to have migrated rather rapidly westward. By September, the center had passed west of 90°W, beyond the region observed, or had decayed; the form of the separation had altered considerably; and the loop had regained a position fairly far to the north.

Comparison with Other Years

Although the period from May through September, 1969, apparently received closer attention than a like period in any other year, there is a considerable body of information on changes in the circulation pattern and its intensity in the eastern Gulf which allows interpretation and some tentative generalizations of the course of events encountered in 1969. The outstanding event of the 1969 observations was the separation of the anticyclone from the loop; first, the cases of separation are discussed; second, the conditions preceding separation; and, third, the conditions following it. Finally, some regularities in the occurrences, size and development of cyclonic and anticyclonic features encountered in 1969 and other years are noted.

Separation

The existence of a separation between the loop and the anticyclone has been documented

*Figure 4-11. Change in the acceleration potential at 250 cl t^{-1} in J kg^{-1} from the first to the second survey of May, 1969. Dashed lines are the acceleration potential lines for the first survey shown in Figure 3-6*a.

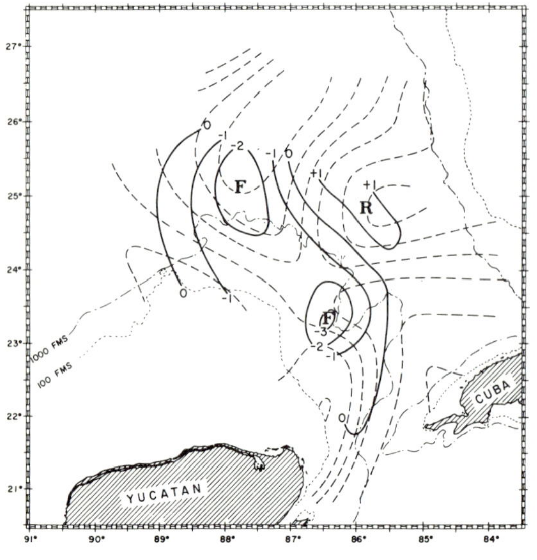

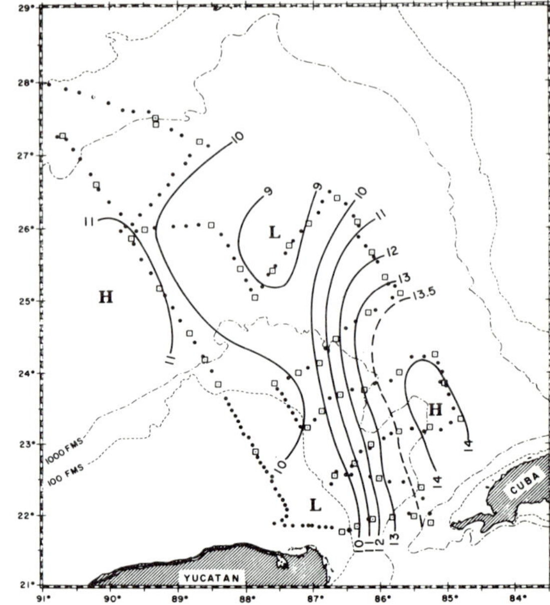

Figure 4-12. Acceleration potential at 250 cl t^{-1} relative to 1000 db in J kg^{-1} (dynamic decimeters) based on observations taken in September, 1969.

from May through August in three cases, that in 1969 described above, one in June, 1967, very clearly delineated by Nowlin et al. (1968), and one in August, 1965, reported by Leipper (1970) and in preliminary form by Cochrane (1966). The locations of each documented case of separation are seen in Figure 4-15 to be rather near the shortest distance from the 100-fm isobath of the Campeche Bank to that of Florida Shelf. Every separation was present to considerable depth and characterized by Shelf Water of low salinity (as Nowlin et al., 1968, have noted for the June, 1967 case).

Conditions Preceding Separation

During 1965, the separation was preceded in July by meanders, one on either side of the loop, if one may judge from the rather widely spaced lines of observation in that month (Leipper, 1970). In May, 1965, however, a cruise conducted by the author in the Yucatan Current region indicated that the Yucatan meander had not yet de-

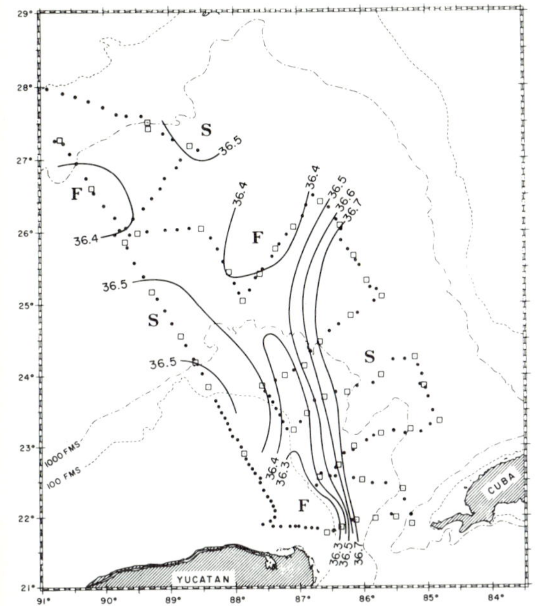

Figure 4-13. Salinity at 250 cl t^{-1} based on observations taken in September, 1969.

veloped. On the other hand, during 1967 (in January and April), rather inadequate lines made by *Alaminos* provide indications of a meander, or cold ridge, extending east from the Campeche Bank. Despite the lack of similarity in conditions preceding the documented cases, one may note that all of the separations were confirmed in the period from May to August and appeared to result from the growth and coalescence of a Florida and a Yucatan meander.

In spring observational series, meanders have been encountered rather frequently, although, because the meanders may be even less than 60 nautical miles across, they may have been missed in some cruises. Between May and August, a meander or cold ridge has been found in the West Florida Current (east limb of the loop) in every year from 1966 to and including 1970, except 1967 when a separation was already present when the earliest observational series of the year reached the Florida Shelf. Figure 4-16 shows where each of the latter group of Florida meanders was found. A Florida meander was also indicated by GEK observations made in May, 1960. In fact, there are no years with observations in which a Florida meander has not been found at some time from May to August. Even in the cruises of May, 1968 (Cochrane, 1969), and June, 1966 (Hubertz, 1967), before a strong meander was clearly present, there was some bulging of the West Florida Current away from the Florida Shelf.

The Yucatan Current, the west side of the loop, seems to undergo seasonal variation of speed, at least in its upper layers. Resultant currents based on ship-drift data given in U.S. Navy Hydrographic Office Miscellaneous Charts 10690-1 to -12 indicate, as Cochrane (1965) points out, a maximum speed in June and a minimum in November. Leipper (1970), on the basis of such seasonal variations and of bathythermogram series over somewhat more than one year, proposes a "spring intrusion" of the loop northward in the Gulf. Although the direct evidence of its occurrence is not extensive, a spring intrusion while the current is intensifying is plausible.

Figure 4-14. Successive positions of the 100-m contour and the anticyclonic center in the 22°C surface for observations taken in 1969.

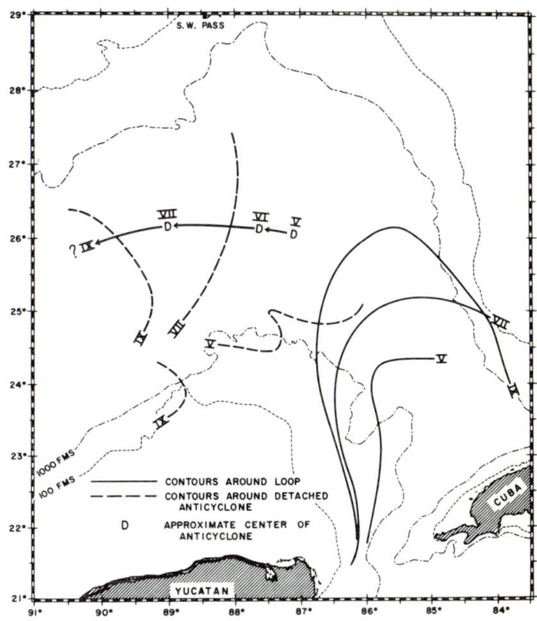

For a Florida meander to form in the locations where they have been found (Figure 4-16), the loop must extend somewhat north of these locations. The spring intrusion proposed by Leipper would lead to the loop's extending fairly well to the north of the meander position by the period from May until August. Possibly, the meander begins to form when the West Florida Current (the east side of the loop) impinges against the edge of the Florida Shelf so as to develop a "topographic" wave of the type described by Warren (1963). The effect of topography on the Yucatan Current is discussed in Chapter 8.

Although the formation of a Florida meander may even be a yearly occurrence, the detachment of an anticyclone is apparently not, as has already been noted. In each of the three years in which a separation was documented, the Yucatan Current was found somewhat east of its normal position. In July, 1965, June, 1967, and May, 1969, the direction of the current at Yucatan Strait was toward the north or somewhat east of north rather than west of north, as it has been when a separation has not occurred. Prior to separation, the Yucatan meander was definitely present in May, 1969, and very probably present in July, 1965, and April, 1967. On the other hand, in August, 1966 (Leipper, 1970) and August, 1968 (Schneider, 1969), when the marked Florida meanders shown in Figure 4-16 were present, the Yucatan Current core lay rather close to the eastern edge of Campeche Bank. Although the loop was constricted, a separation had not formed. It appears that the development of a Yucatan meander is a necessary step in producing a separation.

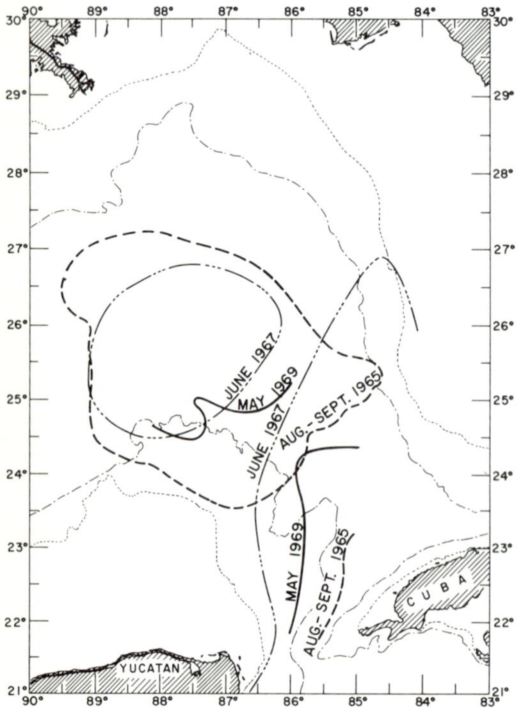

Figure 4-15. Locations where a separation between the Loop Current and an anticyclone has been documented, as indicated by positions of the 100-fm contour of the 22°C surface for August-September, 1965, June, 1967, and May, 1969.

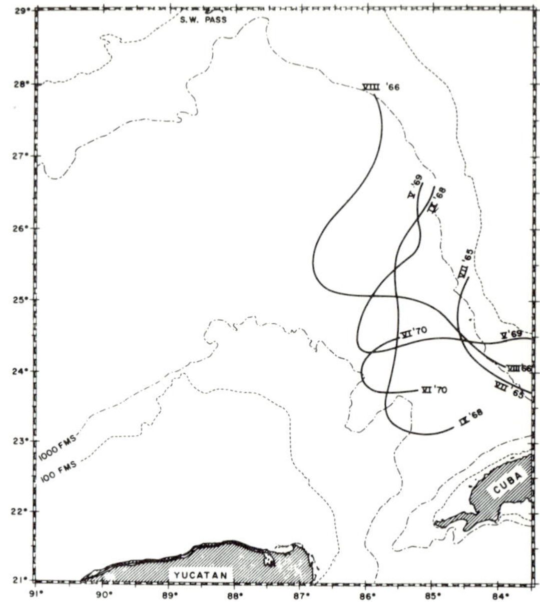

Figure 4-16. Locations where a marked meander has been found in the West Florida Current from May through August.

Possibly, a Yucatan meander starts as a topographic wave. If so, its origin may lie in the Cayman Sea where the Yucatan Current first forms as a western boundary current. Thus, the location, direction and speed at the core of the Caribbean Current where it first reaches the western boundary become important. (See Chapter 8.)

Just how the Yucatan and Florida meanders may grow together is not clear. Possibly, the anticyclonic shear zone set up in the constriction between the two meanders is unstable. The rather strong anticyclonic shear zone indicated between the meanders in the first reconnaissance of May, 1969, was replaced with remarkable rapidity by a cyclonic shear zone.

Conditions Following Separation

The three documented cases of separation show considerable differences after separation. In September, 1965 (Leipper, 1970; see Chapter 5 of this volume, Leipper et al., 1971), the detached eddy broke up and its parts apparently moved to the north and west. The movement was smaller than that which occurred between May and September, 1969. The September and November, 1965, *Alaminos* cruises (Leipper, 1968) considered together indicate broadening of the separation. By November, the loop had pushed somewhat to the east against the Florida Shelf and somewhat farther to the north than it was in August and September. The northern part of the loop was, however, far to the south of its September, 1969, position.

In the August, 1967 cruise (Leipper, 1968), the only cruise available from the period after the separation found in June, the loop had receded far to the south, the separation had broadened enormously, and the detached anticyclone, if it is correctly identified, had moved to the southwest so that its center was just north of Campeche Bank near 88° 30′W and had decreased in intensity.

The tendency of detached anticyclones to move westward suggests that they may account for the anticyclones found occasionally in the central Gulf and often as dominant circulation features in the west. However, further observation of anticyclones detached from the loop are clearly required to establish their relationship to the anticyclones which are encountered farther west.

Cyclonic and Anticyclonic Features

Some aspects of the cyclonic and anticyclonic features found in 1969 have also been noted in other years. While the number of cases is too small to permit generalization, the common aspects of these features seem worthy of note.

Clearly marked cyclonic current rings have been found very seldom, if at all, in the Gulf of Mexico. Broad, amorphous cyclonic regions, on the other hand, have often been encountered, usually extending out from the shelf edge (e.g., Leipper, 1970). Cyclonic shear zones such as the separations between loop and anticyclone are not uncommon, as was noted above. Cyclonic meanders are apparently a yearly occurrence along the Florida Shelf and have often been encountered elsewhere. The shear zones have sometimes been of small scale and the meanders usually have been so. The meanders move or change as rapidly as any feature detected by the ordinary shipboard observations. Clearly, the meanders cannot be described adequately without rapid, detailed coverage. (The necessity of dense networks of observations has been pointed out by Nowlin et al., 1968.)

In contrast to the small, cyclonic features, the large, anticyclonic features are usually rather slow in movement and development. Because of this, there appears to be little question that such large features as the detached anticyclonic rings and the central portions of the loop can be identified over several months. This is surely the case in the 1969 series discussed above and in the sequence of current patterns reviewed by Leipper (1970). However, although the central portions of a large anticyclonic feature can easily be identified, it is probably seldom possible to delineate its peripheral form without distortion on the basis of a single-ship survey because of changes which may occur during the survey in the small cyclonic features of the periphery.

Although small anticyclonic features may occur at scales below those normally detected by commonly used spacing and types of observations from aboard ships, anticyclonic features with the scale of the Florida and Yucatan meanders have not often been found. An intense anticyclonic shear zone has been encountered only once, in the first May, 1969, survey. The occurrence of such a feature is apparently rare or of short duration.

Acknowledgment

This work was supported by the Office of Naval Research under Contract Nonr 2119(04) through the Texas A&M Research Foundation.

References

Cochrane, John D. 1965. *The Yucatan Current.* In Unpubl. Rept. of Dept. of Oceanogr., Texas A&M University, Ref. 65-17T:20-27.

______1966. *The Yucatan Current.* In Unpubl. Rept. of Dept. of Oceanogr., Texas A&M University, Ref. 66-23T:14-25.

Fuglister, F. C. and Worthington, L. V. 1951. Some results of a multiple ship survey of the Gulf Stream. *Tellus*, 3(1):1-14.

Hubertz, Jon M. 1967. *A study of the Loop Current in the Eastern Gulf of Mexico.* Master's Thesis, Texas A&M University.

Leipper, Dale F. 1968. *Hydrographic station data, Gulf of Mexico, August-November STD, 1965-1967.* Unpubl. Rept. of Dept. of Oceanography, Texas A&M University, Ref. 68-14T.

______1970. A sequence of current patterns in the Gulf of Mexico. *J. Geophys. Res.*, 75(3):637-658.

Montgomery, R. B. 1937. A suggested method for representing gradient flow in isentropic surfaces. *Bull. Amer. Meteorol. Soc.*, 18:210-212.

______1938. Circulation in the upper layers of southern North Atlantic deduced with use of isentropic analysis. *Pap. Phys. Oceanogr. and Meteorol.*, 6(2).

______and Wooster, Warren S. 1954. Thermosteric anomaly and the analysis of serial oceanographic data. *Deep-Sea Res.*, 2(1):63-70.

______and Stroup, E. D. 1962. *Equatorial waters and currents at 150°W in July-August 1952.* John S. Hopkins Oceanographic Study No. 1.

Nowlin, W. D., Jr., and McLellan, H. J. 1962. A characterization of the Gulf of Mexico waters in winter. *J. Mar. Res.*, 25(1):29-59.

——, Hubertz, J. M. and Reid, R. O. 1968. A detached eddy in the Gulf of Mexico. *J. Mar. Res.*, 26(2):185-186.

Schneider, Michael J. 1969. *A description of the oceanographic features of the eastern Gulf of Mexico, August 1968.* Master's Thesis, Texas A&M University.

Warren, Bruce A. 1963. Topographic influences on the path of the Gulf Stream. *Tellus*, 15(2): 167-183.

Wüst, Georg. 1964. *Stratification and circulation in the Antillean-Caribbean basins,* Pt. 1. New York: Columbia Univ. Press.

5

A Detached Eddy and Subsequent Changes (1965)

Dale F. Leipper, John D. Cochrane and LCDR John F. Hewitt, USN

Abstract

An isolated eddy which was observed in the Gulf of Mexico in August, 1965, is described. The velocity at the core of the current in the eddy, 113 cm/sec, was comparable to that in the East Gulf Loop Current itself. A month later, following the passage of Hurricane Betsy, the eddy was considerably modified in shape and the volume transport had decreased from 40 to 19 million m^3/sec. The velocities decreased from 113 to 73 cm/sec in the core of the current.

Nowlin, Hubertz and Reid (1968) reported on a cruise in June, 1967, and established the existence of a major eddy which had evidently become detached from the Loop Current in the eastern Gulf of Mexico. Such a phenomenon also occurred in 1965 and has been described in technical reports by Leipper (1970) and Cochrane (1966). In this case, a month after the isolated eddy was observed, another cruise was conducted and a marked change was found to have occurred in the eddy. Hurricane Betsy had passed over the area in the interim.

Three successive cruises conducted in August and September, 1965, defined the isolated eddy and indicated its anticyclonic circulation. Hydrographic stations were made at selected critical positions. BTs were obtained hourly on the first two cruises at the positions shown in Figure 5-1, taken from Leipper (1970). This figure presents the topography of the 22°C isothermal surface, the surface used also by Nowlin et al. to represent partial results of their cruise. The topography shown in Figure 5-1 indicates clearly the separation of a

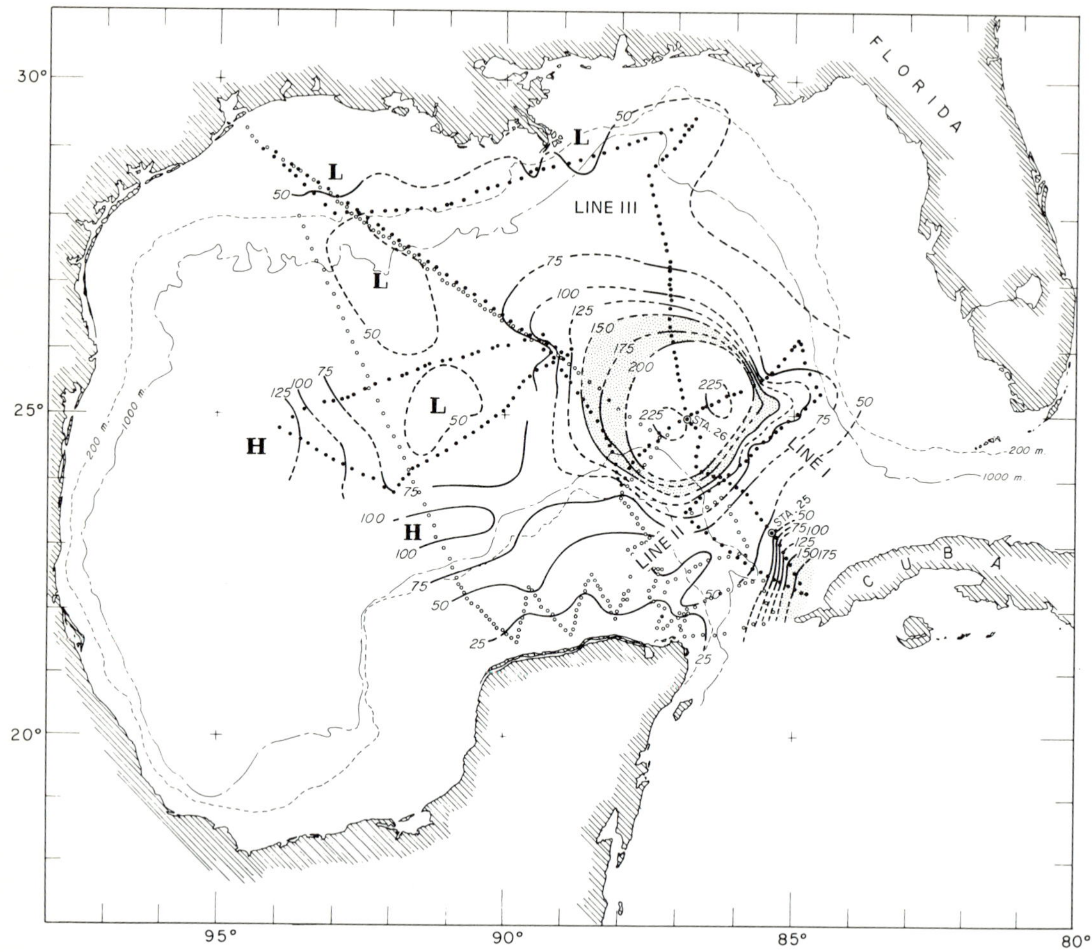

Figure 5-1. Station locations and topography of the 22°C isothermal surface for two cruises, 10-24 Aug. and 31 Aug. - 8 Sept. 1965.

major eddy from the Yucatan Current. Interpretation of the isothermal surface topography in terms of the current pattern in the Gulf is discussed by Ichiye (1962) and Leipper (1970).

Considerable additional evidence was obtained supporting the separation of the eddy from the Loop Current near Cuba. On the first cruise, as shown in Figure 5-1 by solid black dots, two lines of observations cut northwestward from Cuba across the Loop Current and show that current to be between St. 25 and the island. One of these lines extends farther and crosses the detached eddy to the northwest, centered near St. 26. On these lines, between the loop around Cuba and the eddy, the 22°C surface rises to less than 50 m, indicating that the cold ridge defining the left side (to an observer looking downstream) of the Loop Current has been reached.

The thermal structure along Line III, the most nearly north-south line in Figure 5-1, from near Panama City, Florida, to the west tip of Cuba and passing through Sts. 25 and 26 is given in Figure

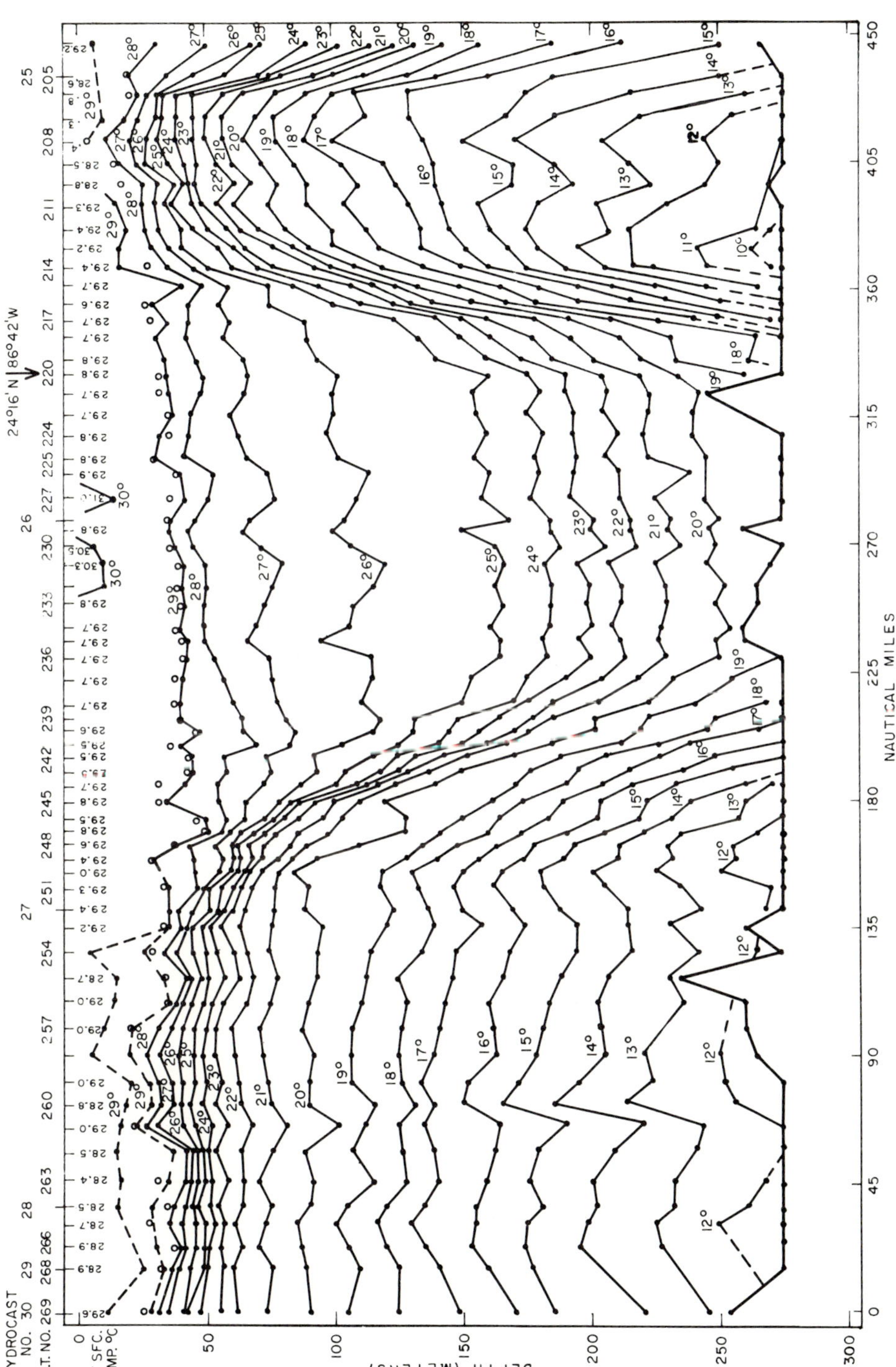

Figure 5-2. North to south temperature section (°C) across the detached eddy and the Loop Current along Line III, 20-22 Aug. 1965.

5-2. The separation between the southwesterly flowing portion of the eddy and the northeasterly flow along the coast of Cuba is shown by the rise and fall in isotherms as they reach and pass a minimum depth at BTs 207 and 208, about 35 nautical miles from the Cuban coast. The BT sections were plotted as the cruise progressed.

Since the separation of the eddy from the Loop Current was suspected, a decision was made to test the separation by making a line of observations which would hopefully fall between the two features. This test line was that marked Line I in Figure 5-1, running from northeast to southwest. The line turned out to be on the north edge of the cold ridge separating the two circulations. Line I extended fully across the deep basin north of the Yucatan Strait. The relatively flat temperature structure observed along this line (Figure 5-3) precludes the possibility of any significant flow of water across it. The 22°C isotherm dips only some 20 m, whereas in a crossing of the loop it dips more than 125 m, as in Figure 5-2. Experience on many cruises has demonstrated, as first mentioned by Ichiye (1962), that the thermal structure is a dependable indication of the presence of strong flows in the Gulf.

From August 31 to September 8, beginning a week after the cruise just discussed, the Western portion of the Yucatan Strait region was surveyed. Locations of BT observations for this cruise are also shown in Figure 5-1 (open circles). These observations indicated, since the 22°C surface was less than 50 m deep, that the cold ridge between the eddy and the loop flow extended well onto the Yucatan Shelf at that time and that the loop and the eddy flow were not connected in this area. Also, systematic observations were made with the GEK. These are shown in Figure 5-4 as reproduced from Cochrane (1966). These direct measurements of current confirm the strong southwesterly flow in the southern portion of the eddy (near top of Figure 5-4) and the unusual northeasterly flow from the Yucatan Strait in the Loop Current. The area between the eddy and the loop shows only weak and variable flow.

St. 25 of the first cruise, taken between the eddy and the loop at the position shown on Figure 5-1, was typical non-Caribbean water, the type called left-hand water by Leipper (1970) and in the Gulf found only outside the East Gulf Loop. This station position is one which in our experience is always encircled by any unbroken loop current extending farther north and which has Caribbean water present in such cases. The present data thus indicate the loop current to be broken. The maximum salinity at St. 25 was at 75 m depth and was less than 36.5 per mil. The temperature was lower than at St. 26 by 1-8°C at all depths above 1000 m. This fact and the low maximum salinity show that Caribbean water was not present at the time of the August, 1965, cruise. On the other hand, St. 26 (see Figure 5-1), which was in the center of the eddy, had a maximum salinity of 36.82 per mil which occurred at about 200 m. It also has the high temperature characteristic of water inside the East Gulf Loop or in its derivatives. A comparison of temperature and salinity values at these stations is shown in Figure 5-5.

Finally, direct current estimates for the first cruise were made independently by *Alaminos* Captain Lewis Newton. Ship speed was about 10 knots. Since most current crossings were at right angles to the current, the set was in the direction of flow and the drift should have been nearly equal to the speed of the current. On the August cruise along Line III (Figure 5-1) from Florida toward Cuba, Captain Newton's calculations showed a drift of 63 cm/sec as the ship crossed the northern side of the eddy and 96 cm/sec in the opposite direction as it crossed the southern side. These drifts persisted 9 hours and 7 hours, respectively, and their reliabilities were 60 and 80%. (The estimate of reliability depended, among other things, on the performance of the Loran.) Geostrophic computations for these same two crossings were 65 cm/sec where the drift was 63 cm/sec and 120 cm/sec where the drift was 96 cm/sec. The three other crossings of the eddy gave geostrophic values between 123 and 129 cm/sec. The

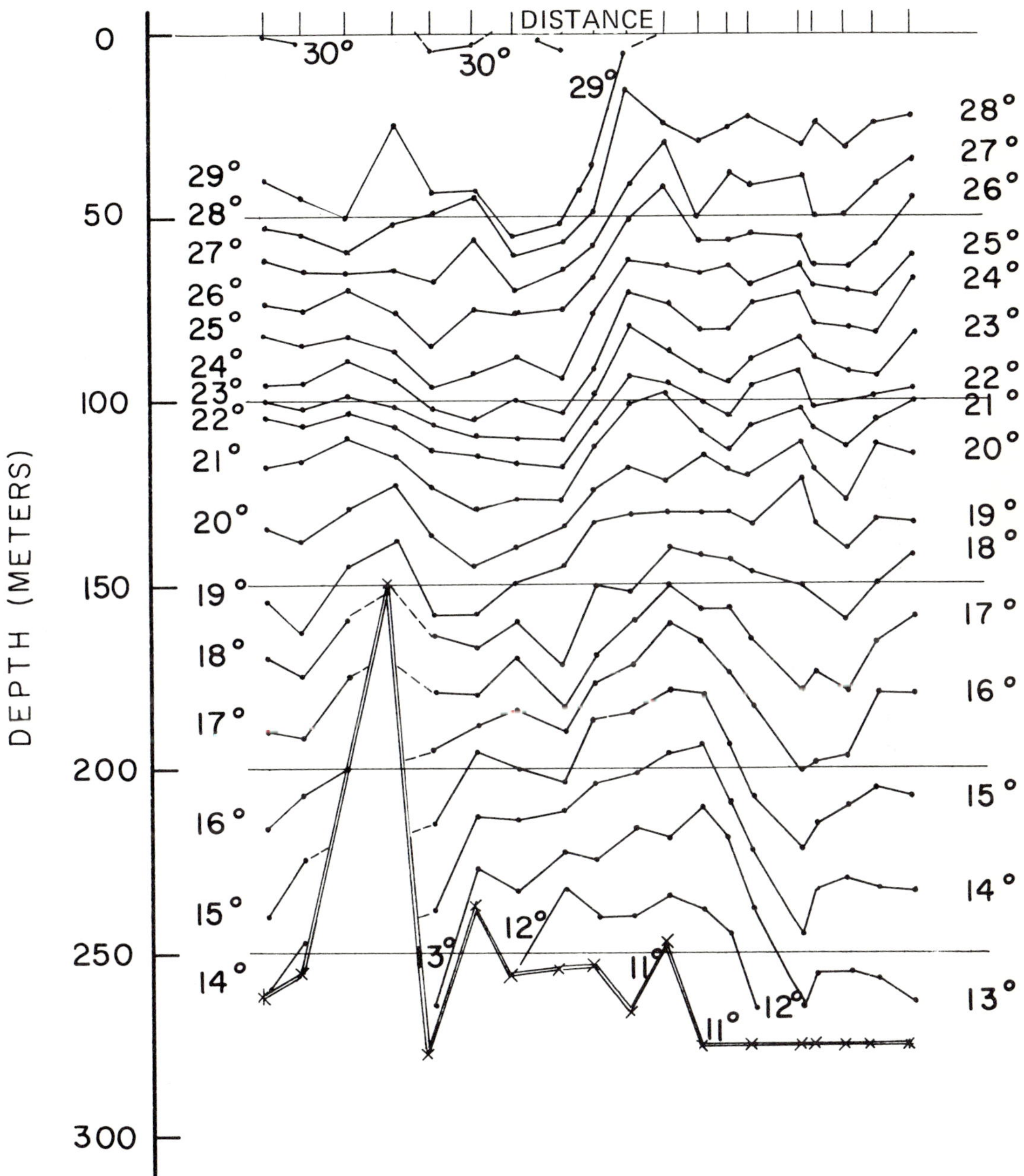

Figure 5-3. Southwest to northeast temperature section (°C), along Line I indicating no significant flow across the deep basin between Yucatan and Florida (Aug. 1965).

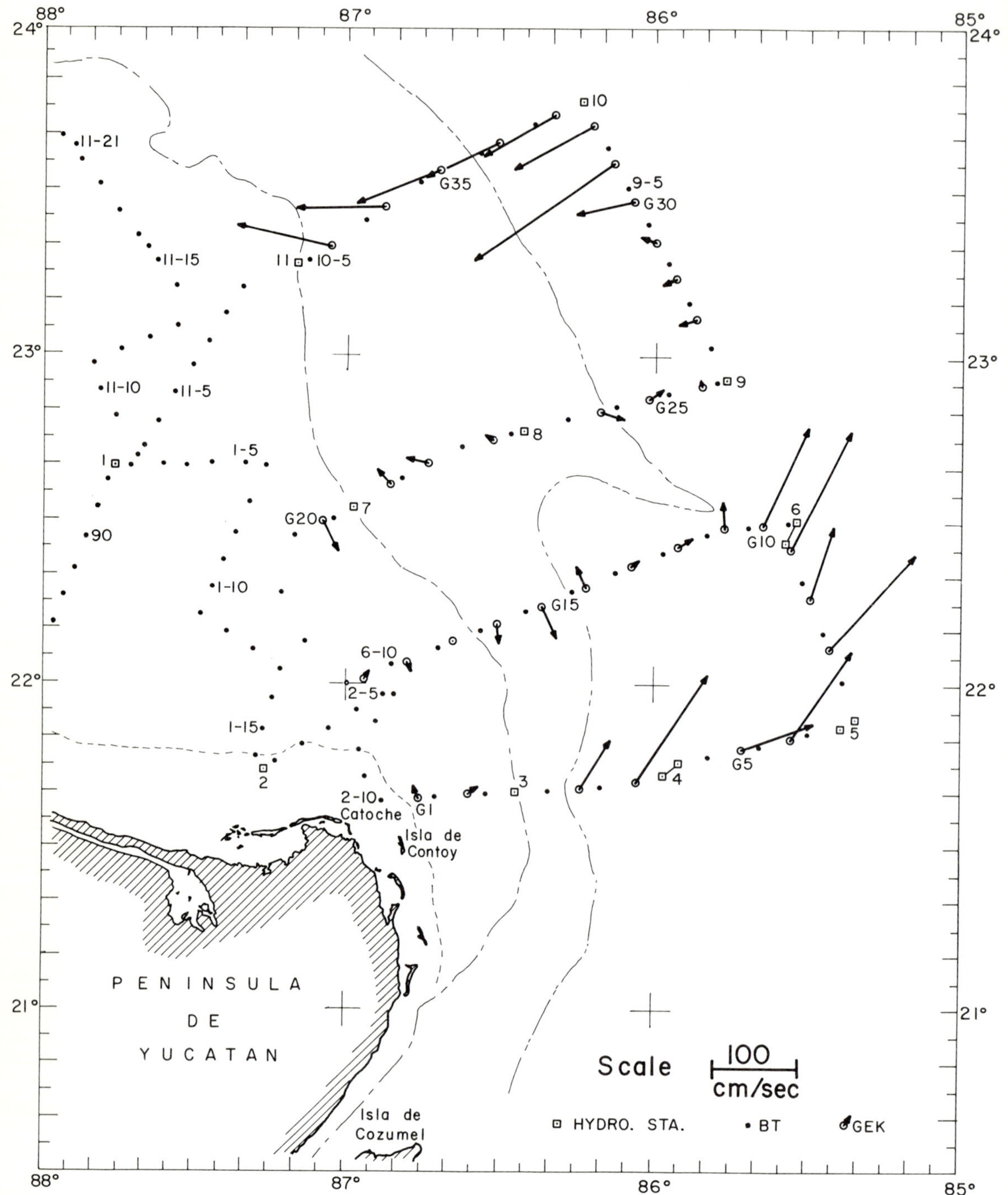

Figure 5-4. Surface current vectors from GEK measurements, 31 Aug. - 8 Sept. 1965.

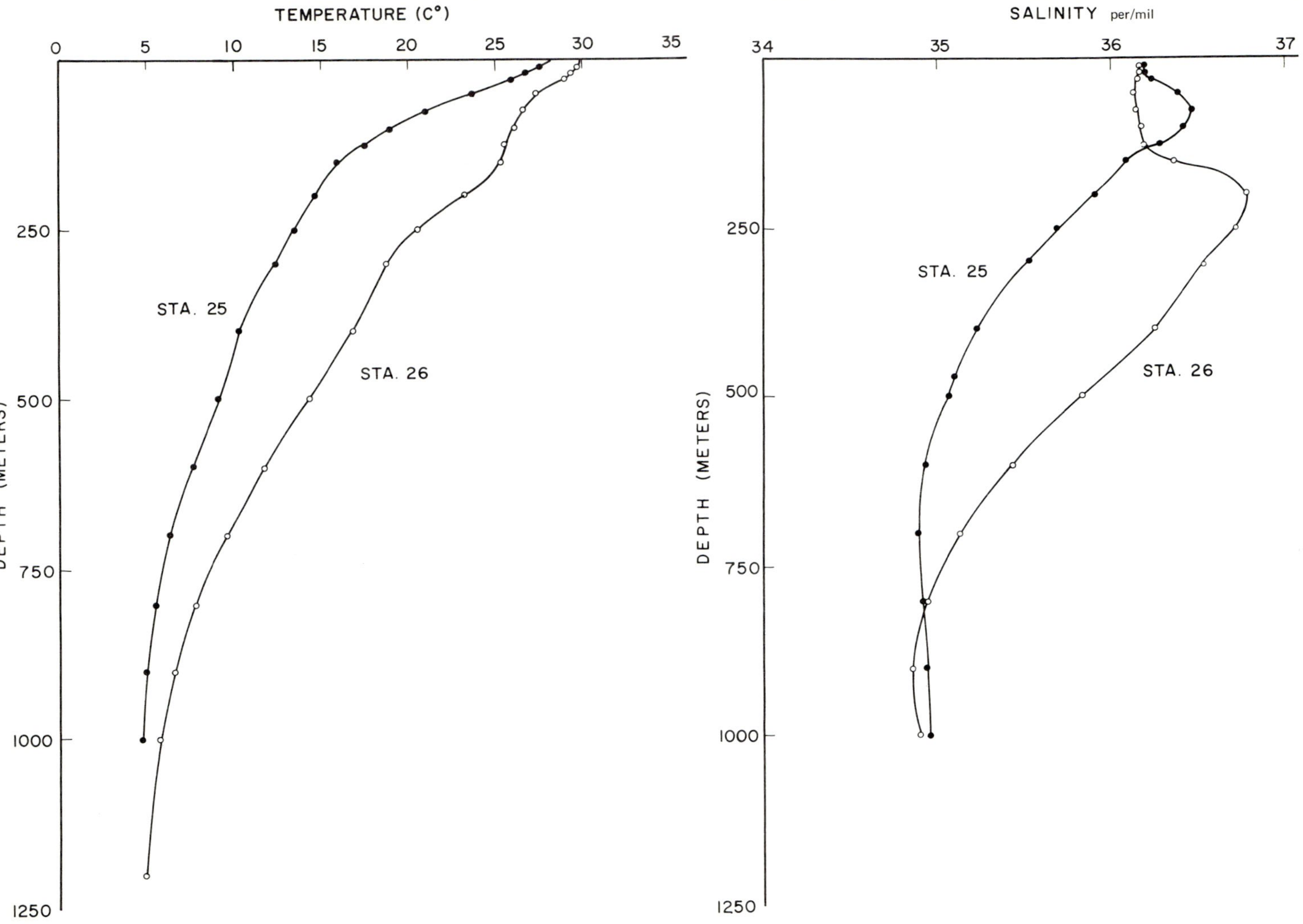

Figure 5-5. Comparison of (a) temperature, and (b) salinity values at Stations 25 and 26, Aug. 10-24, 1965.

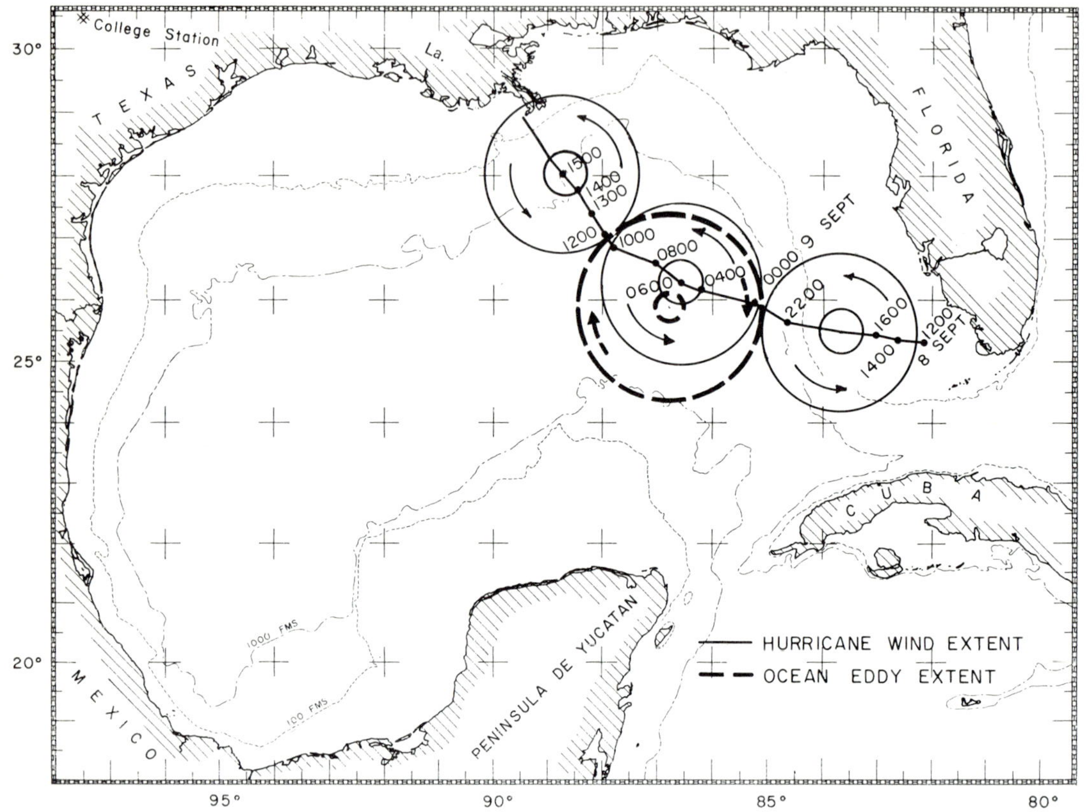

Figure 5-6. Path of Hurricane Betsy and size of the hurricane force wind area over the detached ocean eddy.

low value of 65 at the one northern crossing apparently was related to a broadening of the flow there.

As the ship traversed the center of the eddy along Line III Captain Newton observed a drift of only 13 cm/sec, reliability 50%, over a 16-hour period.

In the two crossings of the Loop Current near Cuba the drift was calculated to be 40 cm/sec for 9 hours, reliability 90%, on the western leg (Line II) and more than 103 cm/sec for 4 hours, 11 minutes, reliability 70%, on Line III. On St. 25 on the northern edge of the loop a drift of 91 cm/sec was calculated with a set toward the northeast over a 4-hour period, reliability 50%. Geostrophic velocity across the loop was 121 cm/sec but was questionable since the data were not complete.

On Line I, positioned hopefully to fall between the loop and the eddy, the Captain, as he left the Florida Shelf and headed southwest, estimated a drift of 52 cm/sec with a set to the left for the first 4 hours, reliability 50%. However, for the 14 hours needed to complete this line he calculated that there was no drift whatsoever and estimated the calculation to have an 80% reliability.

Turning toward Cuba from the Campeche Bank along Line II, Captain Newton estimated a drift of 26 cm/sec with a set to the right for 9½ hours, reliability 50%, before he entered the Loop Current which then set him in the opposite direction.

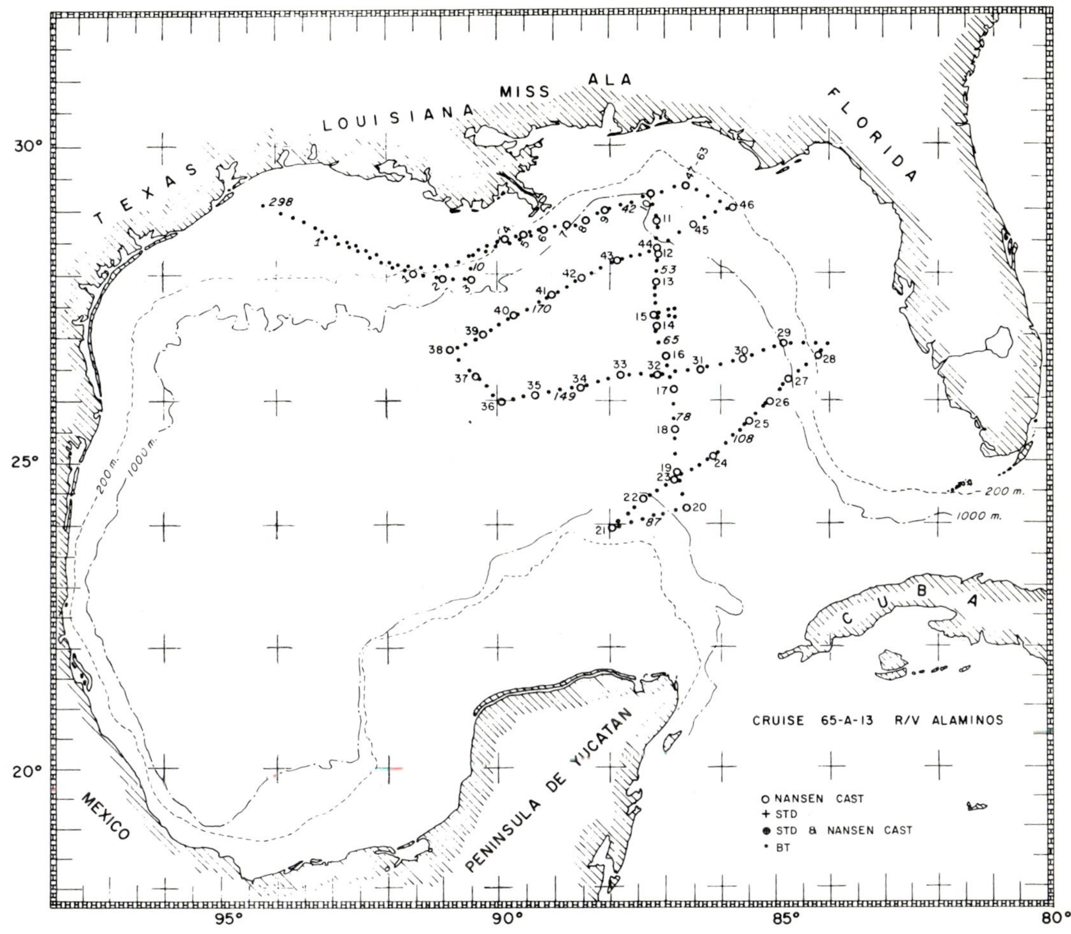

Figure 5-7. Station locations after Hurricane Betsy, Sept. 12-24,1965.

These various lines of evidence indicate clearly that the eddy was detached from the loop. Observations were close together. They were not made in the usual survey grid pattern but were planned in the light of what was known about the area to give maximum information with minimum ship time. Since many of the observation lines were run directly across the current and perpendicular to the flow, a minimum of time was used in each crossing. Although additional hydrographic stations would have been desirable to supplement the BT data more fully, the position and character of the flow were accurately determined.

When Hurricane Betsy (on September 8) crossed the area just described, the original ship plans were canceled and the *Alaminos* became available for 12 days. The path of the storm is shown in Figure 5-6. The ship days were used to repeat the lines of the cruise shown in Figure 5-1 crossed by the hurricane, i.e., the line from Galveston to Panama City, the pertinent northern portion of Line III and the northeast to southwest

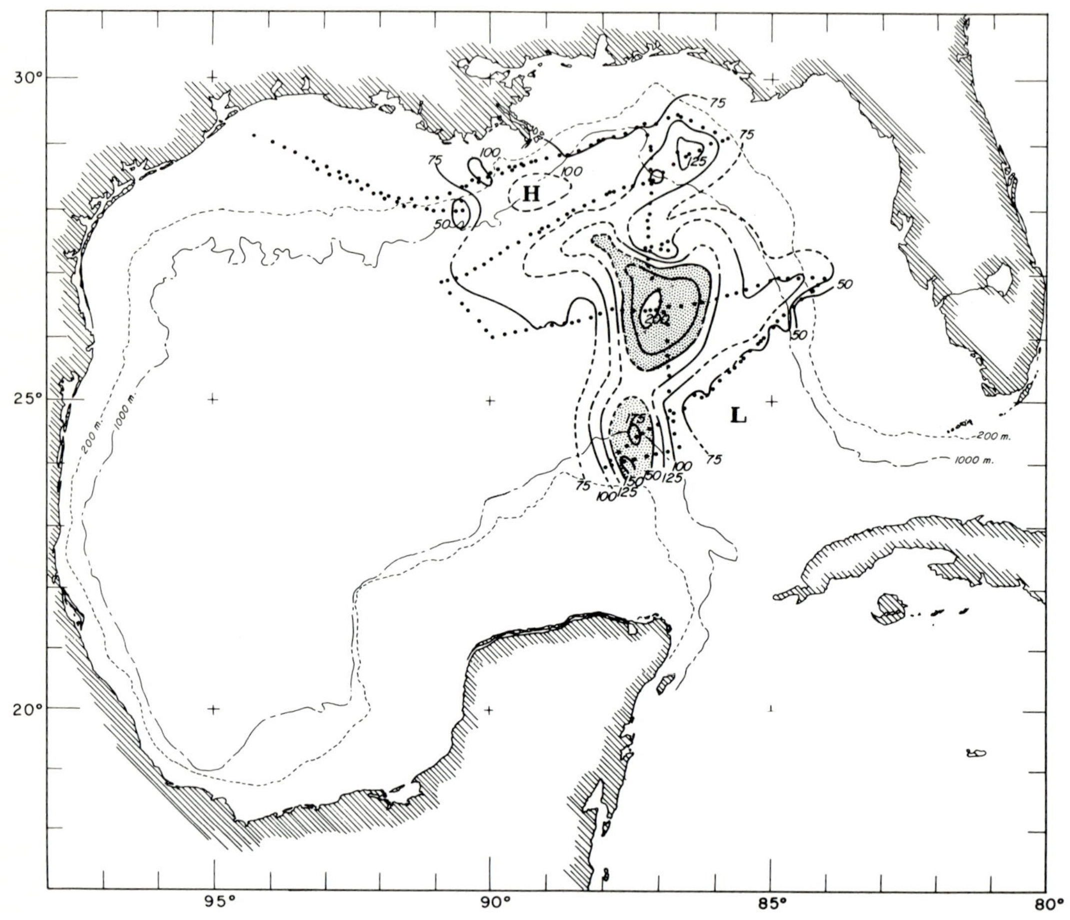

Figure 5-8. Topography of the 22°C isothermal surface, Sept. 12-24, 1965.

line through the center of isolated eddy, Figure 5-1. Two additional northeast to southwest lines were run to more fully cover the hurricane area, Figure 5-7.

At one position the hurricane was centered very close to the center of the isolated eddy, a strong cyclonic circulation in the atmosphere overlying a strong anticyclonic ocean circulation. Figure 5-6 shows the relative sizes of the eddy of the area of winds greater than 75 mph at this time.

A measurement of ocean heat loss to Hurricane Betsy based on the above cruises showed a total loss of 5×10^{18} gm cal from the Gulf and an average loss of 3750 calories per square centimeter in the areas crossed by hurricane winds (Whitaker, 1967). The beginning of the upwelling process which takes place in hurricanes (Leipper, 1967) was noticeable on the temperature sections but did not greatly alter the temperature structure because of the rapid passage of this particular storm.

In the post-hurricane survey September 12-24 the well-defined eddy of Figure 5-1 had split into two smaller eddies, one centered 100 km south and one 140 km north-northeast of the initial eddy, Figure 5-8. The two new eddies were surrounded by a flow of some 7 million m^3/sec and the volume transport was 12 million m^3/sec in each eddy, making a total flow of 19 million m^3/sec. This compares to a calculated average volume transport of 40 million m^3/sec in the initial eddy in August.

The geostrophic velocity averaged 113 cm/sec on the five crossings of the initial eddy and 73 cm/sec on seven crossings of the split eddy. At the sea surface the range of the dynamic height anomaly referred to 1000 m was 1.0555-1.8739 dyn m for the initial eddy and 1.0305-1.6331 for the eddies observed after the hurricane.

Although the small number of hydrographic stations on the initial surveys in August made the geostrophic portion of that analysis difficult, there were sufficient data of various kinds to describe the isolated eddy situation. The post-hurricane cruise was much more complete in the area of interest and the description of rapid change in the eddy which was observed may be useful in future work on circulation in the Gulf.

References

Cochrane, John D. 1966. The Yucatan Current. Report of the Dept. of Oceanography, Texas A&M University, Ref. 66-33T, 14-28.

Ichiye, T. 1962. Circulation and water mass distribution in the Gulf of Mexico. *Geofisica Internacional*, 2(3):47-76

Leipper, Dale F. (1967) Observed ocean conditions and Hurricane Hilda. *Jour. Atm. Sci.*, 24(2):182-196.

——. 1970. A sequence of current patterns in the Gulf of Mexico. *Jour. Geophy. Res.*, 75(3):637-657. (From 1967 Technical Report)

Nowlin, W.D., Jr., Hubertz, J.M. and Reid, R.O. 1968. A detached eddy in the Gulf of Mexico. *Jour. Mar. Res.*, 26:185-186.

Whitaker, A.D. 1967. Quantitative Determination of Heat Transfer from Sea to Air During Passage of Hurricane Betsy. M.S. Thesis, Texas A & M University.

6

Contrasting Summer Circulation Patterns for the Eastern Gulf

W. D. Nowlin, Jr. and J. M. Hubertz

Abstract

The results of two oceanographic surveys of the eastern Gulf of Mexico in June, 1966, and June, 1967, illustrate two contrasting summer circulation patterns of the area and provide the first detailed description of an anticyclonic ring detached from the Loop Current. This ring was observed in 1967 along with part of an older ring which appears to have moved westward. The transport in the upper 1350 m of the principal ring, as well as the Loop Current in 1966 and 1967, is at least 30 x 10^6 m^3/sec. The potential incipient formation of an eddy is noted in 1966 as a meander of the Loop Current.

Water mass analysis indicates that the intermediate and upper waters bounded by the Loop Current and rings are the same as those found in the northwest Caribbean. Water not bounded by these circulations has lesser salinity values at the salinity maximum than the Caribbean water and appears to be formed from the Caribbean mass by vertical mixing along the western edge of the Yucatan Current and over the Campeche Bank.

Vertical kinematic sections normal to the ring and Loop Current indicate that the geostrophic speed distributions, relative to 1350 db, observed in 1967 have the same general features as those observed in 1966.

Introduction

The name Yucatan Current is given to that portion of the Gulf Stream system which flows approximately northward through Yucatan Strait from the Caribbean Sea into the Gulf of Mexico. Within the Gulf this flow turns clockwise and then flows eastward between Cuba and the Florida Keys into the Straits of Florida, where it is known as the Florida Current. The clockwise-flowing stream often extends as a loop-like feature far into the central eastern Gulf, and for that reason it is referred to as the Loop Current of the eastern Gulf.

The circulation of the Gulf of Mexico is dominated by the Yucatan Current and its downstream extension, the Loop Current. Chapter 1 gives a discussion of the general circulation and property distributions in the winter. (See also Nowlin and McLellan, 1967.)

Although definitive evidence was lacking, such data as were available led to speculations (Austin, 1955) that eddies or rings distinct from the Loop Current exist in the Gulf and, more recently, that the Loop Current sheds rings within the Gulf (Leipper, 1967 and 1970). Even though the exact nature of the variability in circulation of the eastern Gulf was unclear, by the early 1960s Cochrane (e.g., 1965, 1966) had established the general character of the seasonal variations in the Yucatan Current.

To obtain quasi-synoptic descriptions of the eastern Gulf circulation during late spring which were as detailed as was feasible, we planned cruises aboard the *Alaminos* for June, 1966 and 1967. By these surveys we hoped (1) to establish a late spring picture which, though more detailed, might be contrasted with the winter situation observed by Nowlin and McLellan (1967) in 1962; (2) to provide environmental data needed in developing and testing numerical circulation models of the Gulf under study at Texas A&M University (e.g., Jacobs and Nowlin, 1968; Wert, 1968 [Chapter 11]; Paskausky, 1969 [Chapter 10]); and (3) to provide some measure of the degree to which the Loop Current repeated itself in the same season of successive years. With regard to this latter point, Cochrane had regularly observed in spring (May) and fall (October-November) seasonal patterns to the Yucatan Current, in Yucatan Strait and along the eastern edge of the Campeche Bank, which appeared to repeat from year to year. We rather expected a similarly regular pattern for the Loop Current.

Since Cochrane commonly observed the Yucatan Current in May, our plan was to concentrate principally on obtaining June observations of the Loop Current downstream from Cochrane's study area. Most of the deep water area of the eastern Gulf was surveyed as rapidly as possible to obtain a combination of traditional Nansen casts, salinity-temperature-depth (STD) profiles, BT observations and GEK surface current estimates. The locations of stations occupied during 1966 on *Alaminos* Cruise 66-A-8 and during 1967 on *Alaminos* Cruise 67-A-4 are shown in Figure 6-1. Planned also were direct current measurements from a vertical array of buoyed meters moored in deep water. No data were obtained from the system due to mechanical difficulties.

The current regimes observed on Cruises 66-A-8 and 67-A-4 proved to be totally different. In 1966, we observed a well-developed Loop Current which extended far into the northeastern Gulf; in 1967, we observed that a large anticyclonic ring of current had become detached from the Loop Current, which itself extended but a little way into the southeastern Gulf. This chapter documents the results of these two cruises, thereby establishing the relative features of two distinct circulation modes within the eastern Gulf. This is also the first attempt at a detailed description of a detached current ring in the Gulf of Mexico, although a rough description was given as a note (Nowlin et al., 1968) in which this same ring was first reported.

Current Patterns

The contrast between the apparent surface flow in June of 1966 with that of June 1967 was truly striking, even as represented by comparing

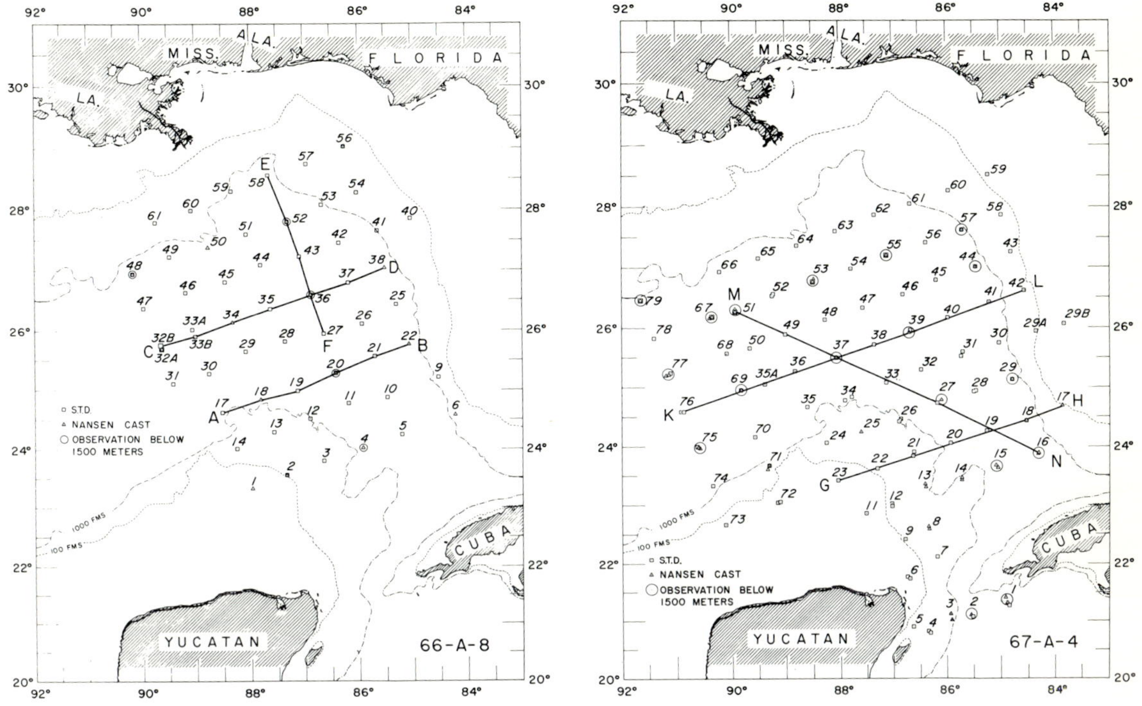

Figure 6-1. Positions of stations occupied during (a) Alaminos *66-A-8 (5-22 June 1966) and (b)* Alaminos *67-A-4 (10 June - 1 July 1967). Locations of vertical sections* AB, CD, EF, GH, KL *and* MN *are shown.*

the surface currents estimated with a GEK. These individual measurements are shown in Figure 6-2. The only correction applied was for spatial distribution of the mean vertical magnetic field intensity. In Figure 6-2*a* Cochrane has kindly permitted the use of two lines of GEK measurements in the Yucatan Strait made on *Alaminos* 66-A-7 in May, 1966, just before Cruise 66-A-8.

During June, 1966, the surface flow of the Yucatan Current and its downstream extension reached north northwestward across the Gulf almost to the continental margin of Mississippi before turning clockwise. It then flowed around the periphery of the Gulf basin toward the Florida Straits, its main current axis remaining offshore of the 2000-m isobath. Thus, the Loop Current was particularly well-defined and extended far into the Gulf. The officers and crew aboard the *Alaminos*, who commonly work in the eastern Gulf of Mexico, claimed not to have previously seen the surface current so strong and broad within the Gulf.

By contrast, GEK measurements from June, 1967, show a northward flowing Yucatan Current which begins to turn clockwise about 24°N and extends into the Gulf only to about 26°N. Based on the southernmost line of observations across the eastern side of the Loop, there appeared to be some clockwise circulation of the surface waters bounded within the Loop Current proper. Northwest of the Loop there was an anticyclonic ring of surface current centered about 25°30′N, 88°W.

The northernmost lines of GEK measurements made during either June are very interesting. The surface currents observed are small but quite variable, especially in direction. Although one really cannot separate spatial from temporal variations, it is tempting to speculate that there might be rela-

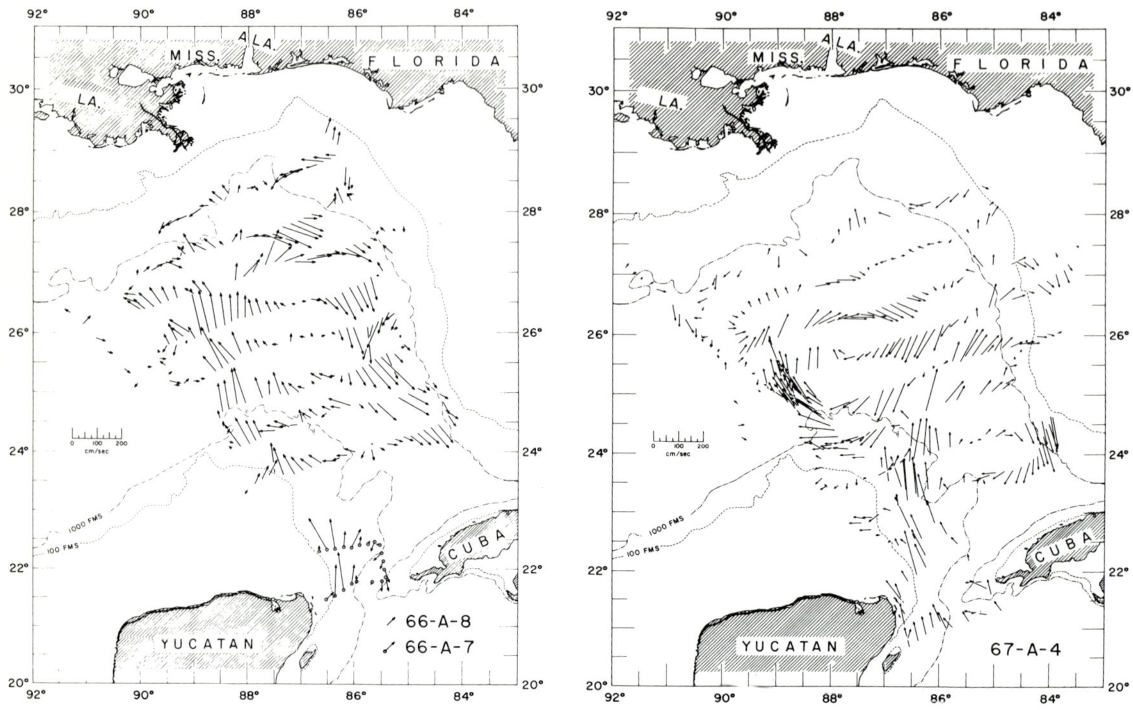

Figure 6-2. GEK surface current observations, Alaminos *(*a*) 66-A-8 and (*b*) 67-A-4. Also shown with 66-A-8 data are some observations made during* Alaminos *66-A-7 in May, 1966.*

tively short-period, regular fluctuations to this flow. If the GEK surface current vector is plotted as a function of time, it rotates in a clockwise direction with a near diurnal period, which is difficult to distinguish from the inertial period (of about 27 hours) at these latitudes. (These observations are discussed further in the Appendix.)

Based on the hydrographic station data and the STD observations, we have calculated the geopotential anomalies of the sea surface relative to 1350 db, as shown in Figure 6-3. As expected, the baroclinic geostrophic surface flows correspond quite closely to the fields of surface flow indicated by the GEK observations in Figure 6-2. There is a notable exception for 1966 however. The geopotential anomalies of the sea surface show a meander or wiggle in the north northwestward flowing current (a trough extends from the west into this current at about 27°N) which is not clearly seen when viewing the GEK surface current observations. It is just such meanders which undoubtedly grow in magnitude to finally separate from the main Loop Current such anticyclonic rings as that observed in June of 1967. The meander pictured might have been limited in its growth unless it moved upstream in the current system and away from the continental shelf. The geopotential anomalies of 1967 likewise show a feature which is not readily apparent in the corresponding GEK field, namely, a small cyclonic gyre northeast of the large anticyclonic ring and centered about 26°30′N, 90°W.

Judging by the internally consistent patterns manifested by these data and others which we present, the circulation features under study must have moved relatively little during the time interval in which the cruises were executed. This is in sharp distinction to the situation observed aboard *Alaminos* during July of 1969 by William Schroeder and Leo Berner (personal com-

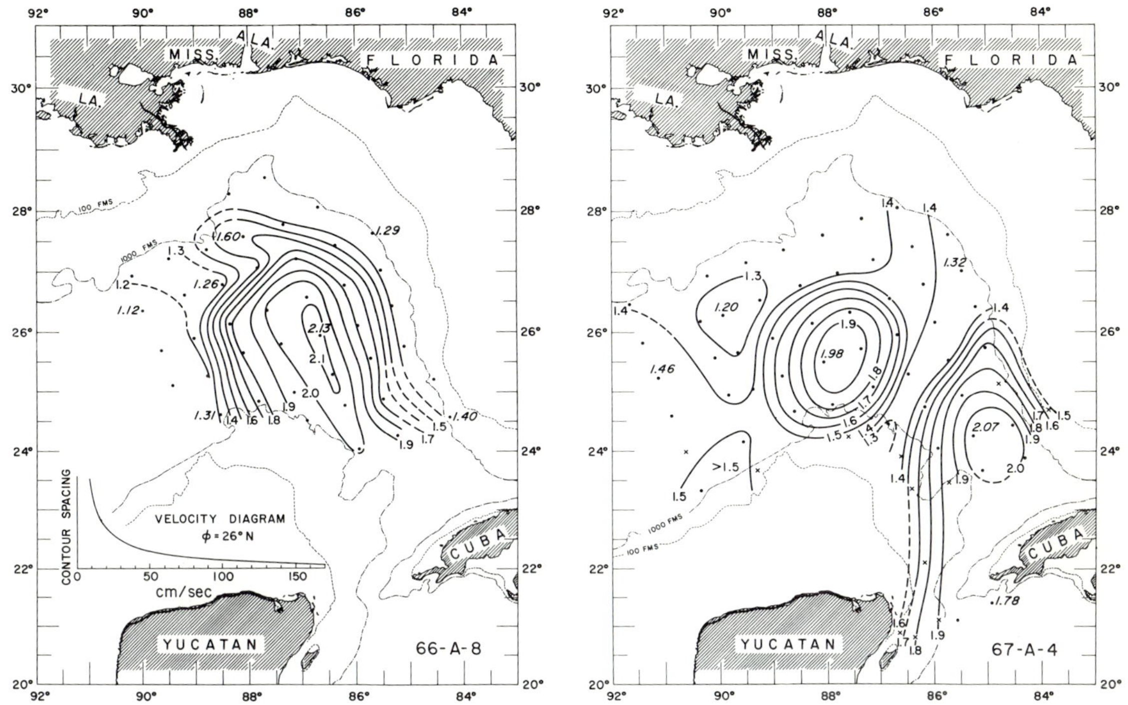

Figure 6-3. Geopotential anomaly of the sea surface relative to the 1350-db surface as inferred from Alaminos *(*a*) 66-A-8 and (*b*) 67-A-4 data.* X's *indicate some extrapolation. Contour interval, 0.1 dynamic meters.*

munication). They observed a ring separated from the Loop Current in the eastern Gulf much as we did in 1967. Their survey crossed the ring twice, with an interval of 10 days between crossings, and showed that its center had moved some 100 nautical miles south southwest during the interval. Compared with the principal ring seen in 1967, the ring they observed was further to the west on both occasions and was less intense.

The geopotential anomalies of the 480-db surface (Figure 6-4) show the same general features as the geopotential anomalies of the sea surface relative to 1350 db with one exception. For 1966, the manifestation of a meander in the current pattern at 480 db was not as distinct as it was in the surface flow pattern.

We chose to relate the distributions of properties to the horizontal current patterns by presenting for both June cruises the depth distributions of the 22°C isothermal surfaces and the horizontal distributions of salinity at the Subtropical Underwater core. (This is the name for the water mass which is characterized within the Gulf by the maximum subsurface values of salinity. Refer to Chapter 1 for more information.) The depths of the 22°C isothermal surfaces (Figure 6-5) were inferred from the BT observations. Data from Cruise 66-A-8 were supplemented with BT data taken by Cochrane on *Alaminos* 66-A-7. The addition of these BT observations to the 1966 picture makes it possible to provide a rather complete geographical coverage of the Loop Current for that year. The location of the main current axis corresponds closely to the position of the maximum slope of the 22°C isothermal surface. The similarity of features in Figures 6-3*a* and 6-5*a* again shows that one may obtain a very good representation of the internal geostrophic flow field simply by studying the temperature distributions alone in these tropical waters.

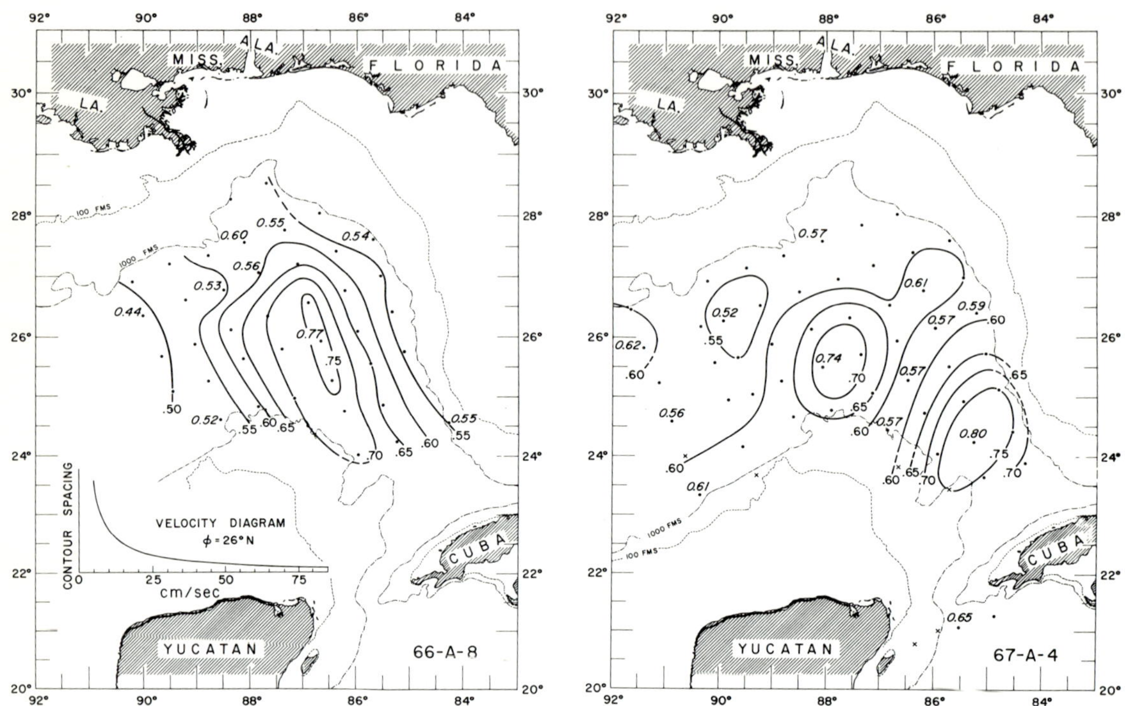

Figure 6-4. Geopotential anomaly of the 480-db surface relative to the 1350-db surface as inferred from Alaminos *(*a*) 66-A-8 and (*b*) 67-A-4 data.* X*'s indicate some extrapolation. Contour interval, 0.05 dynamic meters.*

Salinity at the Subtropical Underwater core observed on *Alaminos* 66-A-8 is not shown here. Salinity values at the core were greater than 36.7 per mil within the Loop Current regime. Sharp salinity gradients observed across the Current corresponded rather well to the greatest slopes of the 22°C isothermal surface. Outside the Loop Current regime the salinity values were generally 36.4-36.5 per mil.

Maximum salinity values at the Subtropical Underwater core in 1967 (Figure 6-6) clearly show that the water within the detached ring is of the same type as that enclosed in the Loop Current. These maximum values were determined by eye from the graphic analog STD records, in contrast to other salinity values which were digitally corrected as explained in the next section. Salinity values within the shear zone separating the Loop Current from the ring have maximum values which are smaller than values commonly found outside the Loop Current regime. Such values are probably due to intense vertical mixing, although they are found in this zone of very strong horizontal surface current shear. Could these regions also have strong vertical current shears?

Examination of Figure 6-6*a* shows two features which might be missed on the basis of the other data. One is a tongue of water with relatively high maximum salinity extending northward along the outer edge of the west Florida Shelf as far as 28°N. Two possible explanations for this distribution are that (1) although we have closed off at about 26°N the geopotential anomaly contours for the sea surface relative to 1350 db (Figure 6-3*b*), the Loop Current in June, 1967, actually has near-surface flow over the western shelf of Florida extending farther north, which is not inconsistent with the GEK pattern (Figure 6-2*b*); or (2) the

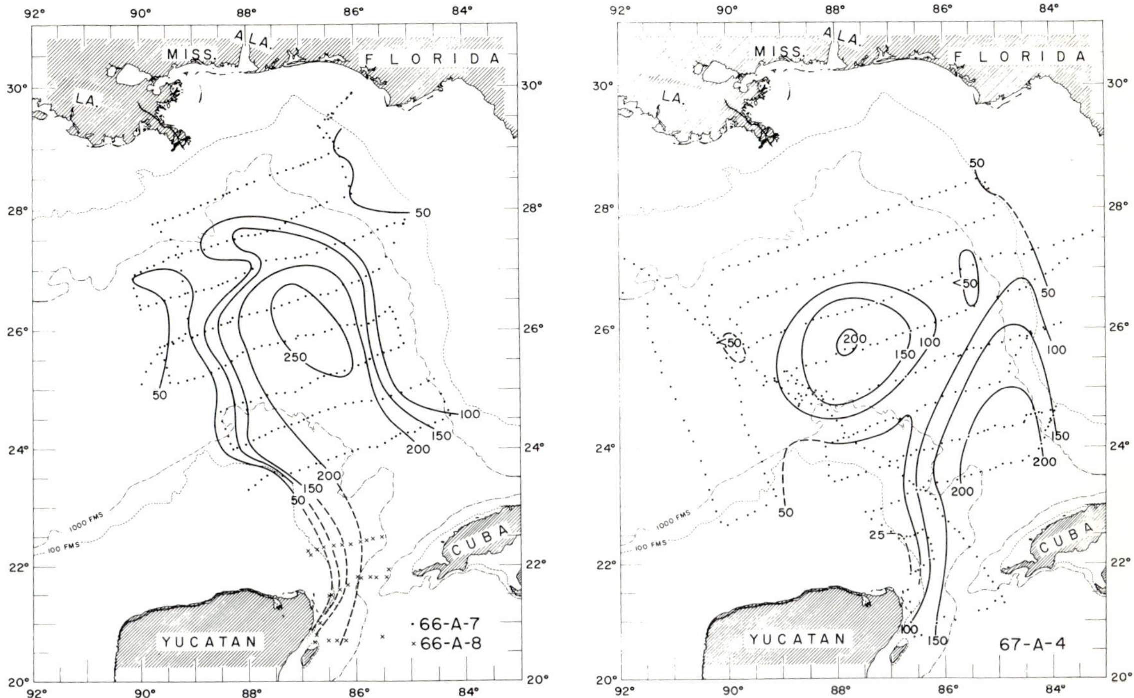

Figure 6-5. Depth contours of the 22° isothermal surface as inferred from BT observations at positions indicated by dots during Alaminos *(*a*) 66-A-8 and (*b*) 67-A-4. The southernmost three lines of observations in (*a*) were made on* Alaminos *66-A-7.*

lowest salinity values in the core around 26°N were missed by our sampling pattern. The second feature unique to the salinity distribution at the Subtropical Underwater core is the high core value (> 36.7 per mil) at St. 78, located near 25°50′N, 91°20′W. This seems to indicate the remnant of another ring which had separated earlier from the Loop Current. Within the region between these two rings in 1967, relatively low salinity values were found at the Subtropical Underwater core at Sts. 51 and 67. These values of less than 36.4 per mil must reflect strong vertical mixing which had occurred between this second ring and the Loop Current.

The contours of depth at which the maximum salinity values were observed in the Subtropical Underwater core are shown in Figure 6-6*b*. The depths of this core layer coincide amazingly well with the depth of the 22°C isothermal surface (Figure 6-5*b*). It appears that the depth distribution of this surface could be used just as well as the thermal distribution to infer the geostrophic circulation in these waters.

The baroclinic component of the geostrophic transport relative to the 1350-db surface has been calculated from the hydrographic station data and the STD observations. Figure 6-7 shows the geostrophic transports in the upper 1350 m; Figure 6-8 shows those in the upper 480 m. These figures were constructed by the following procedure: (1) at each station the geostrophic transport stream function was calculated relative to 1350 db based on a value of the Coriolis parameter appropriate to 26° latitude; (2) for each figure the minimum value of this stream function was subtracted from every value in the field; and (3) the residual values were contoured.

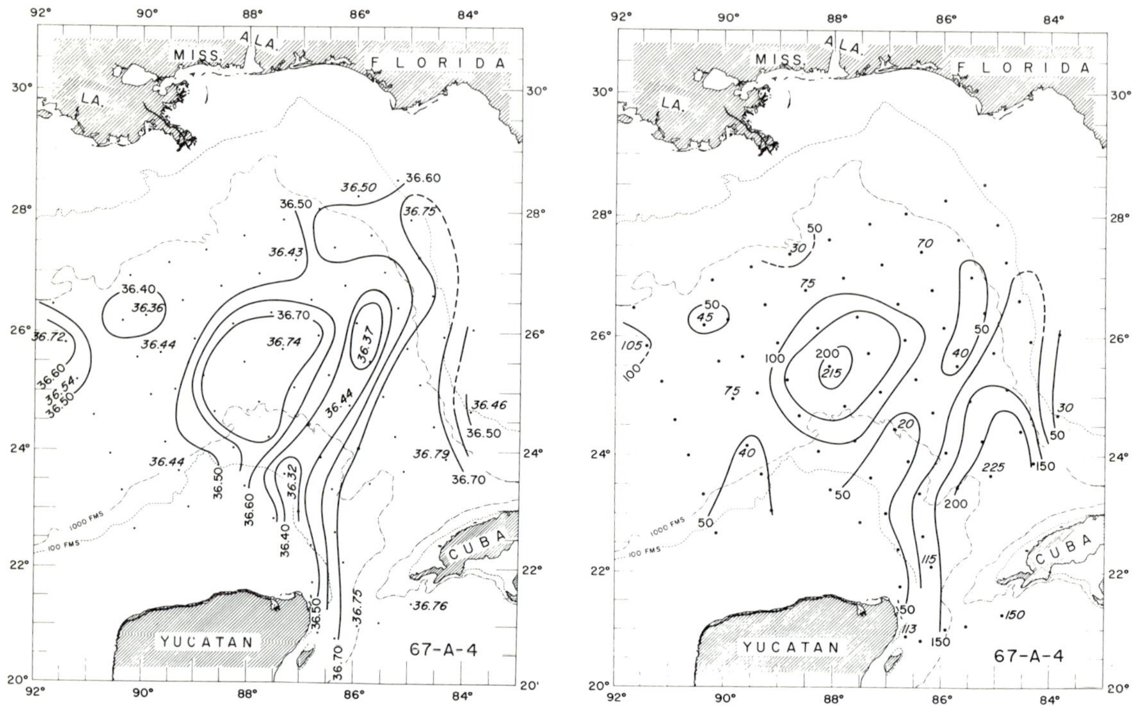

Figure 6-6. Core of Subtropical Underwater, Alaminos *67-A-4: (*a*) salinity (0.1 per mil intervals); (*b*) depth (50-m interval). Sample locations indicated by dots.*

For those portions of the Loop Current observed in 1966 and 1967, as well as for the principal detached ring observed in 1967, the volume transport in the upper 1350 m is at least 30 million m^3/sec, the value now commonly accepted as the probable mean transport in the Florida Current based on the work of Schmitz and Richardson (1968). Transports for the ring current reach 45 million m^3/sec; Loop Current transports exceed 50 million m^3/sec. In Figure 6-7*b* some transport (10 or more million m^3/sec) is seen to be associated with the residual ring in the central Gulf. Geostrophic transport calculations for the upper 480 m, relative to the 1350-db surface, show that at least 25 million m^3/sec of flow take place in the upper 500 m of the water column within the Loop Current.

The circulation and property distributions within the western Gulf of Mexico must vary significantly as rings which have separated from the Loop Current move westward into the region. Based on existing data, the circulation of the western Gulf seems to have an identifiable pattern in winter (see Chapter 1), when rings are not known to form from the Loop Current. Cursory examination of summer circulation patterns (Nowlin, unpublished) indicates that there is no predominant pattern to the western Gulf flow; this circumstance may be due to the sporadic movement into the western Gulf during spring and summer of these current rings which radically change both the current patterns and the distributions of properties.

Temperature-Salinity Relations

Most of the temperature and salinity values for waters within the upper 1200-1500 m were

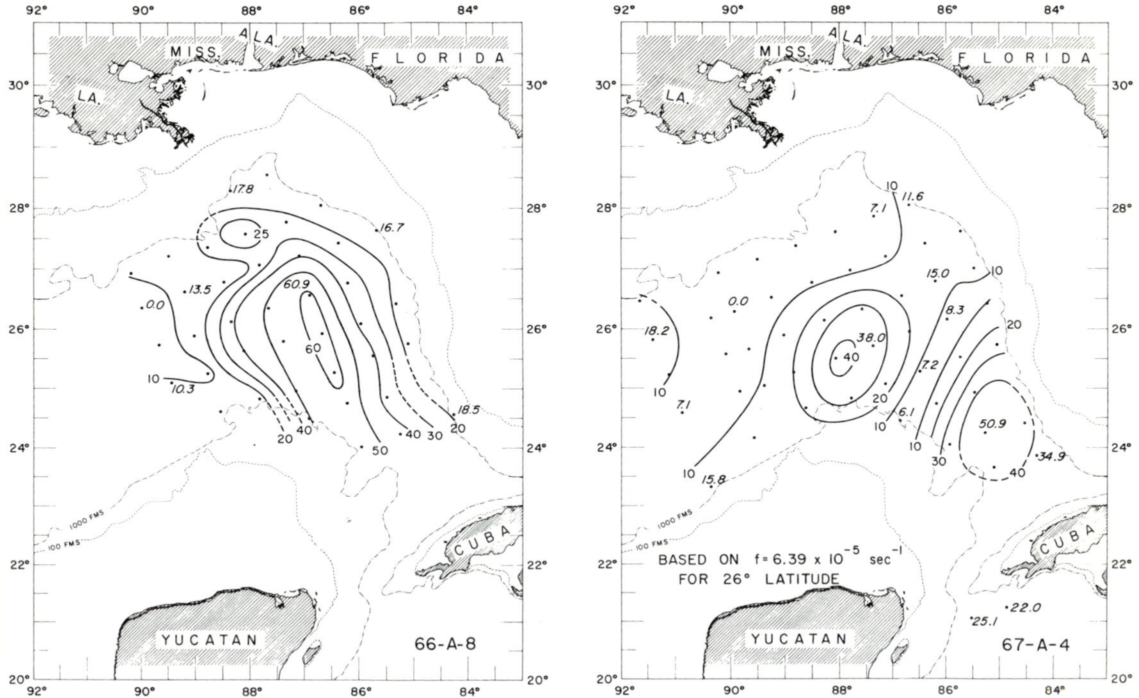

Figure 6-7. Geostrophic transport (10 x 10^6 m^3/sec interval) in upper 1350 m relative to 1350-db surface; Alaminos *(*a*) 66-A-8, (*b*) 67-A-4.*

collected using a Bissett-Berman Model 9006 STD profiler. Values were read from analog records at discrete depth intervals and were digitally corrected in the manner described by Gaul (1966) for errors in apparent salinity induced by differences in the response times of temperature and conductivity sensors. We present for comparison, Figure 6-9, the temperature and salinity values obtained by STD and those obtained by Nansen casts at three stations. In our opinion, the agreement between data derived by these methods is good, especially if one considers that the two types of data were collected on distinct casts separated in time and space, though nominally at the same station location. The reader may notice that the salinity value of the surface Nansen bottle sample at St. 52 (35.57 per mil) is considerably less than the surface salinity value by STD (36.20 per mil). Although the authors find no reason to discard the Nansen cast value, they designate it as questionable.

Here we are interested in the *T-S* relationships of the upper waters. For, as established by Nowlin and McLellan (1967), the *T-S* relationship for the deeper waters with temperatures less than about 16°C is rather uniform throughout the Gulf. Although differences do exist, they are not great enough to be discerned with certainty if one is using a combination of STD and Nansen cast data.

The *T-S* relation characteristic of the near-surface waters in the Caribbean Sea near Yucatan Strait is seen in Figure 6-10. This water mass relation typifies the source water available for upper layer circulation within the Gulf. Based on data from six stations along the center of the water mass bounded by the Loop Current during June, 1966, the *T-S* relation shown in Figure 6-11 characterizes that water bounded by the Loop Current

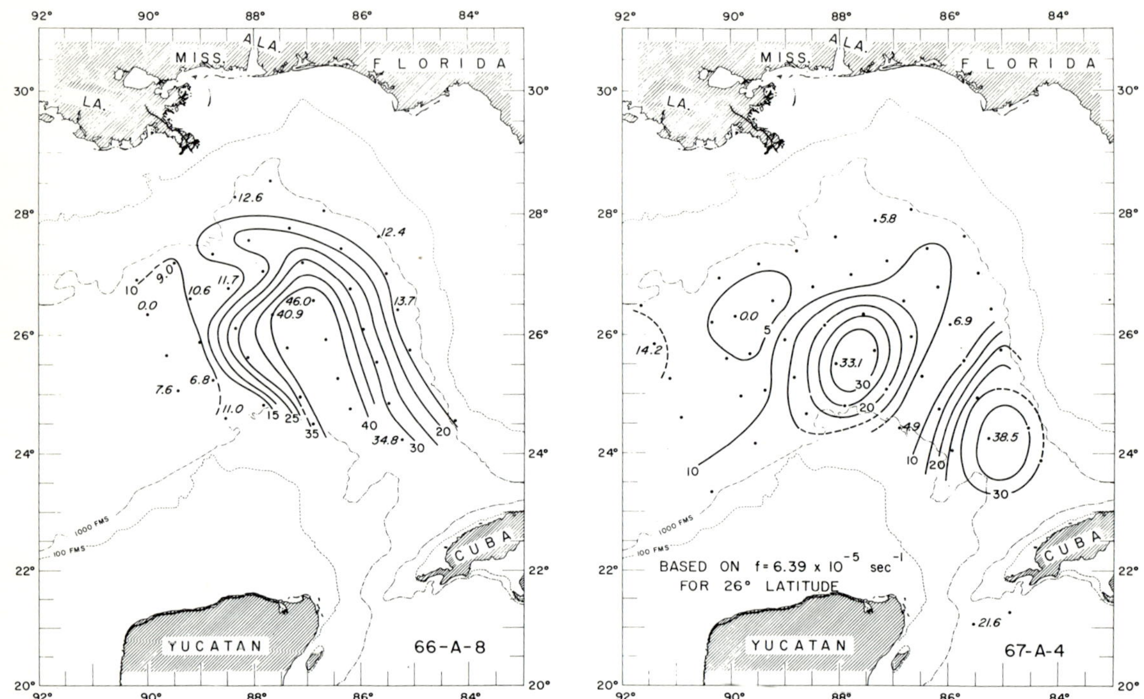

Figure 6-8. Geostrophic transport (5 x 10^6 m^3/sec interval) in upper 480 m relative to 1350-db surface; Alaminos *(*a*) 66-A-8, (*b*) 67-A-4.*

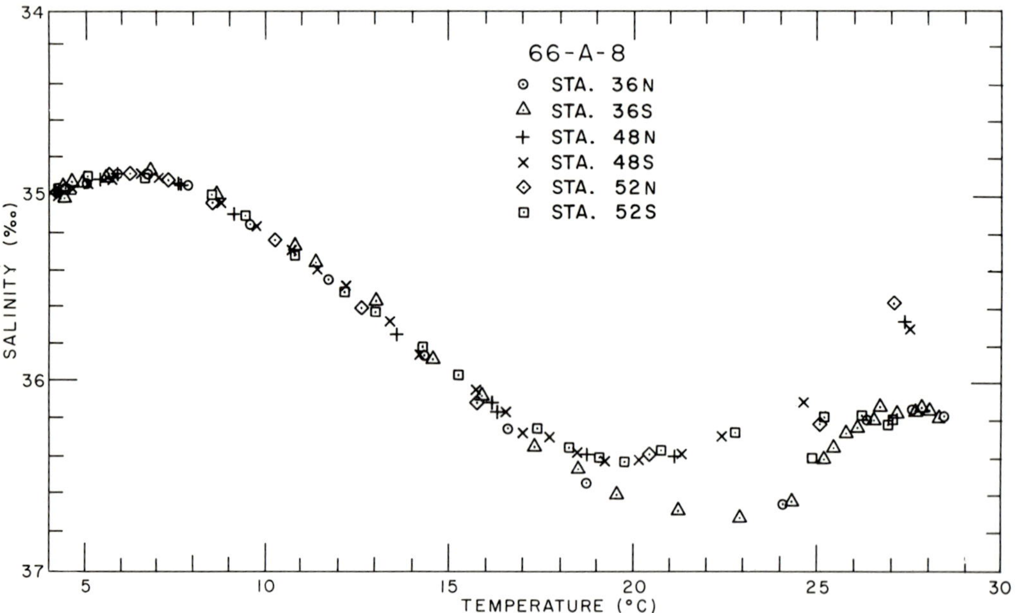

Figure 6-9. Comparison of temperature-salinity data collected by STD (indicated by an S *following station number) and by traditional Nansen casts (indicated by an* N*);* Alaminos *66-A-8, June, 1966.*

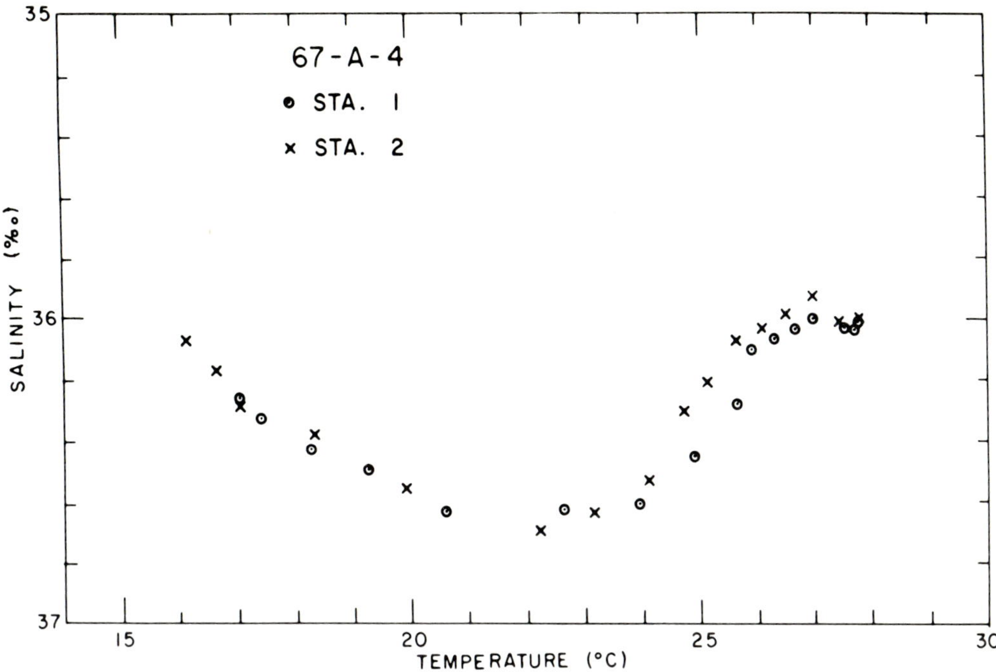

Figure 6-10. Temperature-salinity data from two Caribbean Sea stations near Yucatan Strait; Alaminos *67-A-4, June, 1967.*

within the Gulf. This bounded water mass is seen to have the same *T-S* relation as that water within the northwest Caribbean Sea.

The Gulf waters not bounded by either the Loop Current or one of the current rings evolved therefrom have quite a different *T-S* relation (Figure 6-12). Most of the upper waters of the Gulf are of this modified type, since the Loop Current regime occupies only a part of the eastern Gulf. It could be considered that this is a distinct water mass, and if so, it is likely the only water mass formation of consequence within the Gulf. Having much lower salinity values at the core of the Subtropical Underwater, this water mass is thought to be formed from the Caribbean mass by vertical mixing along the western edge of the Yucatan Current and over the Campeche Bank.

Several stages in the transition from *T-S* relation of unmixed Caribbean mass to that shown in Figure 6-12 can be observed at stations located along a transect across the Loop Current. Figure 6-13 shows this cross-stream variation based on Sts. 28-31 of *Alaminos* 66-A-8. (See Figure 6-1*a* for locations.)

Figure 6-14 shows data which establish the *T-S* relation from Sts. 15, 19 and 29 of *Alaminos* 67-A-4, which were located in the water mass bounded by the limited Loop Current of June, 1967. Clearly this water mass is essentially of unmixed Caribbean water. Moreover, this is also true of the water found at Sts. 34, 37 and 39, which were located in water bounded by the anticyclonic ring observed closest to the Loop Current in June, 1967, and of water found at St. 78 within the second ring indicated by the *Alaminos* 67-A-4 data. The equivalence of water masses within these three areas, when combined with the distinctly different *T-S* relation in the shear zones between rings and Loop Current (Figure 6-15), definitely establishes that these features are indeed rings of current which have separated from the Loop Current, enclosing water masses of the type characteristically bounded by that Loop.

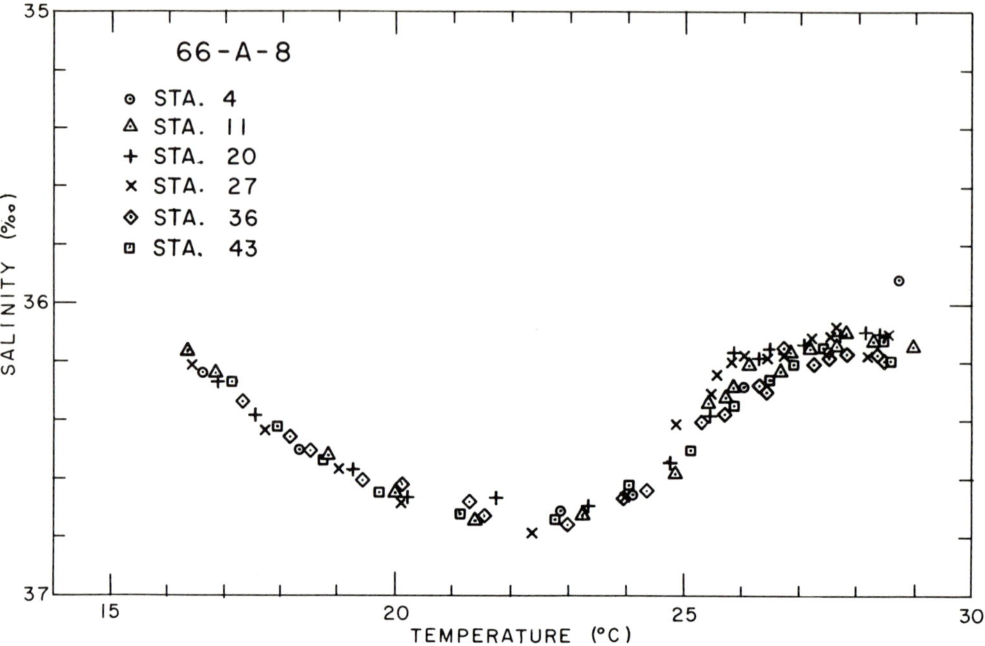

Figure 6-11. Temperature-salinity data from six stations along center of the water mass bounded by the Loop Current; Alaminos *66-A-8, June, 1966.*

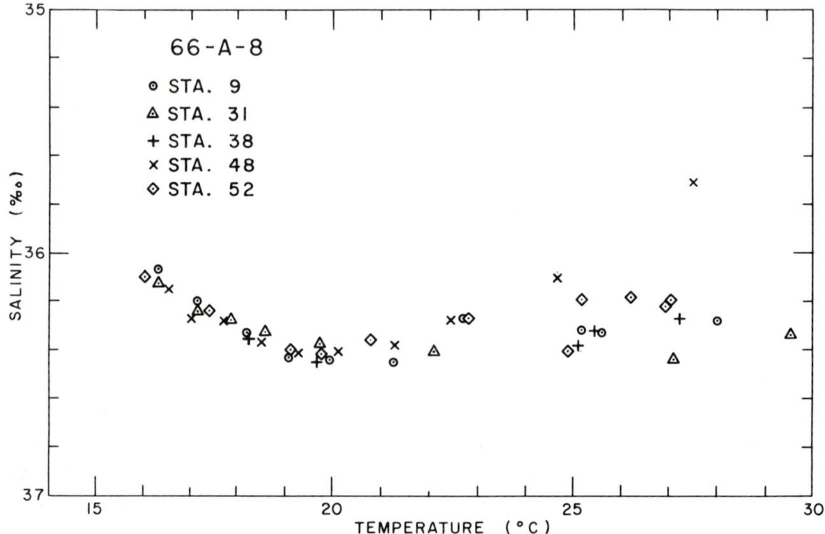

Figure 6-12. Temperature-salinity data from five stations located outside the Loop Current within the Gulf; Alaminos *66-A-8, June, 1966.*

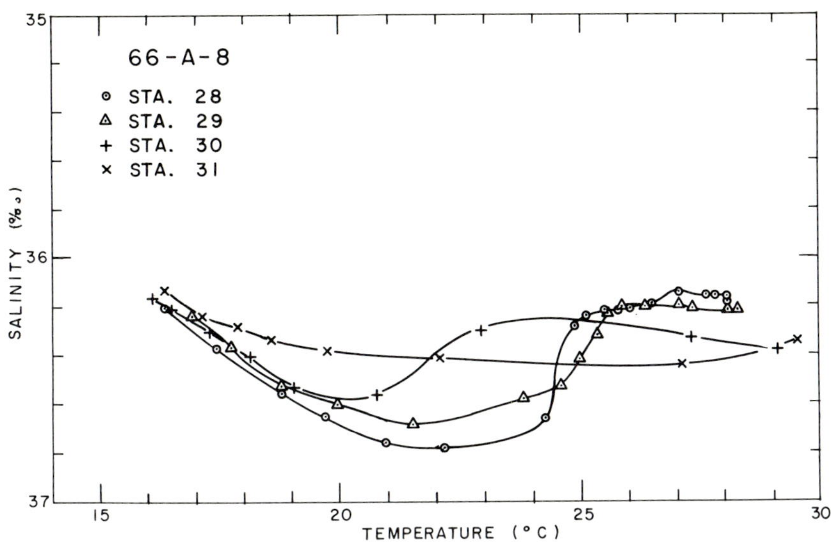

Figure 6-13. Temperature-salinity relations from four stations on a transect across the Loop Current; Alaminos *66-A-8, June, 1966.*

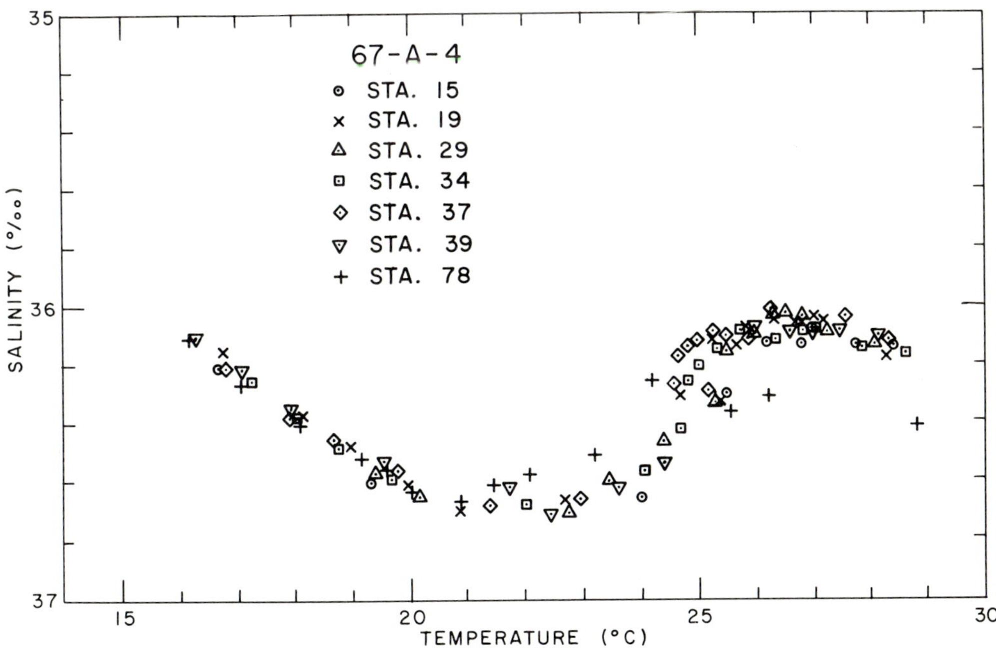

Figure 6-14. Temperature-salinity data from stations within the water masses enclosed by the Loop Current and by two rings; Alaminos *67-A-4, June, 1967.*

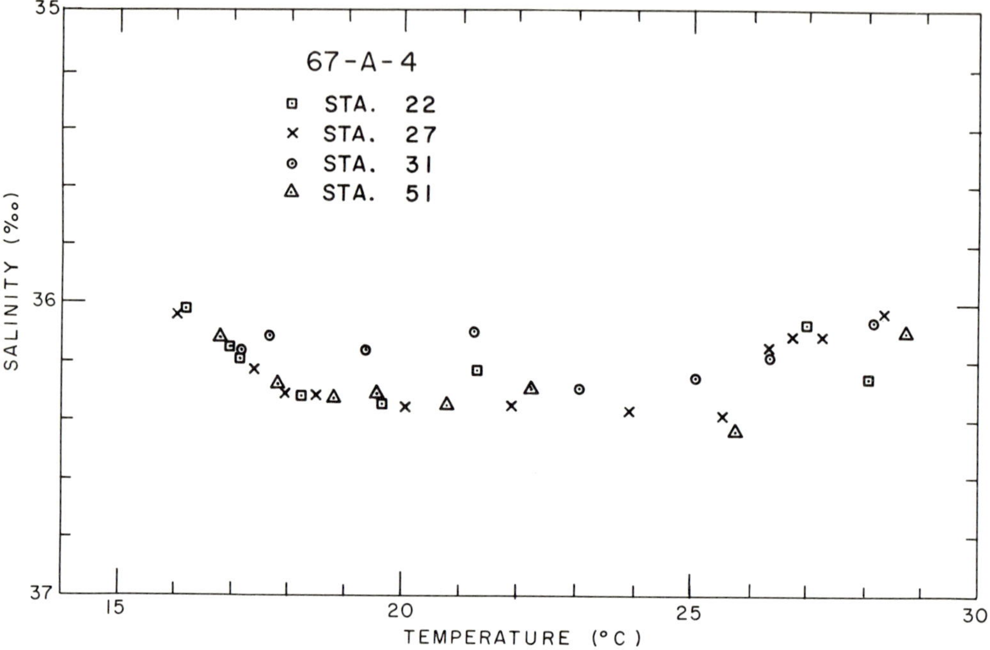

Figure 6-15. Temperature-salinity data from four stations located in the intercurrent shear zones observed during June, 1967; Alaminos *67-A-4.*

As already mentioned, the *T-S* data within the horizontal shear zones in Figure 6-15 show salinity values at the Subtropical Underwater core even less than those commonly found in the upper Gulf waters outside the Loop Current regime (Figure 6-12).

Vertical Kinematic Sections

Rather than indirectly portray the details of the flow in vertical sections by the distributions of temperature, salinity, density or some other property, we chose to present the geostrophic flow relative to 1350 db. For Cruise 66-A-8, Figure 6-16 shows isotachs of baroclinic geostrophic speed normal to Sections *AB, CD* and *EF* (locations in Figure 6-1*a*) computed relative to the 1350-db surface. Sections *AB* and *CD* cross both the inflowing limb (Yucatan Current) of the Loop Current and part of the outflowing limb. Section *EF* crossed the Loop Current at its northernmost limit. Each section is almost normal to the current. The computed geostrophic speed fields in all current crossings are similar. Relative speeds reach 5 cm/sec near 900 m; most waters in which speeds are greater than 25 cm/sec are above 500 m.

For comparison with the baroclinic geostrophic surface speeds relative to 1350 db, the GEK surface current components normal to the sections are also shown in Figure 6-16. Two differences stand out. First, the GEK surface speed estimates are greater than the calculated speeds. This is to be expected since significant flow within the Loop Current exists down to at least 1500 m (Nowlin and McLellan, 1967), and some flow of similar pattern is found even deeper. Second, the maximum downstream surface speeds indicated by the GEK seem displaced to the left (when facing downstream) relative to the position of the surface current core given by the computed geostrophic speeds.

Figure 6-16 also gives the volume transports relative to the reference level for indicated depth intervals between pairs of stations. The direction

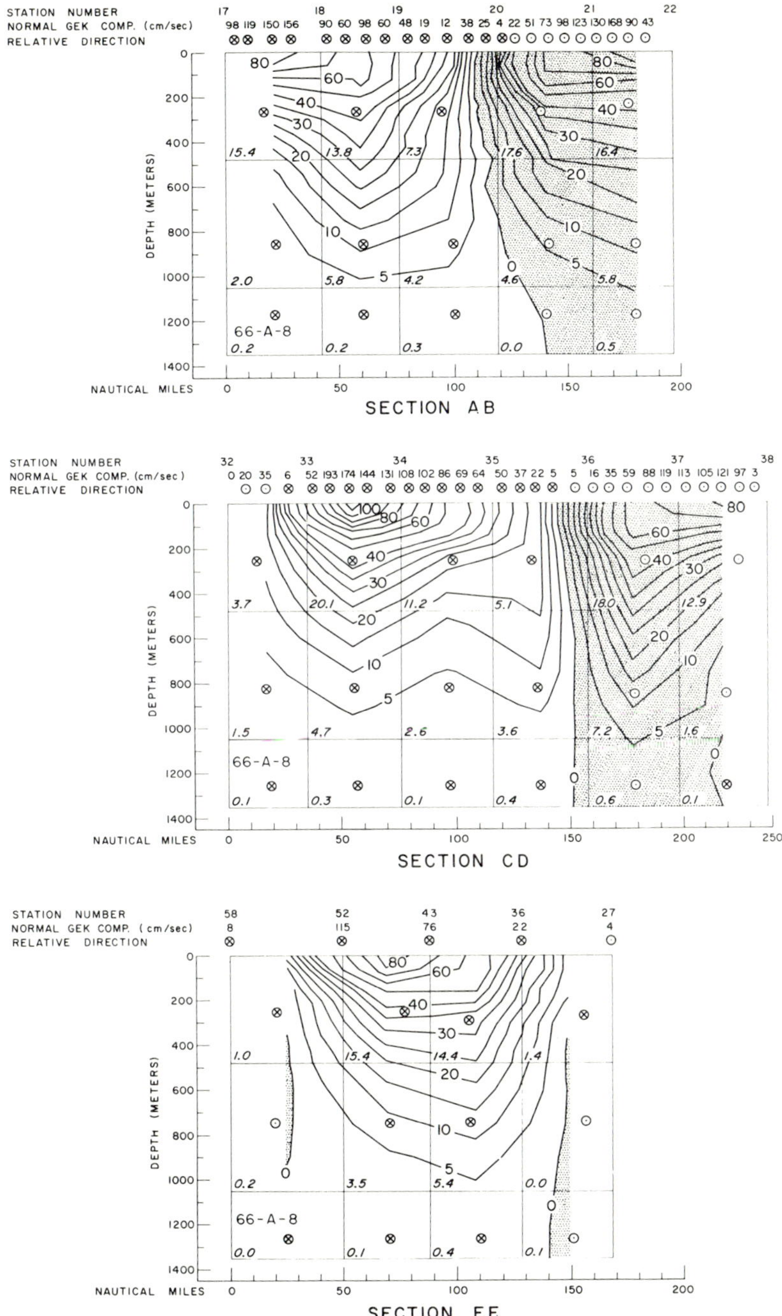

Figure 6-16. Isotachs of geostrophic speed (cm/sec), relative to 1350-db surface, normal to Sections AB, CD *and* EF. *(Locations shown in Figure 6-1*a.*) Also shown are normal components of GEK measurements made along each section and relative volume transport (10^6 m^3/sec) for indicated depth intervals between station pairs. The symbol ⊗ indicates flow into the section as viewed by the reader; ⊙ indicates flow toward the reader. Data are from* Alaminos *66-A-8, June, 1966.*

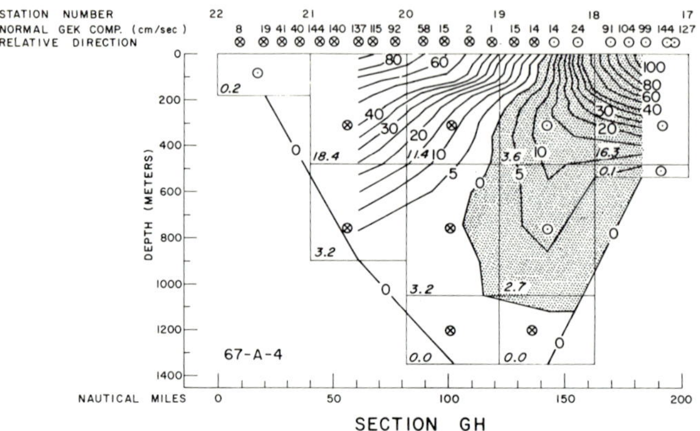

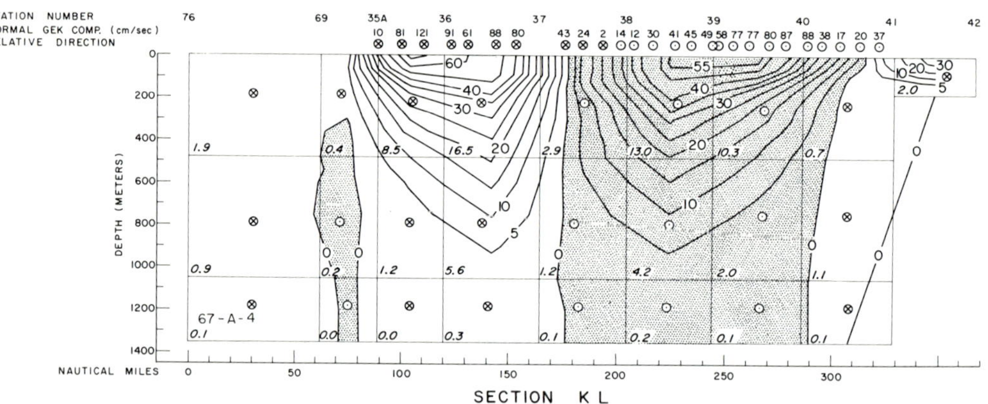

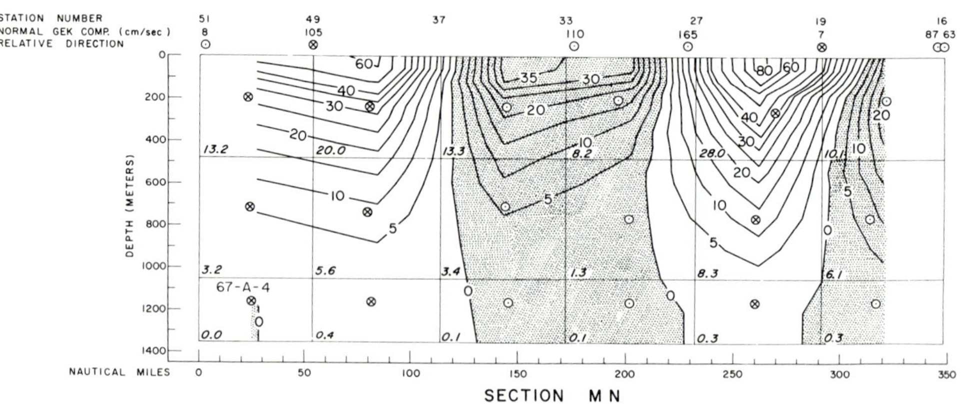

Figure 6-17. Isotachs of geostrophic speed (cm/sec), relative to 1350-db surface, normal to Sections GH, KL *and* MN. *Locations shown in Figure 6-1*b.*) Also shown are normal components of GEK measurements made along each section and relative volume transport (10^6 m^3/sec) for indicated depth intervals between station pairs. The symbol ⊗ indicates flow into the section as viewed by the reader; ⊙ indicates flow toward the reader. Data are from* Alaminos *67-A-4, June, 1967.*

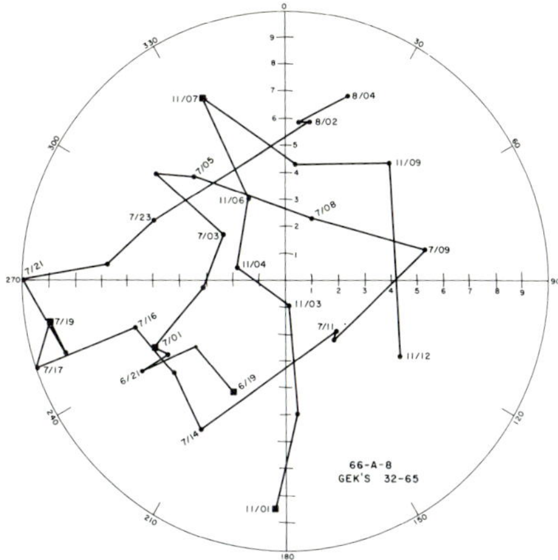

Figure 6-18. Polar plot (magnitudes in 10 cm/sec) of heads of vectors representing GEK surface current estimates 32-65, taken between STs. 61 and 53 on Alaminos *66-A-8 during June, 1966. Day and hour are shown as one and two digit numbers separated by a slash.*

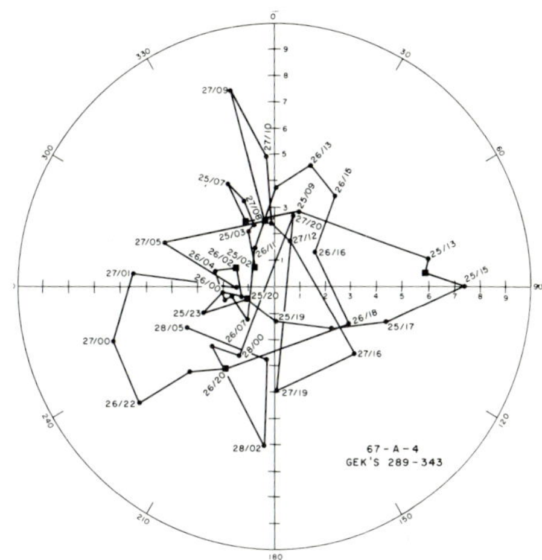

Figure 6-19. Polar plot (magnitudes in 10 cm/sec) of heads of vectors representing GEK surface current estimates 289-343, taken between Sts. 51 and 67 on Alaminos *67-A-4 during June, 1967. Day and hour are shown as one and two digit numbers separated by a slash.*

of transport is indicated by a circled cross if away from the reader or by a circled dot if toward him. The magnitudes of transport (in 10^6 m^3/sec) normal to the section are indicated by numbers in the lower left-hand corners of the rectangles to which the numbers apply.

Figure 6-17 shows vertical sections of relative geostrophic speeds normal to the transects labeled *GH*, *KL* and *MN* in Figure 6-1*b*. Sections *GH* crossed parts of both northward and southward flowing limbs of the Loop Current in June, 1967. Unfortunately, both limbs extended laterally into such shallow water that lateral extrapolation of a reference level to obtain geostrophic speeds seems unjustified. The normal GEK components show that the large apparent shear between the two limbs is mostly an artifact of the wide station spacing.

Section *KL* crossed, at nearly right angles to the flow, both eastern and western extremes of the principal separate ring observed on *Alaminos* 67-A-4. Again, the GEK and computed surface current components show some disagreement which is largely attributable to station spacing and to the fact that these are really different flow components. Isotachs are not shown at the lateral limits of the sections if indicated speeds are quite small or if sampling depths are shallow.

The principal value of Section *MN* is in presenting isotachs normal to the inflowing Yucatan Current, although the section also crossed the separate ring and, between Sts. 16 and 19, part of the outflowing limb of the Loop Current. Note that Section *MN*, like Section *EF*, consists of stations separated rather widely in time as compared with the other sections, on which stations were occupied sequentially.

The reader has probably noticed that the speed distributions in sections normal to the Loop Current and ring observed in 1967 have the same general features as the distribution observed in 1966 (Figure 6-16). Based on these data and personal comparisons with other data from the Loop Current regime, the internal distributions of mass and associated pressure gradients in cross-stream sections as observed on different occasions appear to be basically similar, i.e., no prominent double cores, countercurrents or other such phenomena are apparent.

Appendix

During June of 1966 and 1967, the GEK surface current estimates for the northeastern Gulf outside the Loop Current regime did not seem to indicate a recognizable spatial pattern. If these GEK measurements are treated as though they represent the motion of a given water mass, however, they reveal an intriguing pattern.

Although there is no unambiguous meaning, we have considered the GEK vectors along the northernmost cruise lines as if they represented a time series of horizontal current vectors. We made polar plots of these vectors and then connected the heads of the vectors. Had the GEK surface current estimates been made at the same location, or if the currents at different points over the area of observation are assumed to have moved identically, then the resulting Figures 6-18 and 6-19 would be amenable to physical interpretations.

Figure 6-18 is based on GEK measurements 32-65, taken between Sts. 61 and 53 on *Alaminos* 66-A-8. (The stations on Cruise 66-A-8 were occupied in reverse numerical order; between Sts. 54 and 56, the *Alaminos* put into port because of potential danger from a hurricane traveling northward along the west Florida Shelf.) Interpreted as a time series at one point, these measurements show a clockwise rotation of period about 20 hours superimposed on a westward flow.

Figure 6-19 is based on GEK measurements 289-343, taken between Sts. 51 and 67 on *Alaminos* 67-A-4. Pictured is a surface current with a net westward component and a clockwise rotation of near 24-hour period.

Acknowledgments

We wish to express our thanks to Mrs. Ruby Dee Parker and Mr. Oscar Chancey for their efforts in analysis and plotting of the data and in preparing the final figures. To Mrs. Ina Deel goes our thanks for typing the manuscripts.

All phases of this work, including the ship and technical support, were supported by the Office of Naval Research through contracts Nonr 2119(04) and N00014-68-A-0308-0002 with the Texas A&M Research Foundation.

References

Austin, G.B., Jr. 1955. Some recent oceanographic surveys of the Gulf of Mexico. *Trans. Amer. Geophy. Un.*, 36(5):885-892.

Cochrane, J.D. 1965. The Yucatan Current. In Unpubl. Rept. of Dept. of Oceanogr. and Meteorol., Texas A&M University, Ref. 65-17T:20-27.

________. 1966. The Yucatan Current. In Unpubl. Rept. of Dept. of Oceanogr., Texas A&M University, Ref. 66-23T:14-25.

Gaul, R.D. 1966. Circulation over the continental margin of the northeast Gulf of Mexico. Unpubl. Rept. of Dept. of Oceanogr., Texas A&M University, Ref. 66-18T.

Jacobs, Clifford A. and Nowlin, Worth D., Jr. 1968. A numerical treatment of steady, frictional boundary currents in a homogeneous ocean applied to a two-port basin. Unpubl. Rept. of Dept. of Oceanogr., Texas A&M University, Ref. 68-19T.

Leipper, Dale F. 1967. A sequence of current patterns in the Gulf of Mexico. Unpubl. Rept. of Dept. of Oceanogr., Texas A&M University, Ref. 67-9T.

________. 1970. A sequence of current patterns in the Gulf of Mexico. *J. Geophys. Res.*, 75(3):637-658.

Nowlin, W.D., Jr., and McLellan, H.J. 1967. A characterization of the Gulf of Mexico waters in winter. *J. Marine Res.*, 25(1):29-59.

________. Hubertz, J.M. and Reid, R.O. 1968. A detached eddy in the Gulf of Mexico. *J. Marine Res.*, 26(2):185-6.

Paskausky, David F. 1969. A barotropic prognostic numerical model of the circulation in the Gulf of Mexico. Unpubl. Ph.D. dissertation, Texas A&M University.

Schmitz, W.J., Jr., and Richardson, W.S. 1968. On the transport of the Florida Current. *Deep-Sea Res.*, 15(6):679-693.

Wert, Richard T. 1968. A frictional model of a two-port unbounded ocean basin. Unpubl. Rept. of Dept. of Oceanogr., Texas A&M University, Ref. 68-4T.

7

Objective Analysis of Oceanic Surface Currents

J. M. Hubertz, R. O. Reid, A. Garcia

Abstract

Physical oceanographic surveys of the eastern Gulf of Mexico were made in June, 1966, and June, 1967, with the R/V *Alaminos*. Hourly surface GEK measurements were made during both cruises. Treating measurements from each cruise as synoptic, these results are used to approximate the nondivergent part of the surface velocity field, which is displayed in terms of a stream function.

The method used to obtain the stream function is a numerical relaxation of a form of Poisson's equation. Solutions were obtained for two types of boundary conditions. The gradient of the stream function normal to the boundary was specified in one instance (Neumann-type). In the second case the value of the stream function itself was specified along the boundary in such a way that the net flow through the boundaries was zero (Dirichlet-type).

The data from 1967 are used in an extension of the method which considers the transports within a near-surface layer bounded above by the sea surface and below by a surface of constant potential density.

Introduction

This is part of a study of the major circulation patterns in the Gulf of Mexico. The purpose here is to demonstrate a technique for obtaining a description of the nondivergent part of the surface velocity field using a quasi-synoptic, discrete set of surface current velocities.

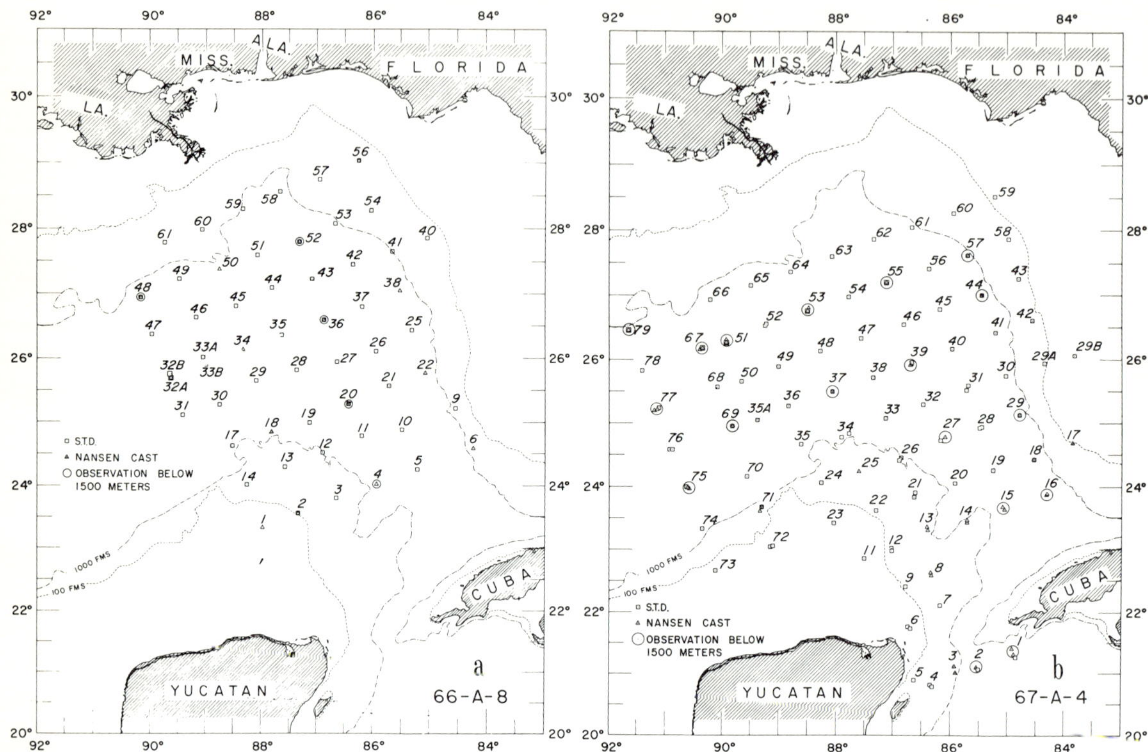

Figure 7-1. Ships track and station locations for Cruises (a) 66-A-8 (5-22 June 1966) after Hubertz (1967) and (b) 67-A-4 (1 June-3 July 1967) of R/V Alaminos *of Texas A&M University.*

Pritchard (1948) proposed an objective method of calculating a stream function for the field of mean surface currents for a portion of the tropical Atlantic and Pacific. This was based on the evaluation of a line integral involving the velocity components. However, since the field of surface currents is not divergenceless, the direct evaluation of the stream function in terms of a line integral is not unique but depends on the path of integration. In his application, an average of stream function values for at least two different paths of integration was obtained.

If the observed flow is nondivergent, the pattern of streamlines should agree with that pattern found from the dynamic topography. Applying the geostrophic assumption to the hydrographic station data implies that the flow is nearly nondivergent. The pattern should also be similar to that of the resultant GEK vectors, with the degree of similarity again depending on the nondivergence of the actual surface circulation.

In addition to the determination of values of the velocity stream function, values of a transport function were also obtained. The values give the transport in a surface layer bounded above by the sea surface and below by a surface of constant potential density. A surface of constant σ_t = 24.0 gm/1 is used as the lower boundary, assuming that σ_t is equivalent to potential density near the surface. The maximum depth of this layer is 128 m, while its minimum is 15 m. If the components of velocity normal to isopycnals are neglected, the transport in this surface layer should be nondivergent.

Cruises 66-A-8 and 67-A-4 of the *Alaminos* during 1966 and 1967 both followed a rectangular grid pattern of hydrographic stations approximately 40 nautical miles apart in the eastern Gulf, Figure 7-1.

Surface current measurements using the GEK were taken hourly, giving about four measurements between stations. These provided a lattice of discrete points at which were obtained components of velocity normal and parallel to the grid pattern.

Normal and parallel components were averaged between stations, and the average values were assigned to a point midway between stations. At each of these points, the vertical component of vorticity was then used to obtain values for the surface velocity. The vertical component of vorticity, ζ, is defined as

$$\zeta \equiv \vec{k} \cdot \nabla x \vec{V} = \partial v/\partial x - \partial u/\partial y,$$

where u and v are respectively the components of velocity in the x- and y-directions.

Theory

In general, any two-dimensional vector field, $\vec{V}(x, y)$, can be represented in terms of two scaler potentials, ϕ and Ψ, as follows (Lamb, 1945):

$$\vec{V} = -\nabla \Phi + \vec{k} \times \nabla \Psi$$

where $\vec{k}$ is a unit vector normal to the x, y plane. The function Φ is related to the divergence of $\vec{V}$ by $\nabla^2\Phi = -\nabla \cdot \vec{V}$, and Ψ is related to the curl of $\vec{V}$ by

$$\nabla^2 \Psi = \vec{k} \cdot \nabla \times \vec{V}. \qquad (7\text{-}1)$$

The part of the flow associated with Φ is divergent and irrotational while that associated with Ψ is nondivergent and rotational. Thus, Ψ controls the rotational part of the flow field, and the locus of points of equal value of Ψ corresponds to a streamline of the rotational flow only.

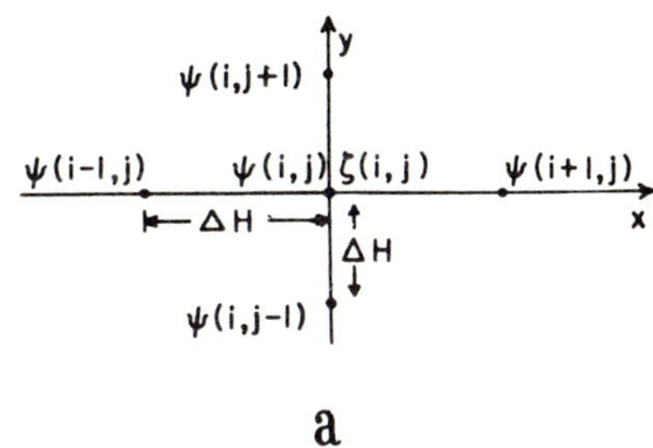

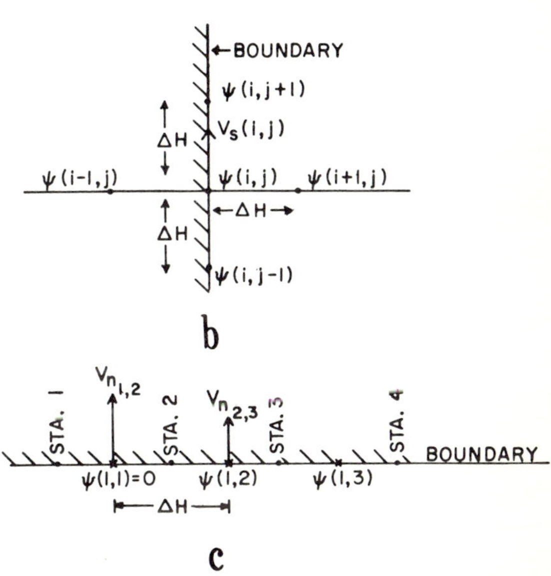

Figure 7.2. (a) Five point finite difference scheme for Poisson's equation. (b) Finite difference scheme for Neumann boundary conditions. (c) Finite difference scheme for Dirichlet boundary conditions.

In principle one can employ the Poisson relation (7-1) to find Ψ, evaluating curl $\vec{V}$ from the given field of $\vec{V}$ and using Neumann-type boundary conditions which specify the normal gradient of Ψ along a boundary from a knowledge of the tangential velocity

$$V_s = \partial\Psi / \partial n \cdot \qquad (7\text{-}2)$$

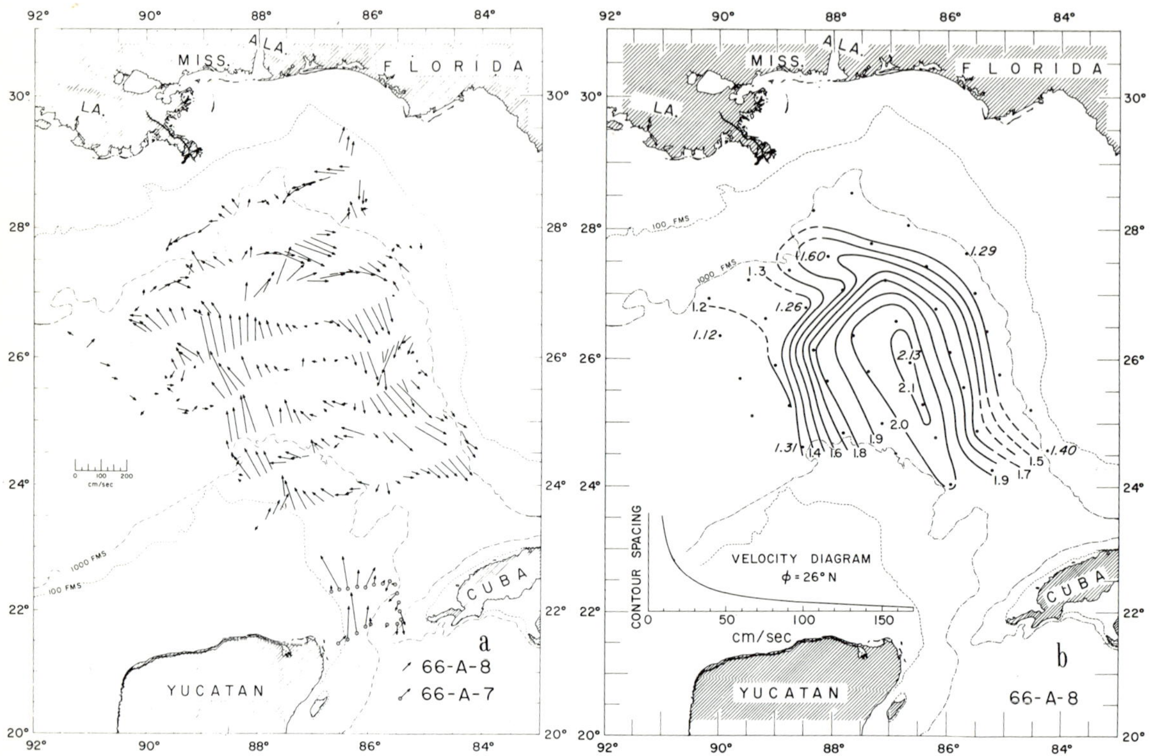

Figure 7-3. GEK vectors and the dynamic topography (dynamic meters) of the sea surface with reference to the 1350-db surface for Alaminos *Cruise 66-A-8 (after Hubertz 1967).*

The tangential velocity used for this boundary condition is that component of the observed value at each point along a boundary. Alternatively one can use Dirichlet-type boundary conditions which specify directly the value of Ψ on a boundary

$$\Psi = \int_{O}^{S} V_n \, ds.$$

In this case the normal component of the observed velocity is adjusted along the outermost lines of stations defining the rectangular grid, so the net flow through this boundary is zero. The method of adjustment is explained later.

Numerical Representations

The finite difference analog of (7-1) may be written as the approximation

$$\Psi(i, j+1) + \Psi(i-1, j) + \Psi(i, j-1) + \Psi(i+1, j) - 4\Psi(i, j) \approx \zeta(i, j)(\Delta H)^2, \quad (7\text{-}3)$$

where ΔH is the distance between grid points. The finite difference scheme for (7-3) is illustrated in Figure7-2*a*.

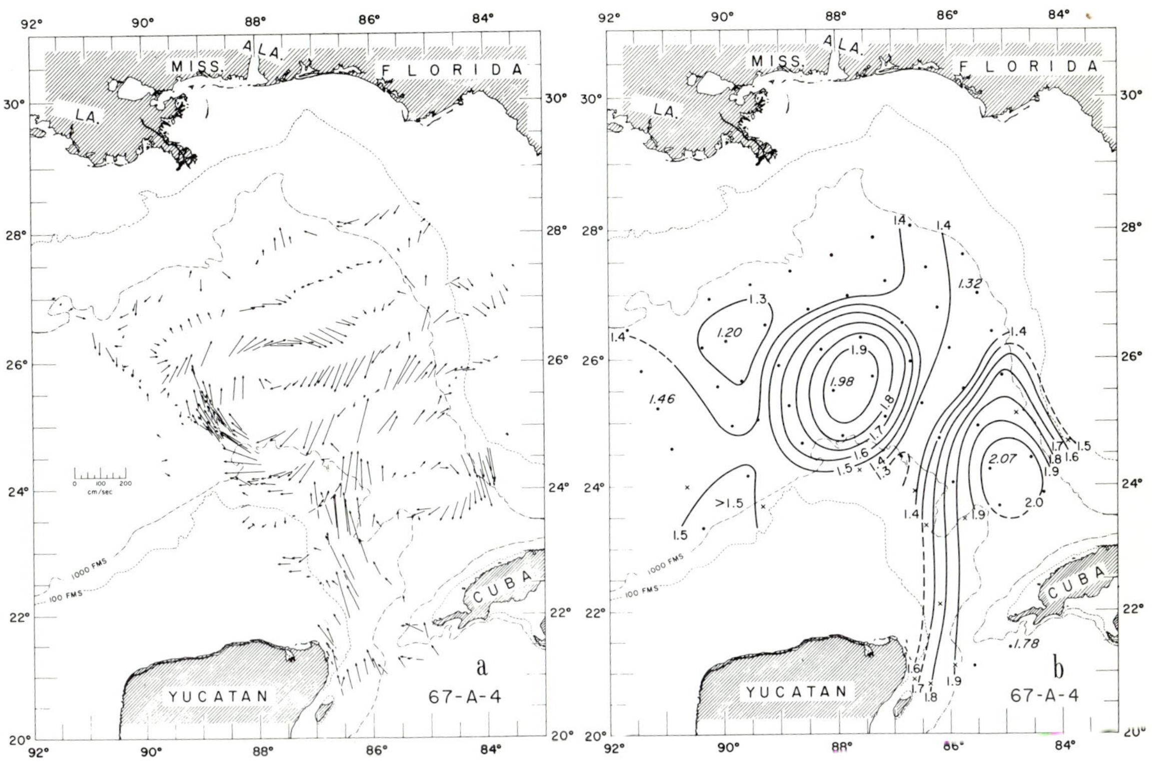

Figure 7-4. GEK vectors and the dynamic topography (dynamic meters) of the sea surface with reference to the 1350-db surface for Alaminos *Cruise 67-A-4.*

The finite difference analog of the vorticity is

$$\zeta\,(i,j) = 1/(2\Delta H) \quad \left\{ [v\,(i+1,j) - v\,(i-1,j)] - [u\,(i,j+1) - u\,(i,j-1)] \right\}. \quad (7\text{-}4)$$

Centered finite difference analogs of (7-2) may be written as

$$\Psi\,(i,j+1) - \Psi\,(i,j-1) \approx -\,2\,u\,(i,j)\,\Delta H \quad (7\text{-}5)$$

and

$$\Psi\,(i+1,j) - \Psi(i-1,j) \approx 2\,v\,(i,j)\Delta H \quad (7\text{-}6)$$

for respectively x- and y- directed boundaries.

Relaxation Technique for Evaluating Ψ

The evaluation of Ψ from (7-3) with the appropriate boundary conditions was carried out by the simultaneous relaxation technique (Thompson, 1961). This is an iterative technique in which a series of successive approximations are made, each approximation hopefully being better than the preceding one. The amount by which the approximation fails to satisfy (7-3) is termed the residual $R\,(i,j)$. If the absolute value of $R\,(i,j)$ falls below some preselected small value at all points, then (7-3) is said to be satisfied for one particular value of $\Psi\,(i,j)$. The values of $\zeta\,(i,j)$ used were obtained from (7-4).

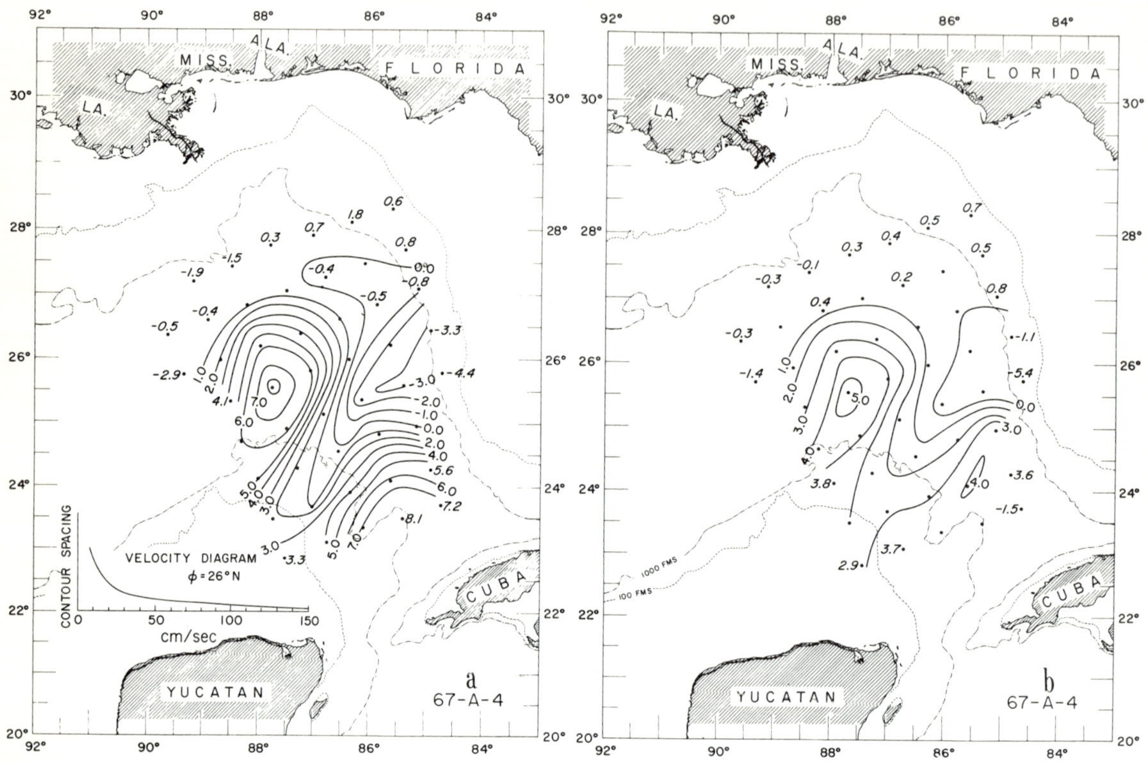

*Figure 7-5. Contours of the velocity stream function (10^8 cm^2/sec) using (*a*) Neumann and (*b*) Dirichlet boundary conditions for* Alaminos *Cruise 66-A-8.*

Specifically, the residual is defined by

$$\Psi^N(i, j+1) + \Psi^N(i-l, j) + \Psi^N(i, j-1) + \Psi^N(i+l, j) - 4\Psi^N(i, j) - (\Delta H)^2\,\zeta\,(i, j) \equiv R^N(i, j), \qquad (7\text{-}7)$$

where the superscript N refers to the value after the nth iteration. The best recursion relation (Thompson, 1961, p. 95) for Poisson's equation is

$$\Psi^{N+1}(i, j) = \Psi^N(i, j) + (1/4)R^N(i, j), \qquad (7\text{-}8)$$

where R (i, j) is evaluated from (7-7). It should be noted that the value of the coefficient of $R\ (i, j)$ can be varied, but ¼ is the optimum value for convergence of this system of equations.

Equations (7-7) and (7-8) are used directly for evaluation of Ψ at all interior points of the grid system. When Neumann-type boundary conditions are used, we specify a fictitious point outside the boundary, as illustrated in Figure 7-2*b* for a left lateral boundary. At that exterior point the value of Ψ^N is obtained from the appropriate form of (7-5) or (7-6). Then (7-7) and (7-8) can be applied to obtain $\Psi^N(i, j)$ on the boundary.

In applying this relaxation technique an initial value of zero is assigned to each interior or boundary point. When Dirichlet-type conditions are used however, the value of the stream function on the boundary is specified, as illustrated for a northern boundary in Figure 7-2*c*.

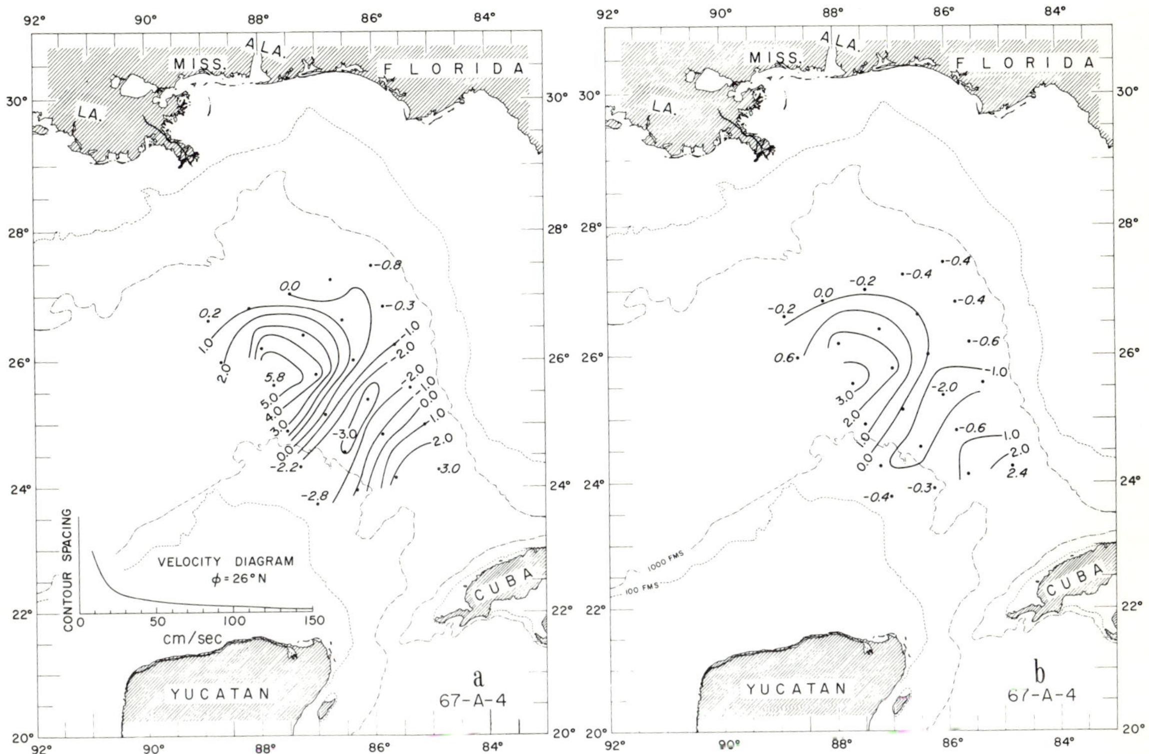

Figure 7-6. Contours of (a) the velocity stream function (10^8 cm^2/sec) and (b) the transport stream function in Sverdrups for the interior region only of Alaminos *Cruise 67-A-4. A value of $\Psi = 0$ is assumed for the boundary.*

Application of Dirichlet Boundary Conditions

When applying Dirichlet boundary conditions it is necessary to specify a value of the stream function at each grid point on the boundary. The values can be determined from a knowledge of the component of velocity normal to the boundary.

For the initial grid point we arbitrarily set Ψ (1, 1) = 0. Referring to Figure 7-2*c*, we wish to obtain $\Delta\Psi$ between grid points (1, 1) and (1, 2). Let $V_{n_{1,2}}$ be the average of the velocity components normal to the boundary, determined from the GEK measurements taken between stations 1 and 2. Similarly $V_{n_{2,3}}$ denotes the average of measurements taken between stations 2 and 3. The value of Ψ at station 2 can now be written, $\Psi_2 = (\bar{V}_{n_{1,2}} + \bar{V}_{n_{2,3}})\ \Delta H/2$. Following a similar procedure a value of Ψ at station 3 is determined as $\Psi_3 = (\bar{V}_{n_{2,3}} + \bar{V}_{n_{3,4}})\ \Delta H/2$. The average of Ψ_2 and Ψ_3 then gives $\Delta\Psi$ such that the value of Ψ at boundary grid point (1, 2) is given by Ψ (1, 2) = Ψ (1, 1,) + $\Delta\Psi$. The quantity $\Delta\Psi$ which we were seeking is now available as $\Delta\Psi = \Psi_{1,2} - \Psi_{1,1}$.

This procedure is used to obtain a value of $\Delta\Psi$ between grid points around the boundary. The sum of these $\Delta\Psi$ values then gives

$$\oint \bar{V}_n\, ds,$$

which should equal zero for a nondivergent region.

Because the region is not completely divergenceless this sum deviated from zero by a quantity we term the residue. The negative of this quantity was divided by the number of boundary grid points and added to the $\Delta\Psi$ value between

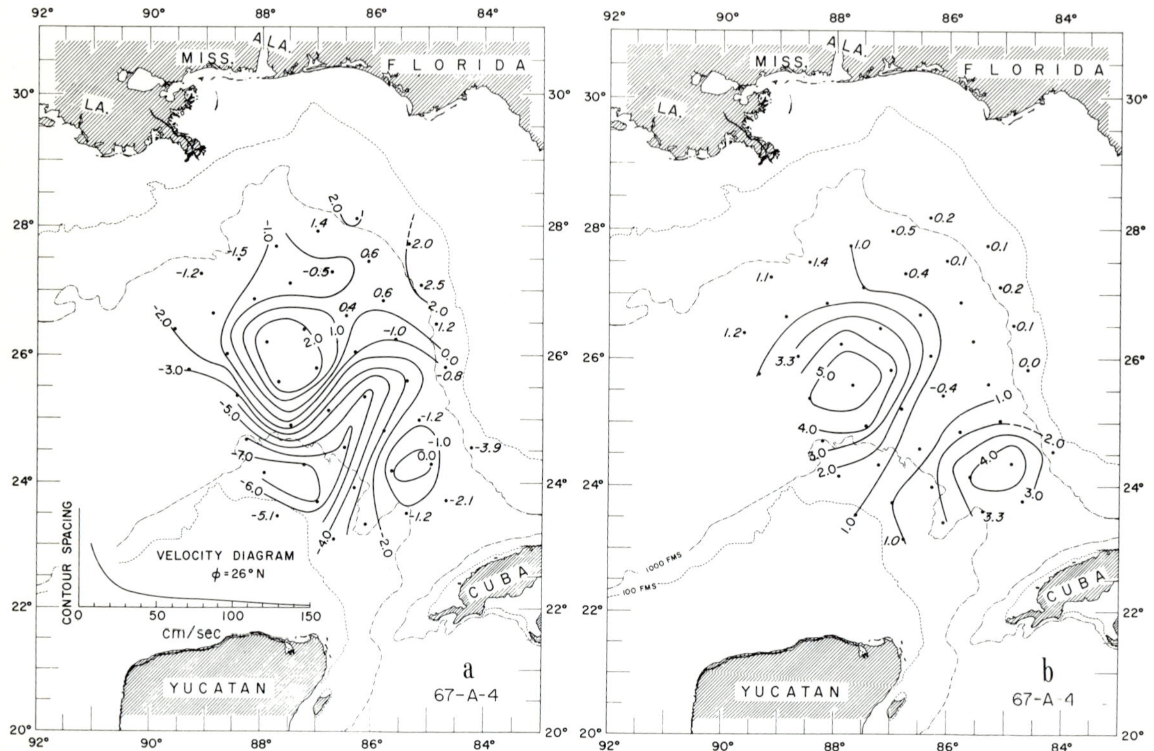

*Figure 7-7. Contours of (*a*) the velocity stream function (10^8 cm^2/sec) and (*b*) the transport stream function in Sverdrups using Neumann boundary conditions for* Alaminos *Cruise 67-A-4.*

each pair of boundary grid points in order to force the adjusted residual to zero. The value of the stream function was then obtained at each boundary point from the adjusted values of $\Delta\Psi$ between grid points.

Extension to a Thin Layer

Consider a layer bounded above the sea surface and below by a surface of uniform σ_t (24.0 gm/1). If we assume that all vertical components of velocity vanish at the upper and lower boundaries, then the vertically integrated, horizontal transport within this layer will be nondivergent. Assuming $\vec{V}$ is independent of depth within this layer, the integrated transport per unit width is given by $h\vec{V}$ where h is the depth to the surface of uniform σ_t. We can define a transport stream function T such that $uh = -\partial T/\partial y$ and $vh = \partial T/\partial x$. It follows that $\nabla^2 T = \partial vh/\partial x - \partial uh/\partial y \equiv Z$, where Z is determined from the observed data. The relaxation scheme to be used in this case consists of the recursion formulae:

$$\begin{aligned} T^N(i, j+1) + T^N(i-1, j) + T^N(i, j-1) \\ + T^N(i+1, j) - 4T^N(i, j) \\ - (\Delta H)^2 Z(i, j) = R^N(i, j) \end{aligned}$$

and

$$\begin{aligned} T^{N+1}(i, j) = T^N(i, j) \\ + (1/4) R^N(i, j). \end{aligned}$$

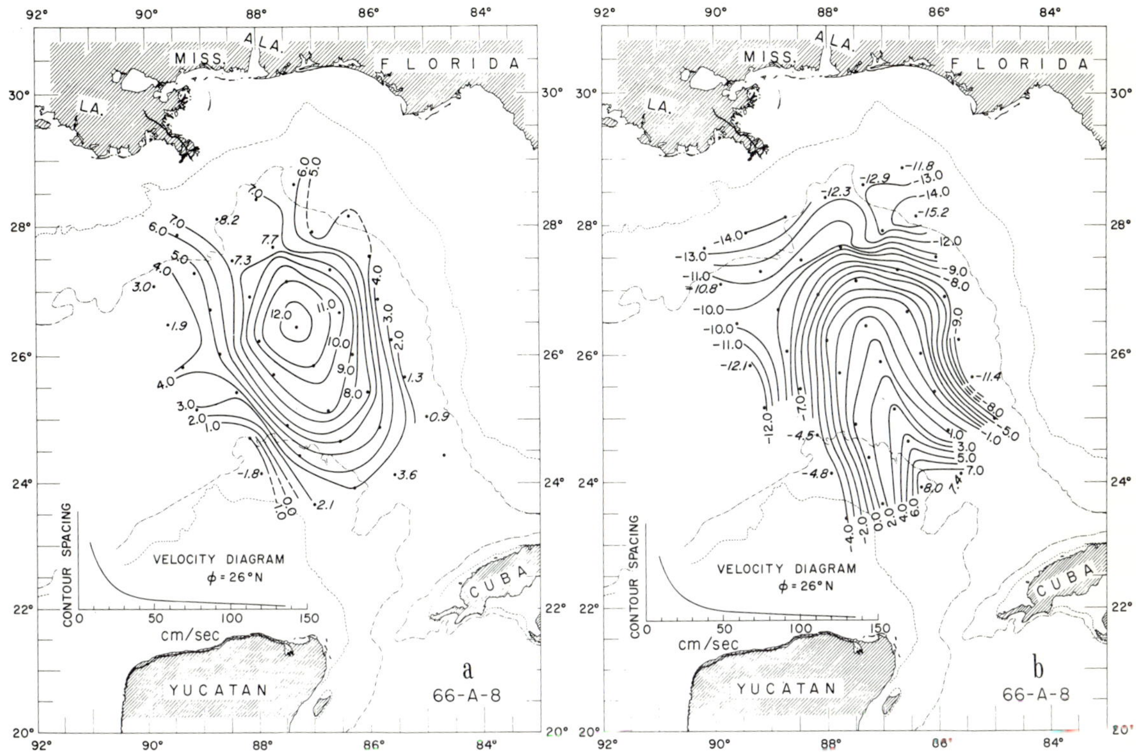

*Figure 7-8. Contours of (*a*) the velocity stream function (10^8 cm^2/sec) and (*b*) the transport stream function is Sverdrups using Dirichlet boundary conditions for* Alaminos *Cruise 67-A-4.*

Results

The data from 1967 covers a larger grid area than that from 1966 and so should be less affected by boundary conditions. We feel the data from Cruise 67-A-4 lends itself to a better illustration of the technique, since the ratio of boundary points to interior points is less than for Cruise 66-A-8. Also, the current regime in 1967 contains an eddy distinct from the main Loop Current, Nowlin et.al. (1968). The presence of an eddy in addition to the Loop Current provides a more complex regime of currents for which the streamlines may be numerically approximated.

Figures 7-3 and 7-4 illustrate the features of the major currents in the eastern Gulf in June of 1966 and 1967. These GEK vectors and dynamic topographies form the basis with which are compared the numerically approximated velocity stream function and transport function.

Contours of the velocity stream function as approximated using Neumann boundary conditions for the June, 1966, situation are presented in Figure 7-5*a*. Comparison with Figure 7-3*b* indicates that this choice of boundary condition does not give a satisfactory result in this instance.

Applying Dirichlet boundary conditions, Figure 7-5*b*, improves the agreement of the pattern of streamlines with the general flow as illustrated in Figure 7-3. Better agreement is found between Figure 7-3*a* and 7-5*b*.

We now focus attention on the results of the 1967 data. Three analyses each were made for the velocity stream function and transport stream function. In the first analysis Ψ and T were approximated for the interior region only; zero values were imposed on the boundaries of the grid. Secondly, Neumann boundary conditions were imposed and finally, Dirichlet boundary conditions.

The results for the interior solution are shown in Figure 7-6. Even though the interior region is relatively small, it still shows parts of the three main features of the circulation. These are the Loop Current, the eddy and the shear zone between them. When boundary conditions are imposed, better definition is obtained on the western side of the eddy and on the southern extension of the shear zone and Loop Current. It is unfortunate that the right lateral boundary did not extend farther south, since it is in this region that we would expect to find the flow of the Loop Current toward the Florida Straits.

The result of applying Neumann boundary conditions appears in Figure 7-7. The separation of the eddy and the Loop Current now has the form of a trough of relatively low values. The final analysis applying Dirichlet boundary conditions is shown in Figure 7-8.

For the various choices of boundary conditions there is some slight variation in the positions of the eddy and the Loop Current, more so for the Loop Current because of its proximity to the boundary and consequent dependence on the boundary conditions. In general, the results of all three analyses are consistent with each other and with the observed pattern in Figure 7-4. Velocities and transports over the region, as calculated from the streamlines, are also reasonable and in agreement with observed values. Comparison of the pattern of velocity streamlines to those of the transport function in each analysis does not show any appreciable difference. The correlation between the velocity stream function and the observed velocity field would indicate that the major flow in the eastern Gulf of Mexico is nondivergent.

References

Hubertz, J.M. 1967. A Study of the Loop Current in the Eastern Gulf of Mexico. Unpublished Masters Thesis, Texas A&M University. Ref. 67-4T.

Lamb, H. 1945. *Hydrodynamics*. New York: Dover Publications, 208-209.

Nowlin W.D., Hubertz, J.M. and Reid, R.O. 1968. A detached eddy in the Gulf of Mexico. *J. Mar. Res.*, 26(2): 185-186.

Pritchard, D.W. 1948. Streamlines from a discrete vector field: with application to ocean currents. *J. Mar. Res.*, 7(3): 296-303.

Thompson, P.D. 1961. *Numerical weather analysis and prediction.* New York: The MacMillan Company, 89-98.

8

The Effect of Topography on the Yucatan Current

Robert L. Molinari and John D. Cochrane

Abstract

The effect of topography on a portion of the Yucatan Current east and northeast of Yucatan Peninsula during May of 1962, 1965 and 1966 is investigated. A graphical method is used in solving the conservation of potential vorticity equation on the basis of actual topography and observed velocity. A numerical method based on a simplified topography is also used as a check on the subjectivity of the graphical method. Through this procedure a path of the current core is obtained and then compared to the observed path. In the three cases considered the calculated paths agree quite well with the observed paths from the Yucatan Strait north to approximately 23°30′N. The current paths all closely follow a particular isobath with meanders of small amplitude. North of 23°30′N a more complicated topographic region is encountered and the calculated paths appear to diverge from the few observations in that region.

Introduction

The Yucatan Current is the prevailing northward flow which extends from the northwestern portion of the Cayman Sea, through the Yucatan Strait, along the east side of the Campeche Bank and sometimes far to the northwest in the Gulf of Mexico (Figure 8-1 shows a map of the area). On various cruises in the region of this current Cochrane (1962, 1963, 1966) has found a marked relationship

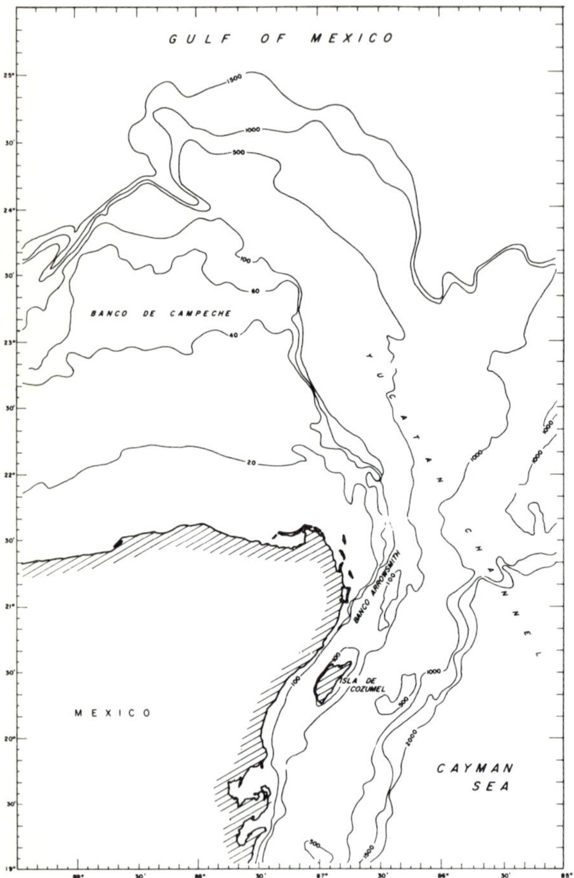

Figure 8-1. Bottom topography in fathoms of the Yucatan Strait region (from U.S. Navy Hydrographic charts BC0904N and BC0905N).

between the core of the current and the underlying contours of bottom topography. This relationship is seen in Table 8-1 (from Cochrane, 1966), which shows the depth associated with the maximum geomagnetic electrokinetograph (GEK) measurement in a particular current crossing. These data indicate a topographic control on the Yucatan Current, especially in the month of May, when the current appears to be strongest.

The majority of recent studies on topographic effects have been extensions of Rossby's (1940) conservation of potential vorticity theory. For instance, Warren (1963) considered a modification of the conservation theory to develop analytical Gulf Stream paths which compared favorably with observed patterns. Holland (1966), using a barotropic model, concluded that the conservation of potential vorticity is quite effective when dealing with a limited region of the ocean containing a current such as the Gulf Stream. Robinson (1965) and Robinson and Niiler (1967 *a,b*) have investigated the basic dynamics of a baroclinic inertial flow which conserved potential vorticity. In all these cases bottom topography causes a meander pattern; however, Hansen (1970) has suggested that topographic meandering cannot alone explain the observed translations of the wave-like features. He concludes that baroclinic instabilities in the Stream are necessary to account for the current path.

Applying Conservation of Potential Vorticity

Rossby (1940) originally investigated the effect of a meridional topographic feature on a zonal wind system, assuming that the flow consisted of "several homogeneous incompressible strata" in which no frictional forces acted. He arrived at the following equation for a column: $d/dt[(\zeta+f)/h] = 0$ where the derivative with respect to time is the total time derivative, ζ is the vertical component of relative vorticity, $f(= 2\Omega \sin\phi)$ the Coriolis parameter and h the vertical thickness of the layer. Thus, at any point along the path of a column of water

$$(\zeta+f)/h = \text{constant}. \qquad (8\text{-}1)$$

The relative vorticity is expressed in natural coordinates as $\zeta = V/R - \partial V/\partial n$ where V is the horizontal speed, R the radius of curvature of the path of a parcel of water taken to be positive if the center of curvature is to the left of the curve and negative if to the right, and n is distance measured normal and to the left of the path. The first term on the right represents the vorticity due to horizontal curvature and the second that due to hori-

Table 8-1. Depth of Ocean Bottom at Yucatan Current Core

Latitude	Date	Current Speed	Depth
21°38′	21 May 1962	196 cm/sec	455 m (249 fm)
22°00′	15 May 1962	165 cm/sec	455 m (249 fm)
22°01′	13 May 1962	175 cm/sec	455 m (249 fm)
22°11′	22 May 1962	197 cm/sec	495 m (271 fm)
22°38′	23 May 1962	220 cm/sec	440 m (241 fm)
22°52′	11 May 1962	166 cm/sec	365 m (200 fm)
23°06′	11 May 1962	201 cm/sec	385 m (211 fm)
23°14′	11 May 1962	134 cm/sec	440 m (241 fm)
23°15′	24 May 1962	230 cm/sec	420 m (230 fm)
21°42′	18 May 1965	180 cm/sec	170 m (93 fm)
22°13′	19 May 1965	178 cm/sec	185 m (101 fm)
23°12′	20 May 1965	160 cm/sec	175 m (96 fm)
19°27′	22 May 1966	111 cm/sec	366 m (200 fm)
21°33′	22 May 1966	162 cm/sec	567 m (310 fm)
22°21′	23 May 1966	147 cm/sec	457 m (250 fm)
21°34′	10 Nov 1962	97 cm/sec	1680 m (919 fm)
22°02′	30 Oct 1962	166 cm/sec	220 m (120 fm)
22°25′	9 Nov 1962	86 cm/sec	295 m (161 fm)
22°29′	8 Nov 1962	124 cm/sec	295 m (161 fm)
22°34′	11 Nov 1962	95 cm/sec	165 m (90 fm)
22°50′	11 Nov 1962	110 cm/sec	330 m (180 fm)
23°05′	12 Nov 1962	104 cm/sec	420 m (230 fm)
21°46′	26 Feb 1965	106 cm/sec	865 m (473 fm)
22°16′	27 Feb 1965	159 cm/sec	350 m (191 fm)
22°41′	27 Feb 1965	176 cm/sec	500 m (273 fm)
23°36′	28 Feb 1965	171 cm/sec	1045 m (572 fm)

zontal shear in the current. In the following development the conservation relationship is to be applied to a column in the core of the current; thus, by definition the shear term is identically zero.

These further assumptions are also made:

1. The ocean is homogeneous.
2. The velocity is constant in depth and downstream.
3. The Coriolis parameter is approximated by $f = f_o + \beta y$, where f_o is a constant and β is the change of Coriolis parameter with y also a constant.
4. An initial inflection point exists in the current such that $\zeta = 0$, at a latitude $f = f_o$, and depth $h = h_o$.

Assumptions 1 and 2 eliminate all baroclinic phenomena and consequently the layer depth is the depth to the ocean floor. With these assumptions equation 8-1 can be written as

$$R = V/[(f_o/h_o)h - f]\,. \qquad (8\text{-}2)$$

The graphical approach uses equation 8-2 to construct the trajectory of a column of water as a series of arcs. The radius of curvature of each arc is calculated from depths determined directly on a contoured map of the area. For each cruise considered, the depth h_o is determined from the depths where the GEKs indicate maximum velocity (Table 8-1). The Coriolis parameter f_o is arbitrarily taken at a latitude of 20°. The initial point of the trajectory is taken where the maximum GEK velocity is observed in the Yucatan Straits. At this point R is calculated from values of f, h and V and used to construct an arc whose direction is that of the GEK velocity. When the arc reaches a depth approximately 50 fathoms different from its starting position, 100 fathoms if the slope is great, the procedure is repeated with the new arc drawn tangent to the old by visual fitting. Thus, an approximate path of the current core is pieced together by these arcs of varying length.

A second method is used as a check on the subjectivity of selecting arc lengths. A numerical solution is obtained by substituting the differential expression for the radius of curvature into equation 8-2 to obtain

$$d^2x/dy^2 = [1+(dx/dy)^2]^{3/2} \times [(h/h_o)f_o - f]/V$$

where x is taken as positive to the east and y positive to the north. The independent variable is taken to be y since the trajectories extend primarily to the north. The differential equation is solved by means of a Runge-Kutta-Nystrom approximation as given by Kreyszig (1962). This finite difference technique is a forward interpolation scheme which uses the information at one point to leapfrog to the next.

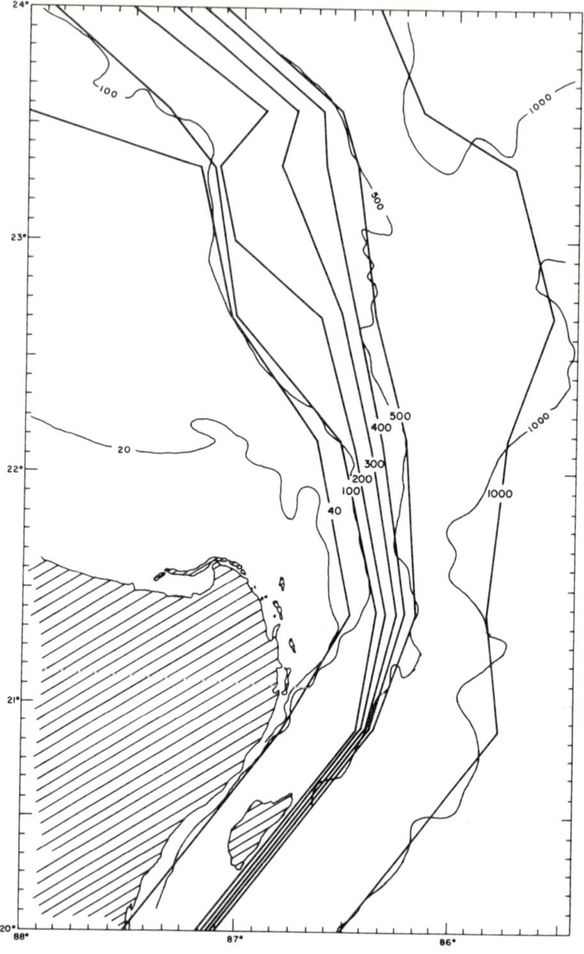

Figure 8-2. Bottom contour lines in fathoms used in approximating the depth for the numerical method of determining the path of the current core.

The actual topography is approximated by a series of straight lines to permit a computer solution. The contours are densest in the region where the graphical method is applied. These are given in Figure 8-2. The space increment used in the numerical approximation is taken as 1/2 mile. The initial parameters of depth, angle and velocity used in the graphical approach are input into the numerical scheme to begin the differentiation. At each successive 1/2 mile step the required depth is

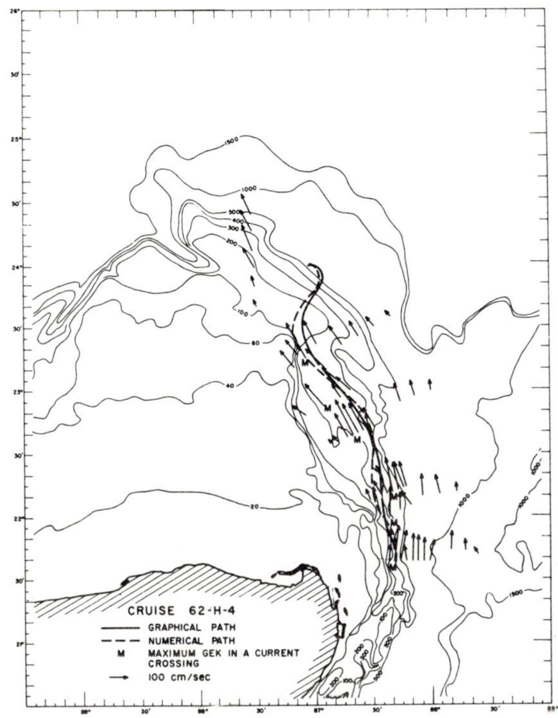

Figure 8-3. Current core path indicated by graphical and numerical methods, and the surface current vectors based on GEK observations for May, 1962 (topography is given in fathoms).

Figure 8-4. Current core path indicated by graphical and numerical methods, and the surface current vectors based on GEK observations for May 1965 (topography is given in fathoms.)

determined by interpolation in the x-direction between the two adjacent contour lines.

Results

The area for which the computations of the current core are made includes the western boundary of the Cayman Sea and the southwest Gulf of Mexico. Particular heed is paid to the region between 21°N and 24°N where the current data are most numerous. Figure 8-1 shows a map of the bottom tography of this region and its surroundings.

The graphical method and the numerical method were applied to data from three of the May cruises listed in Table 8-1, those of 1962, 1965 and 1966. A series of trials were considered in which the initial conditions were kept constant but the velocities were varied. The computed paths given in Figures 8-3, 8-4 and 8-5 are for those velocity trials which most closely resemble the observed current cores. The graphical and the numerical paths do not contain any appreciable subjective bias.

Texas A&M Cruise 62-H-4 (1962) contains the most extensive GEK coverage. The agreement between the actual and the observed paths is very good for a velocity of 75 cm/sec and depth at initial inflection of 250 fm. The similarity with both the position and direction of maximum GEKs extends for approximately 2° of latitude and encompasses two wavelengths of the me-

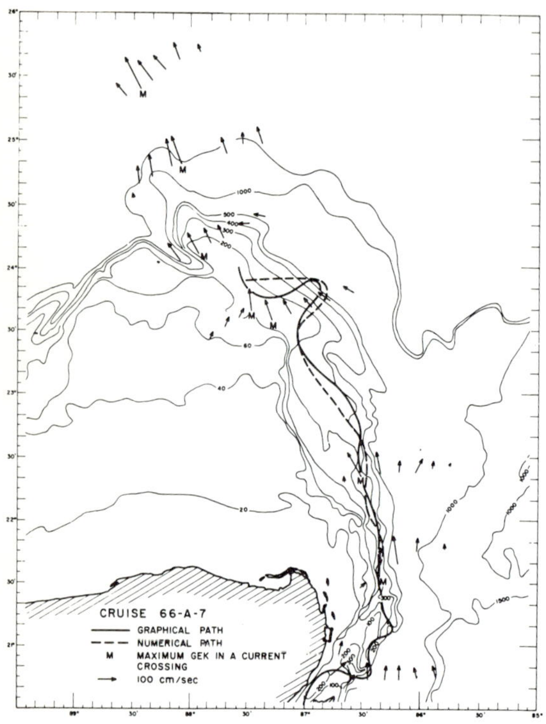

Figure 8-5. Current core path indicated by graphical and numerical methods, and the surface current vectors based on GEK observations for May, 1966, south of 23°N and for June, 1966, north of 23°N (topography is given in fathoms).

andering current. The computed paths are carried only a short distance beyond the westward turn in contours to 24°N where meandering becomes more intense. This intensification occurs in all three cases and will be discussed later.

The calculated paths for Texas A&M Cruise 65-A-7 given in Figure 8-4 again exhibit good agreement with the observed flow. A velocity of 50 cm/sec and a depth at initial inflection of 200 fm are used. The current core oscillates through only one and one-half meanders in essentially the same region as the 1962 cruise. The numerical path does not extend as far north as the graphical one because the depth limit employed in the computer program is reached. This occurred just short of the region of more intense meandering and therefore a more complicated topography is not introduced.

Figure 8-5 shows Cruise 66-A-7 May, 1966, in the southern area. At about 23°30′N the GEK pattern of the June, 1966, cruise is introduced. This cruise entered the region about three weeks later than the May cruise. Although considerable data exist to the south of the initial positions used in 62-H-4 and 65-A-7 the starting point of the computed paths is taken in the same area as the previous cruises. North of this area, agreement with the May cruise is good for a velocity of 50 cm/sec and depth of 250 fm. The first GEK line of the June, 1966, cruise indicates a current core slightly to the west of the calculated core, but this is in the region of intense meandering in the computed paths.

Because of the data present south of the Yucatan Straits the graphical path was projected back for the 1966 cruise using the same procedures as used in tracing the path downstream. Figure 8-5 indicates that similarities exist in the calculated and observed cores. In particular, the calculated path comes around the west side of Cozumel Island, a situation suggested by the GEK observations in this region.

The computed current cores are determined approximately to 24° latitude. North of this region the observed current leaves the constraint of the continental slope to flow into the Gulf of Mexico basin (Figures 8-3 and 8-5). The numerical and graphical paths cannot duplicate this behavior because implicit in the application of Rossby's theory to a homogeneous flow is a restoring force which will return the current to the depth of inflection. A detailed study of this phenomena is beyond the scope of the present investigation, but it is noted that Greenspan (1963) and Pedlosky (1965) have considered the problem of an inertial current separation from the boundary.

Agreement between the calculated and observed paths points out the current pattern similarities of the three cruises considered. The actual current cores all remained in shallow water as they followed the isobaths to the northern part of

Campeche Bank. The calculated paths following this trend also have meanders of similar dimensions, wavelengths between 50 and 60 nautical miles and amplitudes usually less than 5 nautical miles.

Discussion

The similarity between the calculated paths of the current core and the point GEK current measurements suggests that a topographic control caused by the conservation of potential vorticity existed on a portion of the Yucatan Current in May of 1962, 1965 and 1966. The large meanders found north of 23°30′N may mark the limit of the applicability of this theory. The 1962 and 1965 data are insufficient to accurately determine the core path, while the 1966 data indicate a more or less straight current path.

However, in June of 1967 a cut-off eddy was found in the Gulf of Mexico (Nowlin, Hubertz and Reid, 1968). The main body of the Yucatan Current veered east at about 23°30′N instead of continuing north in the Gulf. The eddy's southern edge was just north of the eastward turning of the current. The intense meanderings of the present models could represent a stage in the detachment of this eddy from the Yucatan Current. When the meanders become very elongated and narrow, the developing eddy could separate from the main flow.

Acknowledgment

This research was supported by the Office of Naval Research under Contract Nonr 2119(04) with the Texas A&M Research Foundation.

References

Cochrane, J.D. 1962. Investigation of the Yucatan Current. In *Oceanography and Meteorology of the Gulf of Mexico.* Texas A&M Dept. of Oceano. Ref. 62-14A:5-10.

______.1963. Yucatan Current. In *Oceanography and Meteorology of the Gulf of Mexico.* Texas A&M Dept. of Oceano. Ref. 63-18A:6-11.

______.1966. The Yucatan Current. In *Oceanography and Meteorology of the Gulf of Mexico*. Texas A&M Dept. of Oceano. Ref. 66-23T:14-32.

Greenspan, A.R. 1963. A note concerning topography and inertial currents. *J. Mar. Res.*, 21:147-154.

Hansen, D.V. 1970. Gulf Stream meanders between Cape Hatteras and the Grand Banks. *Deep Sea Res.*, 17:495-512.

Holland, W.R. 1966. *Wind-driven circulation in an ocean with bottom topography.* Doctoral Dissertation, Univ. of California.

Kreyszig, E. 1962. *Advanced engineering mathematics.* New York: John Wiley and Sons.

Niiler, P.P. and Robinson, A.R. 1967. The theory of free inertial jets, II. A numerical experiment for the path of the Gulf Stream. *Tellus*, 19:601-618.

Nowlin, W.D., Jr., Hubertz, J.M. and Reid, R.O. 1968. A detached eddy in the Gulf of Mexico. *J. Mar. Res.*, 26:185-186.

Pedlosky, J. 1965. A necessary condition for the existence of an inertial boundary current in a baroclinic ocean. *J. Mar. Res.*, 23:69-72.

Robinson, A.R. 1965. Three-dimensional model of inertial currents in a variable density ocean. *J. Fluid Mech.*, 21:211-223.

_____, and Niiler, P.P. 1967. The theory of free inertial currents, I. Path and structure. *Tellus*, 19:269-291.

Rossby, C.G. 1940. Planetary flow patterns in the atmosphere. *Quart. J. Roy. Meteor. Soc.*, 66 (Suppl.): 68-87.

Warren, B.A. 1963. Topographic influences on the path of the Gulf Stream. *Tellus*, 15:167-183.

9

A Simple Dynamic Model of the Loop Current

Robert O. Reid

Abstract

When it is well developed in the Gulf of Mexico (north of Yucatan Shelf) the northern portion of the Loop Current is confined entirely to the deep-water region where topographic control is absent. However, the dimensions (northern penetration and width) of the Loop Current can be explained in terms of the variation of Coriolis parameter with latitude and the current speed.

Introduction

In Chapter 8, Molinari and Cochrane demonstrated that in the relatively shallow water of the slope region adjacent to the Yucatan Shelf, the topography exerts a rather strong control on the character of the Yucatan Current. In their treatment, the fluid is regarded as homogeneous. This appears to be justifiable as long as the water depth is comparable to, or at least not much greater than, the depth of the thermocline for the actual water column.

The question now arises as to what happens when the current reaches the northernmost corner of the Yucatan Shelf. At this point the topographic control seems to be lost, for indeed the usual observed patterns of current structure rarely show any tendency for the current to turn westward following the contours along the northern slope of the Yucatan Shelf.

Figure 5 of Chapter 8 shows a typical case where the current runs northwesterly out over the abyssal region of the Gulf. This occurs near the abrupt change in the orientation of the topographic contours. There is some tendency for cyclonic curvature of the streamlines after leaving the 1000-fathom contour, but this is not sufficient to bring the current back into the control of the continental shelf. Instead, if the current has reached this far into the Gulf, it typically forms an anticyclonic loop within the deep part of the Eastern Gulf, ultimately exiting at Florida Strait (see, for example, Figure 1 of Chapter 4).

It is suggested here that the reason for the loss of topographic control at the northeast corner of the Yucatan Shelf is two-fold:

1. The abruptness of the change in orientation of the topography coupled with
2. The fact that when the current reaches a depth greatly exceeding the thermocline depth the topography has little control since the bottom current becomes negligible compared with the current above the thermocline.

Indeed a simplified model of the Loop Current can be constructed by considering an idealized two-layer system with the lower layer at rest, so that bottom topography exerts no control at all. The question we wish to ask is simply this: Can the variation of Coriolis parameter alone cause the current to turn back to the south before it is forced to do so by the northern continental slope in the Gulf? The answer to this seems to be affirmative for typical current speeds.

Theory

For a two-layer frictionless system with the lower layer at rest we have for steady state

$$[\zeta + f(y)]\ /H = \text{constant} \qquad (9\text{-}1)$$

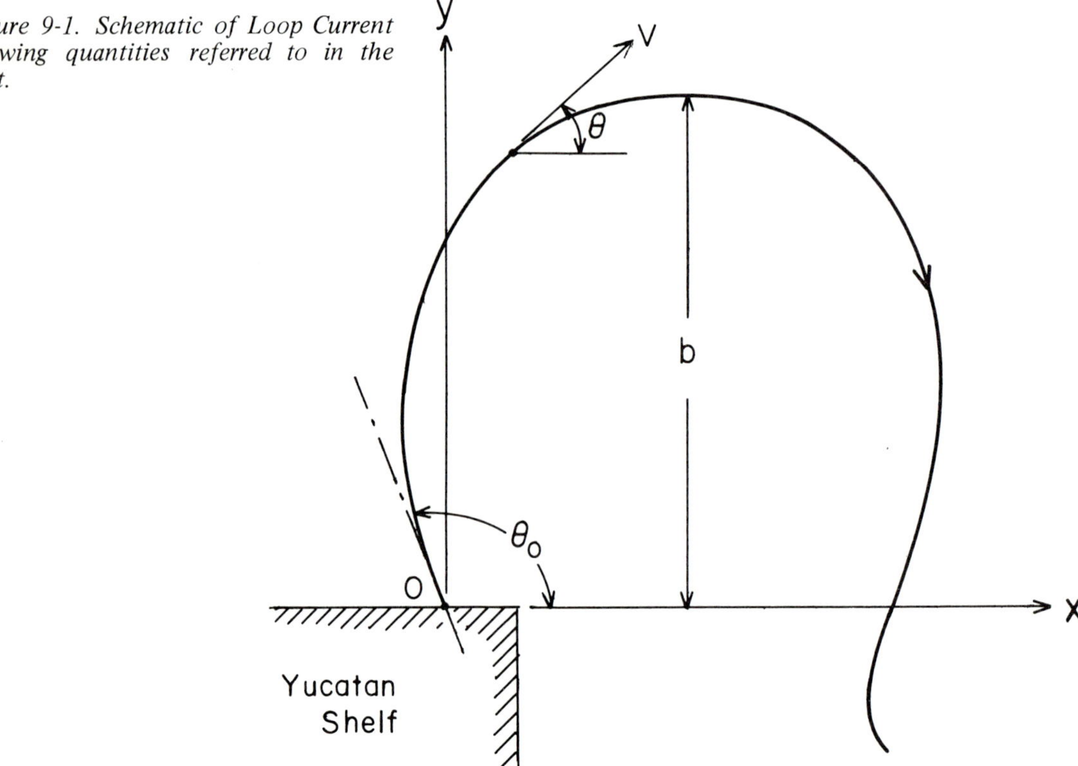

Figure 9-1. Schematic of Loop Current showing quantities referred to in the text.

along a streamline, and

$$\mathbf{V} = (g\epsilon/f)\mathbf{k} \times \nabla H \qquad (9\text{-}2)$$

where H is the upper layer thickness, $\mathbf{V}$ is the horizontal velocity vector for the upper layer, ϵ is the density contrast parameter $(\Delta\rho/\rho)$, $\mathbf{k}$ is the vertical unit vector, g is gravity, f is the Coriolis parameter and ζ is the vertical component of the relative vorticity of the upper layer. We take x,y as horizontal coordinates with x eastward and y northward. Equation (9-1) expresses the conservation of potential vorticity for the upper layer, in which the layer depth H is now a dependent variable. Equation (9-2) is the geostrophic relation, which should be a good approximation as long as the curvature of the streamline is not excessive. Finally, we will approximate f by the usual truncated series expansion $f = f_o + \beta y$ where f_o and β are constants.

Since the flow is geostrophic, a streamline is also an isoline of H. Consequently, (9-1) reduces to

$$\zeta + f_o + \beta y = \text{constant} \qquad (9\text{-}3)$$

along a streamline. If we focus attention on that streamline at or near the core of the current, then

$$\zeta = V(d\theta/ds) \qquad (9\text{-}4)$$

where V is the magnitude of the velocity in the upper layer and $d\theta/ds$ is the curvature of the streamline (θ being the angle between the vector V and the x axis and s the arc distance along the streamline).

For simplicity we will suppose that the core streamline leaves the northern end of the Yucatan Shelf with zero curvature. Moreover, we will take $x = y = 0$ at this point (Figure 9-1). Then the constant in (9-3) is simply f_o, and hence (9-3) and (9-4) combine to give

$$(d\theta/ds) = -(\beta Y/V). \qquad (9\text{-}5)$$

This shows immediately that $d\theta/ds$ is negative for $y > 0$; i.e., the curvature is anticyclonic.

In order to proceed further we will assume that V is constant as an approximation, which implies that the neighboring isolines of H remain nearly equally spaced. Now from the geometry

$$(dy/ds) = \sin\theta.$$

Hence (9-5) transforms to $\sin\theta\,(d\theta/ds) = -(\beta/V)Y$. With V taken as constant, this leads immediately to the following simple integrated form

$$\cos\theta = \cos\theta_o + (\beta/2V)y^2 \qquad (9\text{-}6)$$

where θ_o is the initial value of θ (Figure 9-1).

At the northernmost reach of the streamline ($y = b$) we require $\theta = 0$. Hence 9-6 yields*

$$b = [(2V\text{B})(1 - \cos\theta_o)]^{1/2}.$$

As an example with $V = 50$ cm/sec, $\beta = 2 \times 10^{-13}$ $\text{cm}^{-1}\ \text{sec}^{-1}$ and $\theta_o = 90°$, then $b \doteq 220$ km or about 2.0 degrees of latitude. This seems to be in accord with the northward penetration of observed Loop Current regimes beyond the northern edge of the Yucatan Shelf. Moreover, it can be shown that the east-west range of the Loop is comparable to b as is observed.

Acknowledgments

This work was supported by the Office of Naval Research under contract Nonr 2119 (04).

Reference

Ichiye, T. 1962. Circulation and water mass distribution in the Gulf of Mexico. *Geofisica Internacional,* 2(3):47-76.

*It has recently come to the attention of the author that a similar relation was given by Ichiye (1962, p. 75). In the present development it is stressed that this holds only for the deep-water region and hence the northward penetration b is measured from the northern edge of the Yucatan Shelf.

Section *3*
Numerical Modeling

10

A Barotropic Prognostic Numerical Circulation Model

David F. Paskausky and Robert O. Reid

Abstract

A prognostic vorticity equation for a barotropic numerical model is applied to a basin simulating the Gulf of Mexico. Advection of vorticity, planetary vorticity tendency and frictional torques associated with lateral and bottom stresses are included. Wind stress is not included; instead, the forcing function of this model is the prescribed input flow at one of the two ports (Yucatan Channel). Vorticity is predicted using the old stream-field; then Gauss-Seidel overrelaxation is used to obtain a new stream-field from the predicted vorticity field. An increase from weak to strong western intensification in the input flow over a period of four months approximates the inflow conditions in the Yucatan Channel from late winter to early summer. This variation of western intensification in the input is associated with the subsequent detachment of an anticyclonic eddy, a feature which agrees qualitatively with observed seasonal patterns in the Gulf of Mexico.

Introduction

A numerical prognostic model for the barotropic circulation in a two-port basin with prescribed input is applied with rectangular boundaries but with internal bathymetry approximating that of the Gulf of Mexico (Figure 10-1).

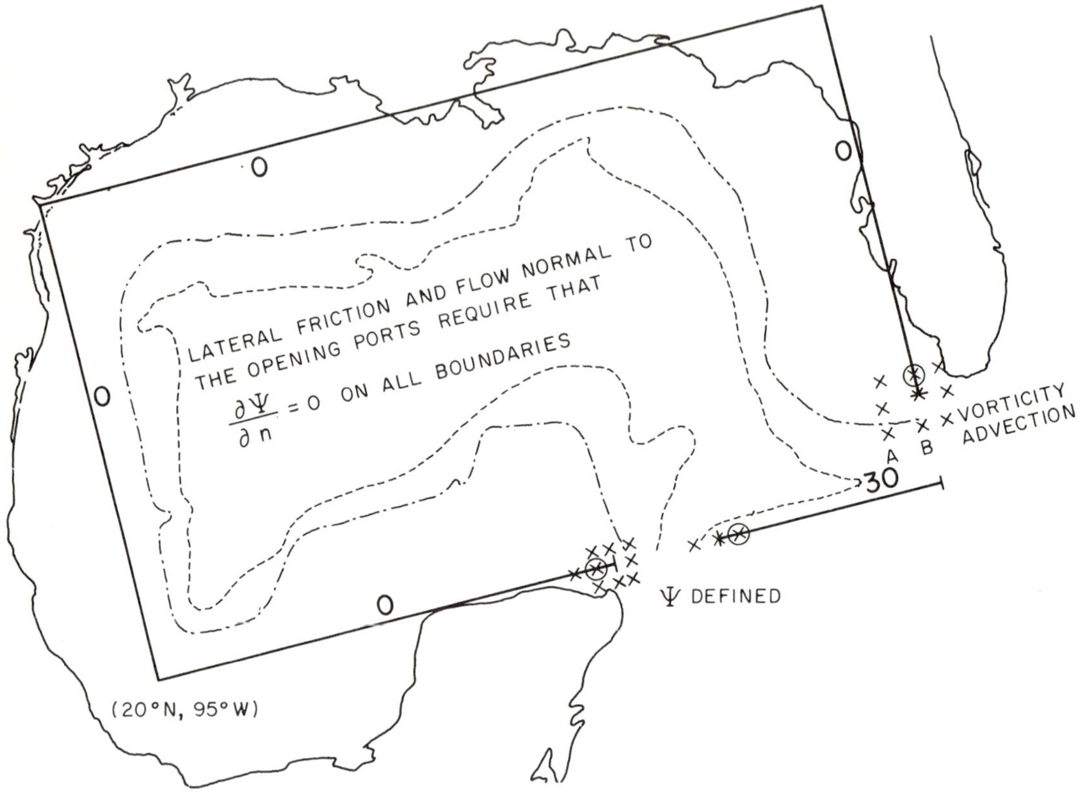

Figure 10-1. Rectangular grid fitted to the Gulf of Mexico. The dashed lines are the 100- and 1000-fathom contours of the model. Boundary conditions are shown.

In this model a numerical analog of the vertically integrated vorticity equation in terms of a volume transport stream function is used. Wind stress is neglected and thus the primary driving force is the flow at the Yucatan Channel. The Gulf of Mexico is assumed to be a basin where a nonlinear barotropic model driven by the inflow and excluding wind and thermohaline effects can approximate the general circulation.

The present study differs from that of Holland (1966) in which the study of topographic effects on a barotropic flow regime were restricted to steady state conditions. Indeed we are interested in time changes of the inertially controlled flow regime in the present case. This study has many points in common with that of Baer et. al. (1968) in which application of a quasi-geostrophic advection model was made to the Gulf. The primary restrictions of the present model are the neglect of wind and of stratification and associated baroclinic shear, such as is included in the numerical oceanic circulation model investigated by Bryan and Cox (1967).

Unfortunately, it was discovered *after* carrying out the numerical experiment reported here that the numerical analog of the vorticity prediction equation, which represents the hub of the problem, did not allow for vorticity tendency due to topographic stretching. Nevertheless, the results show a remarkable qualitative agreement with the seasonal regime of circulation in the Gulf of Mexico.

Assumptions and Restrictions

The principal assumptions of the model and/or restrictions are that:

1. The fluid is homogeneous and the pressure hydrostatic.
2. Inflow and outflow are normal to the cross section of the ports.
3. Transport through the openings is constant (30×10^6 m^3 sec^{-1}) even though the velocity distribution is not.
4. A zero slip condition is employed on all lateral boundaries (excluding ports).
5. Wind effects are neglected.
6. Tidal forces are neglected.
7. The frictional torque at the sea bed is assumed to be proportional to the vorticity.
8. The effect of vertical velocity on the Coriolis acceleration is neglected.
9. The frictional torque associated with lateral mixing is assumed to be proportional to the horizontal Laplacian of the vorticity.
10. At the sea surface the vertical velocity is taken as zero, while at the bottom we require only that the latter be a stream-surface.

Theory

In view of assumption 1, the horizontal pressure gradient is independent of depth; hence, the horizontal velocity vector, $\mathbf{V}_h$, must be independent of depth except in the surface and bottom frictional (Ekman) layers which are regarded as a small fraction of the total depth of fluid, D. Consequently, the vertically averaged equation for horizontal motion is the same as that at any level but with the surface and bottom stress, τ^s and τ^b respectively, entering as equivalent body force terms. This equation with the traditional approximation (h) for the Coriolis acceleration (which neglects that due to vertical motion) is

$$(\partial \mathbf{V}_h/\partial t)\ (\mathbf{V}_h \cdot \nabla_h)\,\mathbf{V}_h - f\mathbf{k} \times \mathbf{V}_h + \nabla_h \lambda = K_h\, \nabla^2{}_h\, \mathbf{V}_h + (\tau^s - \tau^b)/\rho D \qquad (10\text{-}1)$$

where ρ is mass density, $\mathbf{k}$ is a unit vertical vector (Positive upwards), ∇h is the horizontal gradient operator, $\rho\lambda$ is pressure at a fixed level within the fluid, K_h is a horizontal eddy viscosity, f is the Coriolis parameter which is proportional to the sine of the latitude and t is time. Since we are concerned with inertially controlled flow, the wind stress, τ^s is taken as zero.

The variable λ can be eliminated from (10-1) by forming a vector curl operation on (10-1) of which only the vertical component is pertinent; the result is the following vorticity prediction relation:

$$(\partial\zeta/\partial t) + \nabla \cdot [(f+\zeta)\,\mathbf{V}_h] = K_h\, \nabla^2{}_h\, \zeta - \mathbf{k}\cdot \text{curl}\,(\tau^b/\rho D) \qquad (10\text{-}2)$$

where ζ is the vertical component of the fluid's vorticity relative to the earth; i.e., $\zeta = \mathbf{k} \cdot \text{curl}\, \mathbf{V}_h$. The second term on the right side of (10-2), which represents a bed resistance "torque," is approximated by a simple linear expression $\sigma\zeta$ where σ is a coefficient with dimensions of frequency (the variation of D being ignored compared with the variation of τ^b). If we adopt an Ekman boundary layer approximation at the sea bed, then it can be shown that

$$\sigma = (K_v\, f/2)^{1/2}\, D^{-1}$$

where K_v is the vertical eddy viscosity. The vorticity equation consequently becomes

$$(\partial\zeta/\partial t) + \nabla \cdot [(f+\zeta)\,\mathbf{V}_h] = K_h \nabla^2{}_h\, \zeta - \sigma\zeta. \qquad (10\text{-}3)$$

The continuity equation for a column of fluid is, in view of assumption 10,

$$\nabla_h \cdot (D\,\mathbf{V}_h) = 0. \qquad (10\text{-}4)$$

Hence, a volume transport stream function, ψ, exists such that

$$D\mathbf{V}_h = \mathbf{k} \times \nabla_h \psi.$$

Moreover in terms of ψ, the vorticity can be expressed as

$$\zeta = \nabla_h \cdot (D^{-1}\nabla_h\, \psi)\,. \qquad (10\text{-}5)$$

Consequently, if one predicts the field of ζ for given t using (10-3), then ψ can be obtained by relaxation using (10-5) plus appropriate boundary conditions.

An alternative form of (10-3) which displays the separate effects of advection, topography and the earth's curvature more clearly is

$$(\partial\zeta/\partial t) + \overset{(A)}{\mathbf{V}_h \cdot \nabla\zeta} + \overset{(B)}{\mathbf{V}_h \cdot \nabla f} - \overset{(C)}{(f+\zeta)\,\mathbf{V}_h \cdot (\nabla D/D)} = K_h \nabla^2{}_h \zeta - \sigma\zeta,$$

which makes use of relation (10-4). Term A represents the advection of vorticity which is clearly nonlinear; term B is the planetary vorticity tendency for which the magnitude of ∇f is proportional to the earth's curvature as well as the angular speed of the earth; finally, term C represents the effect of stretching or shrinking of the water column associated with motion across the bottom topography.

The model is fourth order spatially in the volume transport stream function, ψ, so two conditions are required on all boundaries for a unique solution, given specific initial conditions. The boundary conditions, consistent with the foregoing restrictions are taken as follows:

1. Solid boundaries: the stream function, ψ, is taken as fixed (0 or 30 x 10^6 m^3 sec^{-1}, see Figure 9-1); the no slip condition requires that $\partial\psi/\partial n = 0$.
2. Input port: the stream function is specified versus distance across the opening at each time step; flow is perpendicular to the port cross section so $\partial\psi/\partial n = 0$.
3. Exit port: vorticity is advected from the interior to the port; flow is perpendicular to the port cross section so $\partial\psi/\partial n = 0$.

Thus, in summary, $\partial\psi/\partial n = 0$ at *all* boundary points, ψ is specified at all boundary points *excluding the exit port* where ψ is found by relaxation from a knowledge of ψ as predicted from a simple outward advection relation (discussed in the following section).

Numerical Algorithm

The finite difference analog of (10-3) for a discrete mesh ($x = J\Delta s$, $y = K\Delta s$, $t = N\Delta t$ where J, K, N are integers) is based on centered differencing in time and space. Grid spacing (Δs) in x (directed 14° north of east) and y (directed 14° west of north) is 20 km. To obtain an analog prediction equation, the DuFort-Frankel scheme (Smith, 1965) is used for the lateral exchange term and the volume transport stream function ψ is introduced to evaluate the velocity in (10-3):

$$\zeta^{N+1}_{J,K} = A_{J,K}\,(B\,\zeta^{N-1}_{J,K} - BT^{N}_{J,K} + XL^{N}_{J,K}) \qquad (10\text{-}6)$$

where

$$A_{J,K} = 1/(1 + 2\,\Delta t\sigma_{J,K} + M),$$

$$M = 4\,\Delta t\,K_h/(\Delta s)^2,$$

$$B = (1 - M),$$

$$XL^{N}_{J,K} = \frac{M}{2}\,(\zeta^{N}_{J+1,K} + \zeta^{N}_{J-1,K} + \zeta^{N}_{J,K+1} + \zeta^{N}_{J,K-1}),$$

$$BT^{N}_{J,K} = [(f_{J,K+1} + \zeta^{N}_{J,K+1})\,(\psi^{N}_{J+1,K+1} - \psi^{N}_{J-1,K+1}) - (f_{J,K-1} + \zeta^{N}_{J,K-1})\,(\psi^{N}_{J+1,K-1} - \psi^{N}_{J-1,K-1}) + (f_{J+1,K} + \zeta^{N}_{J+1,K})\,(\psi^{N}_{J+1,K-1} - \psi^{N}_{J+1,K+1}$$

$$- (f_{J-1,K} + \zeta^{N}_{J-1,K})\,(\psi^{N}_{J-1,K-1} - \psi^{N}_{J-1,K+1})]\,\frac{(\Delta t)}{2D_{J,K}\,(\Delta s)^2}.$$

In the above relations, A and B are frictional damping coefficients whose magnitudes are somewhat less than unity. The quantity $XL^N_{J,K}$ contains that part of the lateral exchange term which is evaluated at time level N in the DuFort-Frankel scheme. The term BT^N_{JK} was taken as the analog of

$\nabla_h \cdot [(f+\zeta)\mathbf{V}_h]$.

It was not until after the extensive computations reported here were carried out that the authors realized that the topographic stretching effect is not implicitly included in the analog of $BT^N_{J,K}$. Accordingly, although a variable depth model was employed which in turn influences the magnitude of the inertial terms and bottom friction, the topographic control which one should find in a barotropic flow is missing for the finite difference analog employed in this work. For this reason the flow acts more like a baroclinic regime, although D is of course stipulated a priori and is not like a dependent layer depth as in a two-layer model. In spite of this shortcoming of the model, we felt that the flow regime which it predicts would be of sufficient interest to be published.

Since a "tilted" coordinate system has been used, the evaluation of f depends upon both x and y:

$$f = 2\,\Omega \sin\phi$$

where

$$\phi = \phi_o + (x \sin\theta + y\cos\theta)/r \qquad (10\text{-}7)$$

where Ω is the angular speed of the earth, r is the mean radius of the earth, θ is the tilt angle of the x-axis with respect to east (14°) and ϕ_O is the latitude of the origin for x, y (20° N).

The grid points used in computing the terms of (10-6) are shown in Figure 10-2. The total mesh is 81 by 43 in the x and y directions, respectively. To use the grid points shown in Figure 10-2, (10-5) is written in expanded form as:

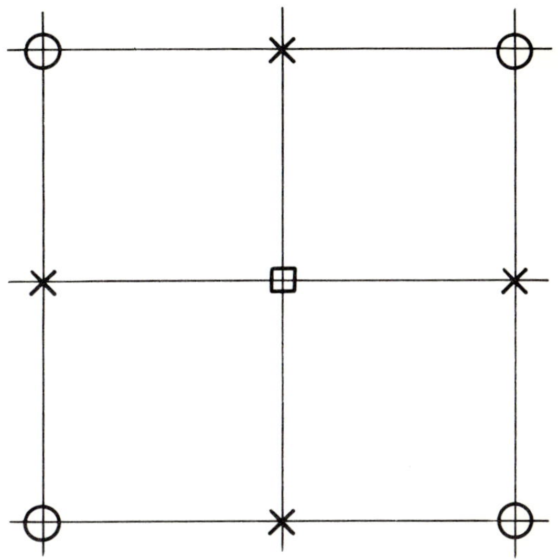

Figure 10-2. Grid points used in computation.

$$\zeta = (\nabla_h^2\psi / D) - (\nabla_h D\, \nabla_h \psi / D^2) \qquad (10\text{-}8)$$

The finite difference analog of (10-8) solved for ψ at the computation point, (J, K), is listed below

$$\psi_{J,K} = \alpha\Big\{ \psi_{J+1,K} + \psi_{J-1,K} + \psi_{J,K+1} + \psi_{J,K-1} - (\Delta s)^2\, \zeta^{N+1}_{J,K}\, D_{J,K}$$
$$- 0.25\,[(D_{J+1,K} - D_{J-1,K})(\psi_{J+1,K} - \psi_{J-1,K}) + (D_{J,K+1} - D_{J,K-1})$$
$$(\psi_{J,K+1} - \psi_{J,K-1})]/D_{J,K}\Big\} + (1-4\alpha)\,\psi'_{J,K} \qquad (10\text{-}9)$$

Table 10-1

Distirbution of Ψ (10^6 m^3 sec^{-1}) for Grid Intervals Across Input Port

Westward Intensification	0	1	2	3	4	5	6	7	8	9	10
Weak	0.0	0.1	0.7	2.0	4.0	7.5	12.7	18.0	23.0	27.0	30.0
Strong	0.0	1.0	5.0	11.0	15.0	19.0	22.0	25.0	27.0	29.0	30.0

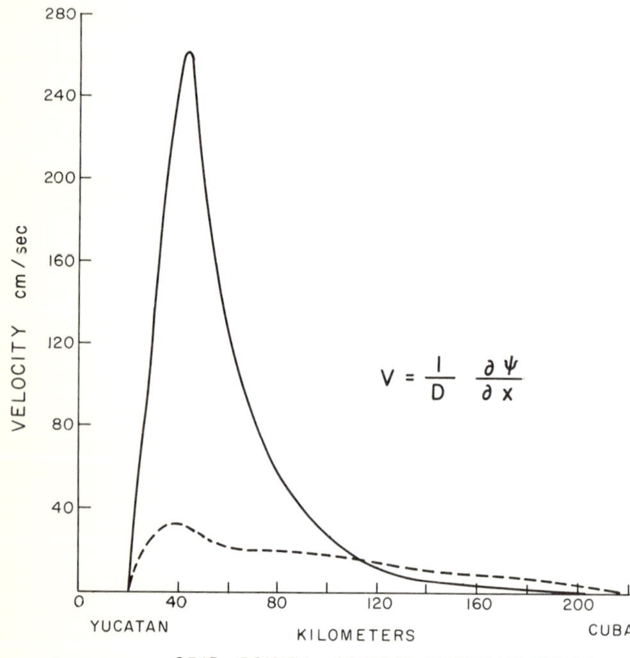

Figure 10-3. Velocity profiles across the Yucatan Strait opening. Solid line represents strong western intensification. Dashed line represents weak western intensification. (V in cm/sec).

where $\psi'_{J,K}$ is the previous value of $\psi_{J,K}$ and α is a relaxation coefficient which varies between 1/4 and 1/2. The ζ field predicted by (10-6) is used in the relaxational equation, (10-9), until the change in ψ at each iteration is less than 0.01 x 10^6 m^3 sec/$^{-1}$. Use of over-relaxation reduced the number of iterations by a factor of 1/5.

Finally, the condition of advection of vorticity at the exit port was taken as follows (see Figure 10-1 for location of points *A,B*):

$$\zeta_B^{N+1} = \gamma(\zeta_A^N + f_A) + (1-\gamma)(\zeta_B^N + f_B) - f_B \qquad (10\text{-}10)$$

where

$$\gamma = u_B^N \, \Delta t / \Delta s ,$$

in which u_B^N is the outward velocity at *B* which is evaluated from the numerical analog of $-D^{-1}\,\partial\psi/\partial y$ •

A value of 20 cm^2 sec^{-1} was chosen for K_V. This gives a range of σ from 10^{-7} sec^{-1} in the deep water to 10^{-5} sec^{-1} in the shallowest water. Charney (1960) used a K_V of 15 cm^2 sec^{-1}, while Arons and Stommel (1956) used about 1 cm^2 sec^{-1} in studies of the planetary deep-water circulation. Our value may be somewhat high.

On the basis of trial runs over periods of 5 to 10 days varying only K_h, the value of 1 x 10^6 cm^2 sec^{-1} was chosen and used in the seasonal variation run.

The time step, Δ*t*, was taken as 5400 sec; this is about the largest value consistent with a stable computing scheme as determined by numerical experimentation.

In starting the computations from an initial field of ψ, the first prediction of ζ is done using a forward time step. From then on the centered

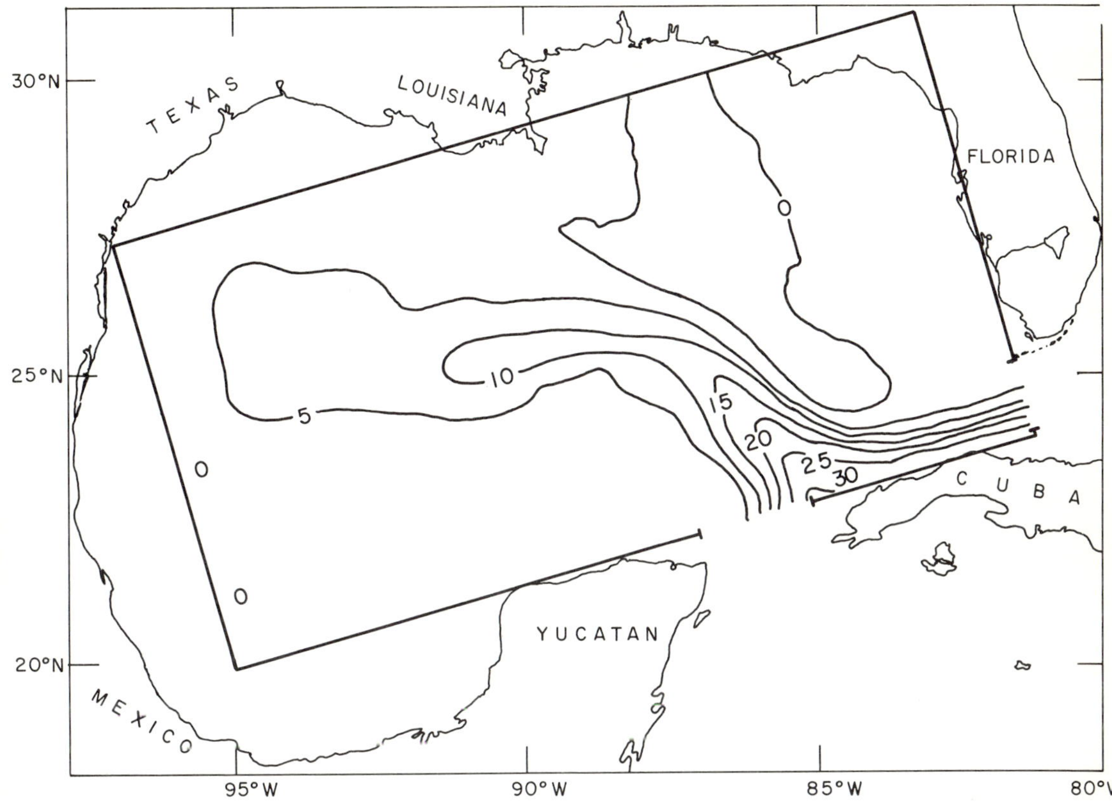

Figure 10-4. Initial quasi-steady state circulation corresponding to a sustained strong westward intensification of inflow at Yucatan channel. Contours of ψ are in units of 10^6 m^3 sec^{-1}.

time step indicated by (10-6) is employed. The procedure for a general time step is to evaluate the new ζ field using (10-6) for each interior point and (10-10) at the exit port. Then evaluate the associated new ψ field by relaxation at all interior points and in the exit port using (10-9). On all other boundary grid points and across the input port, ψ is specified. Next, use the condition that $\partial\psi/\partial n = 0$ to assign a fictitious ψ value one grid interval beyond the boundary, all the way around the boundary. Finally, evaluate the ζ values using the finite difference version of (10-8) at all remaining boundary points other than across the exit port. This completes the evaluation of both ζ and ψ at time level $N + 1$. The whole procedure is then repeated for the next time level.

Results and Discussion

Cochrane's data for Yucatan Channel (personal communication) and Richardson's (1965) findings for the Florida Straits support the idea that there is a significant seasonal variation in the velocity distribution across the Yucatan

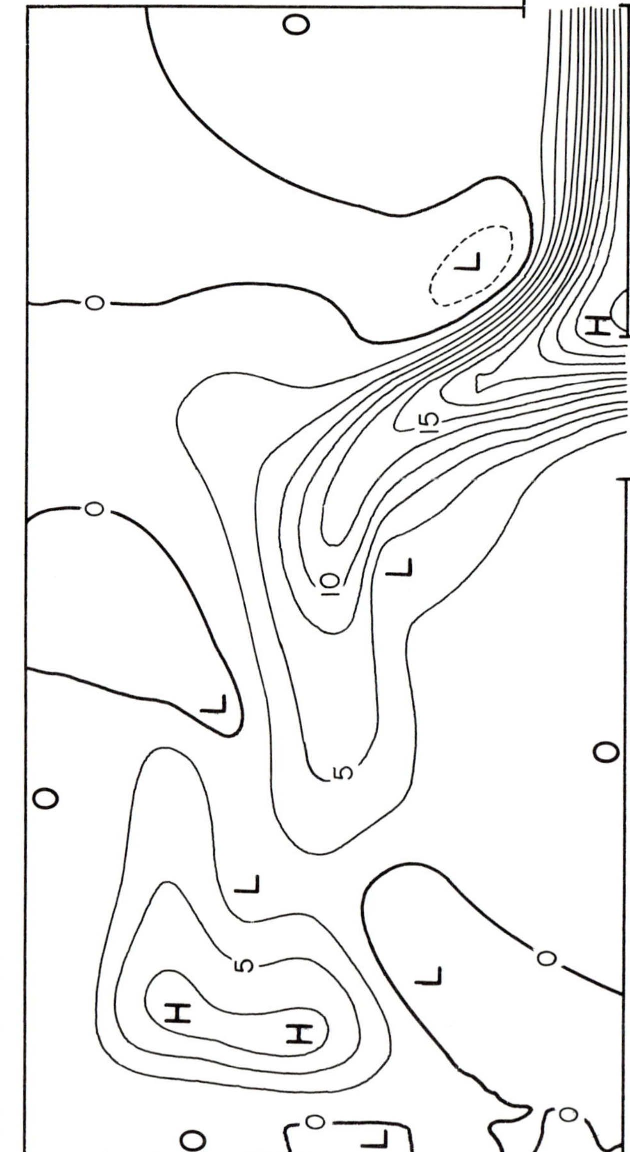

Figure 10-5. Weak western intensification. Contour interval is 2.5 x 10^6 m^3 sec^{-1}. Negative streamlines are dashed. Highs and lows are denoted by H's *and* L's *respectively.*

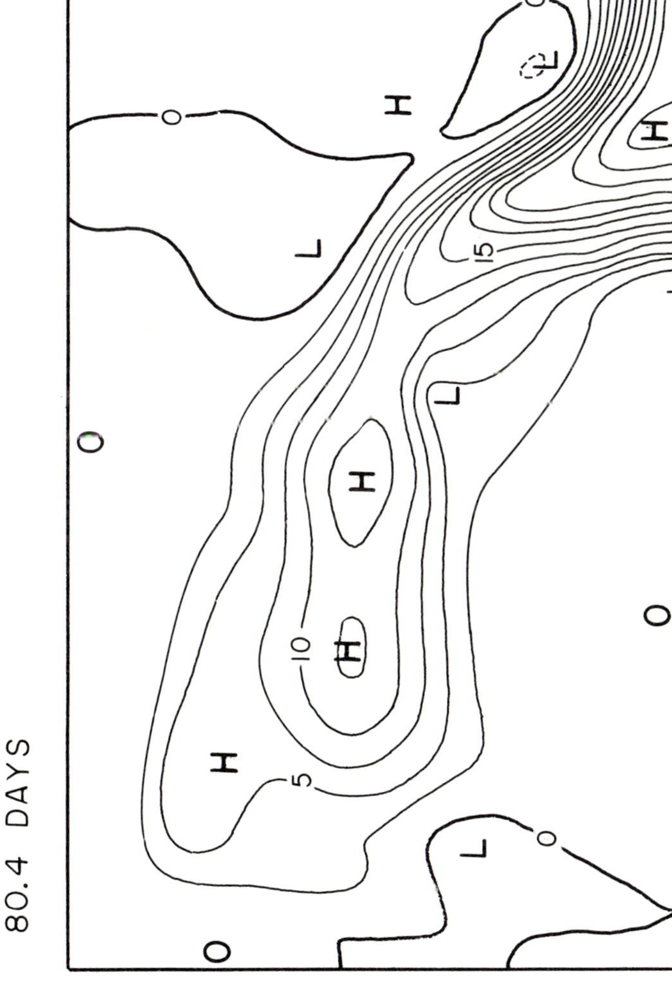

Figure 10-6. Increasing westward intensification. Contour interval is 2.5 x 10^6 m^3 sec^{-1}. Negative streamlines are dashed. Highs and lows are denoted by H's and L's *respectively.*

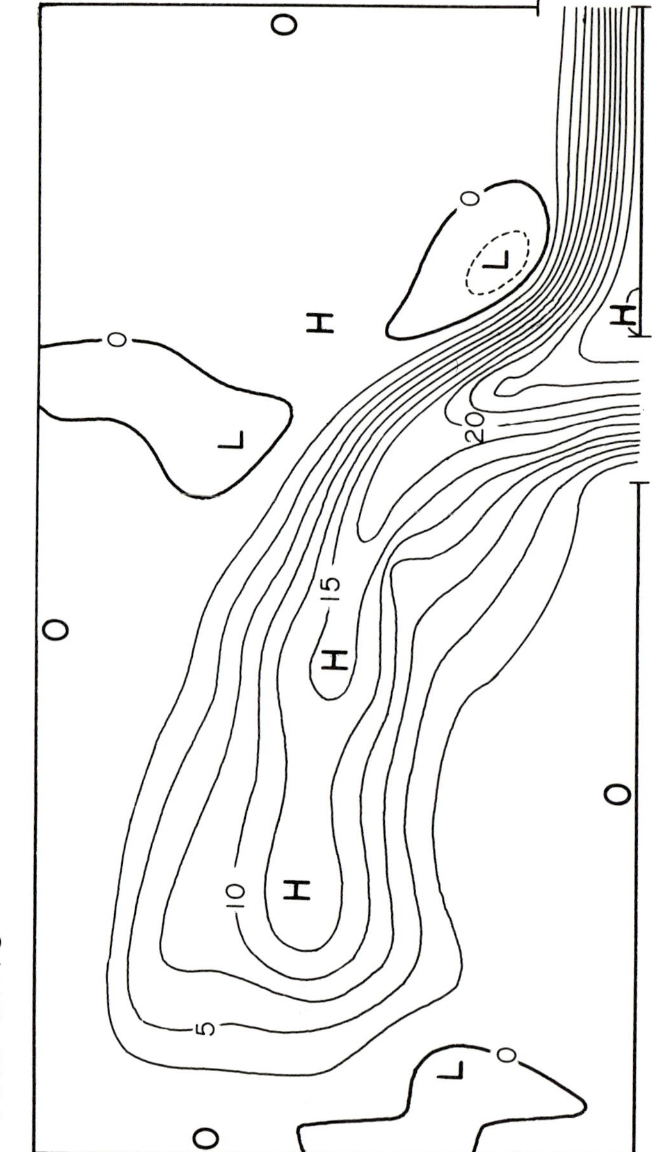

Figure 10-7. Strong westward intensification. Contour interval is 2.5 x 10^6 m^3 sec^{-1}. Negative streamlines are dashed. Highs and lows are denoted by H's *and* L's, *respectively.*

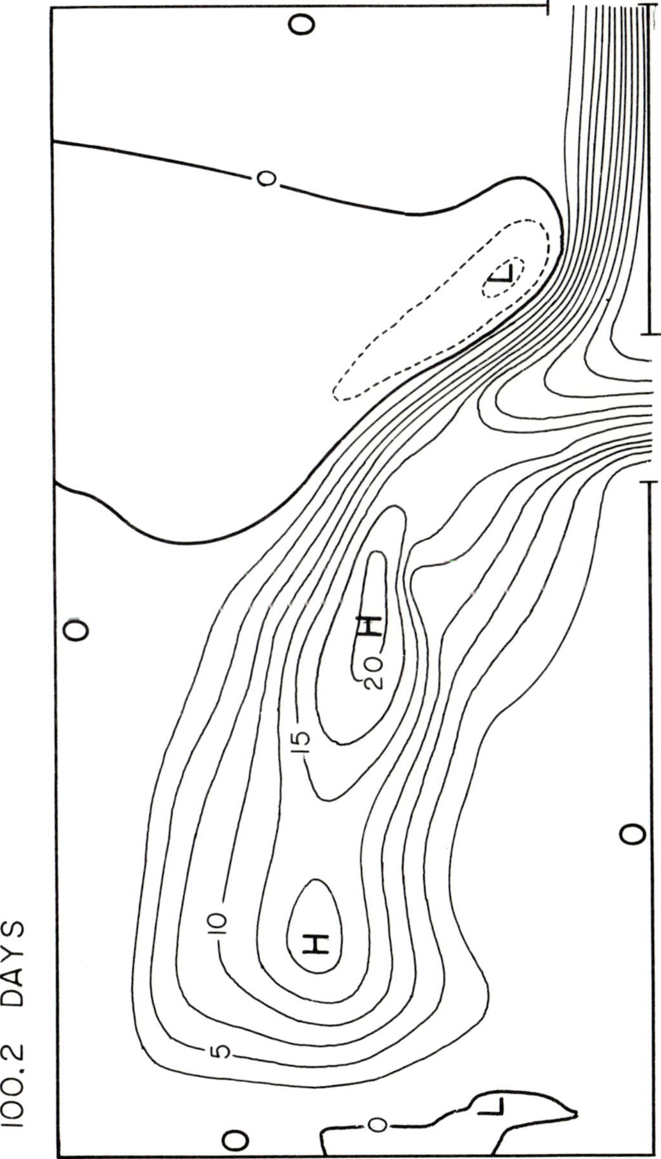

Figure 10-8. Strong westward intensification. Contour interval is 2.5 x 10^6 m^3 sec^{-1}. Negative streamlines are dashed. Highs and lows are denoted by H's and L's, respectively.

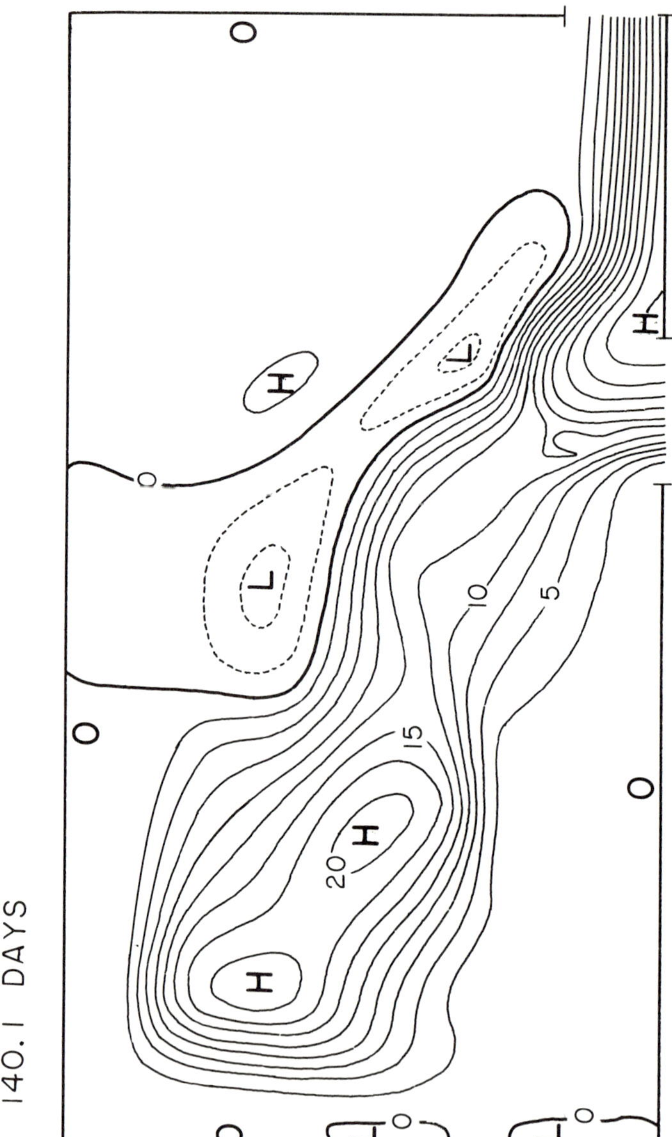

Figure 10-9. Strong westward intensification. Contour interval is $2.5 \times 10^6\ m^3\ sec^{-1}$. Negative streamlines are dashed. Highs and lows are denoted by H's *and* L's *respectively.*

Channel. When westward intensification is weak, most transport is into the Gulf and immediately out through the Florida Straits. With increasing westward intensification in the input, a well-developed loop occurs and an anticyclonic eddy can break off and may migrate to the west. Thus, the distribution of the input flow seems to exert a strong if not major influence on the circulation pattern of the Gulf of Mexico.

Accordingly, in the numerical experiment the distribution of velocity in the Yucatan Strait was varied to represent a seasonal change from early spring to summer. The total transport into (and out of) the basin was held fixed at 30 x 10^6 m^3 sec^{-1}, which is the average value given by Richardson (1965). Figure 10-3 shows the two extreme input distributions of velocity. The associated ψ for the 10 grid intervals across the input port are given in Table 10-1. The relatively shallow depths on the western side of the channel are responsible for the extreme difference in velocities on the western side for the two cases, even though the ψ values differ by only a modest amount on the western side.

The initial circulation pattern was taken as that corresponding to an input of sustained intense westward intensification. The latter was generated by running the model with this constant input until a near steady state was achieved. The resulting initial quasi-steady state of circulation is shown in Figure 10-4.

Starting from this quasi-steady state flow with strong western intensification in the Yucatan Channel, the model was run for 60 days with weak westward intensification in the inflow (Figure 10-3) to allow the flow to settle to a spring circulation pattern. At 60 days the western intensification was slowly increased over a period 30 days to a flow with strong western intensification (Figure 10-3) at 90 days. This strong western intensification was continued for 60 days.

Figures 10-5 to 10-9 are representative of this period. Anticyclonic and cyclonic flow are represented by highs (H) and lows (L) of ψ respectively. Figure 10-5 shows most of the flow in the eastern Gulf with a decaying high in the western Gulf and is similar to the spring circulation presented by Nowlin and McLellan (1967). The spread of the flow to the west in the deeper portion of the Gulf is seen in Figures 10-6 and 10-7. The spawning of a strong high is seen in Figure 10-8, occurring just after the full westward intensification of the inflow has been completed. Figure 10-9 shows that the high has migrated to the west and is weakening.

In an earlier study (Paskausky, 1969) the circulation in a constant depth model essentially followed the boundaries of the basin. Minor variation of the other parameters did not produce significant changes in the general pattern and their changes would not be physically realistic. As Figures 10-5 to 10-9 show, there is a persistent low near Key West, but it is in the deep water. A detailed examination of the topography in the model shows a depression there that may be causing the low.

The circulation pattern of the model agrees qualitatively with observed circulation patterns in the Gulf of Mexico. Seasonal change of total transport and wind were neglected in the model. Inclusion of wind in the form of a hurricane may help cause the low near Key West to be less persistent and might move more of the flow onto the Florida Shelf.

Again the reader should be cautioned that these results were obtained from a numerical model in which the topographic stretching effect on the vorticity was inadvertently omitted. It would be of interest to compare the present results with those for the same input but including the topographic stretching effect. Hopefully this will be carred out in further studies by one of the authors.

Acknowledgements

This work was supported by the Office of Naval Research under contract Nonr 2119 (04). Acknowledgment is made to the National Center for Atmospheric Research for use of its Control Data 6600 computer.

References

Arons, A. and Stommel, H. 1956. A beta-plane analysis of free periods of the second class in meridional and zonal oceans, *Deep Sea Res.*, 4(1):23-31.

Baer, L., Adamo, L. and Adelfang, S. 1968. Experiments in oceanic forecasting for the advective region by numerical modeling, Pt. 2, Gulf of Mexico. *J. Geophys, Res.*, 73(16):5091-5104.

Bryan, K. and Cox, M. 1967. A numerical investigation of the oceanic general circulation. *Tellus,* 19(1):54-80.

Charney, J.G. 1955. On the theory of the wind-driven ocean circulation. *J. of Fluid Mech.*, 12:49-80.

Holland, W.R. 1967. Wind-driven circulation in an ocean with bottom topography. *Tellus,* 19(4):582-600.

Nowlin, W.D., and McLellan, H. 1967. A characterization of the Gulf of Mexico water in winter. *J. of Marine Res.*, 25(1):29-59.

Paskausky, D.F. 1969. A barotropic prognostic numerical model of the circulation in the Gulf of Mexico. Dissertation, Texas A&M University.

Richardson, W. and Schmitz, W. 1965 A technique of the direct measurement of transport with application to the Straits of Florida. *J. of Mar. Res.*, 23(2):172-185.

Smith, G.F. 1965 Numerical solution of partial differential equations. New York: Oxford University Press.

11

A Baroclinic Prognostic Numerical Circulation Model

Richard T. Wert and Robert O. Reid

Abstract

Considered is a two-layer prognostic model of the circulation in the Gulf of Mexico. This two-layer model represents the simplest finite difference approximation to the continuously stratified real ocean. The equations of momentum, considered for each layer, include horizontal and vertical exchange of momentum, Coriolis effect, nonlinear advection of momentum and the effect of topography. In the model, however, the topography is restricted to the lower layer. The associated geostrophic-vorticity equations are formed and the baroclinic and barotropic modes are separated. The baroclinic and barotropic vorticity prediction equations are then put into finite difference form utilizing centered differences with a double step in time. The DuFort-Frankel scheme is used for the lateral diffusion of vorticity terms and the resulting system of finite difference equations is solved using the Gauss-Seidel method with successive overrelaxation.

Laplacian flow is initially specified in the upper layer and the lower layer is considered at rest. The model is spun up to a quasi-steady state which is in turn used as the initial state for the subsequent computations. A 1-year prediction is then made of the baroclinic and barotropic modes of circulation in the Gulf during which time the input through the Yucatan Strait is varied seasonally. During this prediction, an anticyclonic eddy is formed by the Loop Current, partially detaches from the loop and decays. At the end of the prediction, a new eddy is in the process of being developed.

Introduction

Past investigations of oceanic circulation have been concerned primarily with steady-state or mean circulation features. However, as the oceans were observed more closely, what was once thought to be a rather steady circulation is actually a dynamic pattern, with major changes occurring in a time scale of days or weeks. With this realization, some effort has been turned toward the study of time-varying circulation. This study is a second step toward a full-scale model which can predict the time-varying circulation in the Gulf of Mexico in which baroclinic aspects are incorporated via a two-layer approximation. The first step was the barotropic prognostic model by Paskausky (1969), which included a smoothed version of the actual topography of the Gulf. A previous prognostic model of the Gulf, developed by Baer, Adamo and Adelfang (1968) and Hamm and Lesser (1968), differs from the present model as well as from Paskausky's in that only an intermediate layer, the advective region, of the Gulf was considered. Thus, any topographic effect is excluded in the studies of Baer et al.

The Gulf of Mexico is an ideal area to use in developing prognostic circulation models due to its limited size, approximately 1/2% of the world's oceans, and its well-defined boundaries. The Gulf also has a rather well-studied circulation pattern which varies markedly with time. The major feature of this circulation is a loop current, which develops a large anticyclonic eddy that apparently detaches from the main loop. Both this model and the prognostic model developed by Paskausky have exhibited this pattern to some degree.

Formulation of the Model

Assumptions and Restrictions

Considered is a two-port basin containing a two-layer, stably stratified ocean of layer densities ρ and ρ'. Unless otherwise noted, the primed variables will denote the lower layer.

The principal assumptions and/or restrictions used in the developement of the model are

1. Two layers of homogeneous, incompressible fluid exist with constant densities ρ and ρ' with $\rho < \rho'$.
2. Variation of basin depth is allowed only below the layer interface.
3. There is no exchange of mass between layers, through the surface or bottom, or through the boundaries except at the two ports.
4. The model is situated on a beta-plane, and the local horizontal components of the earth's rotation are neglected.
5. Wind effects are neglected.
6. Tidal effects are neglected.
7. Both surface and internal gravity waves are filtered out by use of the geostrophic approximation in the vorticity equations for the system.
8. The vertical transfer of momentum between layers and to the sea bed is proportional to the difference in velocity across the appropriate interface.
9. The horizontal transfer of momentum is expressed in Newton's form.

A surface-referenced, right-handed Cartesian coordinate system is used with x directed eastward, y directed northward and z directed upward.

Assumption 2 allows topography variation only within the lower layer. Therefore, the topography directly affects only the barotropic mode of circulation while having only a secondary effect on the baroclinic mode. This restriction is adopted because the problem of the layer interface intersecting the bottom severely complicates the boundary conditions.

Assumption 3 rules out any mixing at the interface. However, transfer of momentum between layers is allowed.

The beta-plane approximation, assumption 4, is a mathematical abstraction which is commonly applied to models in the low and intermediate latitudes. This approximation allows the Coriolis parameter, f, to be expressed as the linear function $f = f_o + \beta y$ where y is the northward coordinate. A detailed analysis of this approximation is found in the works of Veronis (1963*a*, 1963*b*).

The effects of wind on the model are neglected, assumption 5, because they are negligible as a driving force compared to the inflow through the Yucatan Strait. This assumption makes the model somewhat atypical. However, this term could easily be added if the model were to be applied to an area where the effects of wind are significant.

Tides and other gravity waves are neglected, assumptions 6 and 7, because they do not have a major effect on the long-term circulation.

Equations of Motion

In the following text, all vectors and the operators ∇ and ∇^2 refer to the two-dimensional $x-y$ plane. The only exception is the vertical unit vector, $\hat{k}$. In these equations, the fluid velocity is represented by the vector $\hat{V} = u\,\hat{i} + v\,\hat{j}$. The scalar D represents basin depth for the fluid at rest, while H represents layer thickness. The coefficients for interlayer vertical friction, bottom vertical friction and lateral friction, are σ, σ' and K, respectively. Using these definitions and the assumptions stated previously, the vertically integrated equations of motion for the two-layer system and the associated continuity equations can be written as:

$$\partial\hat{V}/\partial t + (\hat{V}\cdot\nabla)\hat{V} + f\hat{k}\times\hat{V} + g\nabla(H + H' - D) = -\sigma(\hat{V} - \hat{V}')/H + K\nabla^2\hat{V} \qquad (11\text{-}1)$$

$$\partial\hat{V}'/\partial t + (\hat{V}'\cdot\nabla)\hat{V}' + fk\times\hat{V} + g\nabla(\rho H/\rho' + H' - D) = \sigma(\hat{V} - \hat{V}')/H' - \sigma'\hat{V}'/H' + K\nabla^2\hat{V}'; \qquad (11\text{-}2)$$

$$\partial H/\partial t + \nabla\cdot(H\hat{V}) = 0; \qquad (11\text{-}3)$$

$$\partial H'/\partial t + \nabla\cdot(H'\hat{V}') = 0.$$

Defining ζ as the vertical component of vorticity relative to the earth, the standard techniques can be applied to (11-1) and (11-2) to obtain the following vorticity equations:

$$\partial\zeta/\partial t + \nabla\cdot[(f+\zeta)\hat{V}] = T \qquad (11\text{-}4)$$

$$\partial\zeta'/\partial t + \nabla\cdot[(f+\zeta')\hat{V}'] = T' \qquad (11\text{-}5)$$

where

$$T = -\sigma(\zeta - \zeta')/H + K\nabla^2\zeta \qquad (11\text{-}6)$$

and

$$T' = \sigma(\zeta - \zeta')/H' - \sigma'\zeta'/H' + K\nabla^2\zeta' \qquad (11\text{-}7)$$

are the torque terms arising from the effect of friction. Expanding the second term of (11-4) and utilizing (11-3) yields

$$\partial\zeta/\partial t + \hat{V}H\cdot\nabla[(f+\zeta)/H] - (\partial H/\partial t)(f+\zeta)/H = T. \qquad (11\text{-}8)$$

Similarly, (11-5) becomes

$$\partial\zeta'/\partial t + \hat{V}'H'\cdot\nabla[(f+\zeta')/H'] - (\partial H'/\partial t)(f+\zeta')/H' = T'. \qquad (11\text{-}9)$$

The geostrophic approximations for vorticity and velocity, namely

$$\zeta \simeq (g/f)\nabla^2(H + H' - D),$$

$$\zeta' \simeq (g/f)\nabla^2(\rho H/\rho' + H' - D), \qquad (11\text{-}10)$$

$$\hat{V} \simeq (g/f)[\hat{k}\times\nabla(H + H' - D)]$$

and

$$\hat{V}' \simeq (g/f)[\hat{k}\times\nabla(\rho H/\rho' + H' - D)]$$

are now applied to (11-8) and (11-9) which filters out the external and internal gravity waves (Thompson, 1961). The resulting equations are

$$\nabla^2 (\partial H/\partial t + \partial H'/\partial t) + HJ\,[H + H' - D, (f+\zeta)/H] - (f/g)\,[\,(f+\zeta)/H\,]\,(\partial H/\partial t) = fT/g, \qquad (11\text{-}11)$$

$$\nabla^2\,[\,(\rho/\rho')(\partial H/\partial t) + \partial H'/\partial t] + H'J\,[\rho H/\rho' + H' - D, (f+\zeta')/H'] - (f/g)\,[\,(f+\zeta')/H'\,]\,(\partial H/\partial t) = fT'/f. \qquad (11\text{-}12)$$

The Jacobian term of (11-11) comes from

$$\partial\{[\,(f+\zeta)/H\,]\,[\partial\,(H + H' - D)/\partial x\,]\}/\partial y - \partial\{[\,(f+\zeta)/H\,]\,[\partial\,(H + H' - D)/\partial y]\}/\partial x = J\,[H + H' - D, (f+\zeta)/H\,]$$

and similarly for (11-12).

It is now desirable to separate the circulation into the baroclinic and the barotropic modes. The barotropic velocity is that component of the velocity which is uniform throughout the total water column and is equal to the bottom layer velocity. The velocity of the upper layer relative to the lower layer is defined as the baroclinic velocity.

Defining the barotropic height anomaly and a non-dimensional density anomaly as

$$B = \rho H/\rho' + H' - D \qquad (11\text{-}13)$$

and

$$\epsilon = (\rho' - \rho)/\rho',$$

respectively, (11-11) and (11-12) can be rewritten as

$$\nabla^2\,[\partial\,(\epsilon H + B)/\partial t] + HJ\,[\epsilon H + B, (f+\zeta)/H\,] - (f/g)\,[\,(f+\zeta)/H\,]\,(\partial H/\partial t) = fT/g \qquad (11\text{-}14)$$

and

$$\nabla^2\,(\partial B/\partial t) + H'J\,[B, (f+\zeta')/H'] + (f/g)\,[\,(f+\zeta')/H'\,]\,(\partial H/\partial t) = fT'/g. \qquad (11\text{-}15)$$

In the latter equation, the approximation

$$\partial H/\partial t \simeq -\,\partial H'/\partial t$$

has been used in the last term on the left, since $\partial B/\partial t$ is small compared with $\partial H/\partial t$. Subtracting (11-15) from (11-14) yields the baroclinic prediction equation

$$\epsilon\nabla^2\,(\partial H/\partial t) + \epsilon HJ\,[H, (f+\zeta)/H\,] + HJ\,[B, (f+\zeta)/H\,] - H'J\,[B, (f+\zeta')/H'] - (f/g)(\partial H/\partial t)\,[\,(f+\zeta)/H + (f+\zeta')/H'] = f(T - T')/g. \qquad (11\text{-}16)$$

Equation (11-16) along with the barotropic relation (11-15) are the basic prediction equations used in the model.

Numerical Methods

The prediction equations (11-15) and (11-16) in association with (11-6) and (11-7) are put into finite difference form. Centered difference forms are used in both space and time and involve a double time step. One exception to this is that the problem is reduced to a single forward step in time occasionally to suppress spurious oscillations which can develop in the double time step solution over extended runs. The other exception is that the horizontal diffusion terms of (11-6) and (11-7) lead to numerical instability if the normal centered difference scheme is used. Two techniques which can be used to make the system stable are the "lag-scheme" and the DuFort-Frankel scheme (Richtmyer and Morton, 1967). In the lag-scheme, the Laplacian is evaluated at the previous time step rather than the current time step. This method is conditionally stable if the time step meets the criterion

$$\Delta t \leq (\Delta S)^2 / (4K),$$

where ΔS is the grid spacing and Δt is the time step. Although the DuFort-Frankel scheme involves all three time steps in evaluating the Laplacian, it has the advantage of being unconditionally stable for all Δt. Both schemes were tried with no appreciable difference being observed. The DuFort-Frankel scheme was selected so that variations in K would not affect the time step selection. Equations (11-17), (11-18) and (11-19) show how the analog form of the Laplacian is evaluated for the centered difference, lag and DuFort-Frankel schemes, respectively. The superscripts in parentheses indicate time level.

$$\nabla^2 \zeta(i,j) \approx [\zeta^{(2)}(i+1,j) + \zeta^{(2)}(i-1,j) + \zeta^{(2)}(i,j+1) + \zeta^{(2)}(i,j-1) - 4\zeta^{(2)}(i,j)] / (\Delta S)^2, \quad (11\text{-}17)$$

$$\nabla^2 \zeta(i,j) \approx [\zeta^{(1)}(i+1,j) + \zeta^{(1)}(i-1,j) + \zeta^{(1)}(i,j+1) + \zeta^{(1)}(i,j-1) - 4\zeta^{(1)}(i,j)] / (\Delta S)^2 \quad (11\text{-}18)$$

and

$$\nabla^2 \zeta(i,j) \approx [\zeta^{(2)}(i+1,j) + \zeta^{(2)}(i-1,j) + \zeta^{(2)}(i,j+1) + \zeta^{(2)}(i,j-1) - 2\zeta^{(1)}(i,j) - 2\zeta^{(3)}(i,j)] / (\Delta S)^2, \quad (11\text{-}19)$$

where

$$x = i\Delta S \text{ for } i = 1, 2, 3, \cdots$$

$$y = j\Delta S \text{ for } j = 1, 2, 3, \cdots$$

The application of the DuFort-Frankel scheme to the barotropic predicition equation can be accomplished by utilizing (11-10), (11-13) and the finite difference form for the time derivative.

$$\partial\zeta'/\partial t = (g/f)\ \nabla^2 (\partial B/\partial t) \approx [\zeta'^{(3)}(i,j) - \zeta'^{(1)}(i,j)]/(2\Delta t)$$

Equation (11-19) can be rewritten

$$\nabla^2 \zeta(i,j) = [\zeta^{(2)}(i+1,j) + \zeta^{(2)}(i-1,j) + \zeta^{(2)}(i,j+1) + \zeta^{(2)}(i,j-1) - 4\zeta^{(1)}(i,j)] / (\Delta S)^2 - \left\{4\Delta t / [(\Delta S)^2]\right\} [(\zeta^{(3)}(i,j) - \zeta^{(1)}(i,j)] / (2\Delta t), \quad (11\text{-}20)$$

and the first term on the right-hand side of equation (11-20) is defined as the skew form of the Laplacian, $\nabla^2 \zeta$ (skew). Using this definition, the Du Fort-Frankel form for ζ' can be written

$$\nabla^2 \zeta' = \nabla^2 \zeta'^{\,(\text{skew})} - \left\{4g\Delta t / \left[f(\Delta S)^2\right]\right\} \nabla^2 (\partial B/\partial t),$$

and similarly for ζ.

Using the previously described finite difference operators ∇^2 and $\nabla^{2\,(\text{skew})}$ along with the following finite difference analog of the Jacobian

$$J_a[P(i,j), Q(i,j)] = \left\{1 / \left[4(\Delta S)^2\right]\right\}$$

$$\{Q(i,j+1)[P(i+1,j+1) - P(i-1,j+1)] - Q(i,j-1)[P(i+1,j-1) - P(i-1,j-1)] - Q(i+1,j)[P(i+1,j+1) - P(i+1,j-1)] + Q(i-1,j)[P(i-1,j+1) - P(i-1,j-1)]\} \quad (11\text{-}21)$$

the finite difference forms of (11-15) and (11-16) can be written

$$\left\{1 + 4K\Delta t/\left[(\Delta S)^2\right]\right\}\nabla^2 \dot{B}(i,j) + H'(i,j)J_a\,[B(i,j), Z'(i,j)] + [f(j)/g)]\,Z'(i,j)\dot{H}(i,j) = [f(j)/g]\,T_a'(i,j), \quad (11\text{-}22)$$

$$\epsilon\left\{1 + 4K\Delta t/\left[(\Delta S)^2\right]\right\}\nabla^2 \dot{H}(i,j) + \epsilon H(i,j)J_a\,[H(i,j), Z(i,j)] + H(i,j)J_a\,[B(i,j), Z(i,j)] - H'(i,j)J_a\,[B(i,j), Z'(i,j)] - [f(j)/g]\,\dot{H}(i,j)\,[Z(i,j)] + Z'(i,j)] = [f(j)/g]\,[T_a(i,j) - T_a'(i,j)] \quad (11\text{-}23)$$

where

$$\dot{B}(i,j) = [B^{(3)}(i,j) - B^{(1)}(i,j)]/(2\nabla t), \quad (11\text{-}24)$$

$$\dot{H}(i,j) = [H^{(3)}(i,j) - H^{(1)}(i,j)]/(2\nabla t) \quad (11\text{-}25)$$

$$T_a(i,j) = -[\sigma/H(i,j)]\,[\zeta(i,j) - \zeta'(i,j)] + K[\nabla^2\zeta(i,j)]^{(skew)}, \quad (11\text{-}26)$$

$$T_a'(i,j) = [\sigma/H'(i,j)]\,[\zeta(i,j) - \zeta'(i,j)] - [\sigma'/H'(i,j)]\,\zeta'(i,j) + K[\nabla^2\zeta'(i,j)]^{(skew)}, \quad (11\text{-}27)$$

$$Z(i,j) = [f(j) + \zeta(i,j)]\,/\,H(i,j), \quad (11\text{-}28)$$

$$Z'(i,j) = [f(j) + \zeta'(i,j)]\,/\,H'(i,j). \quad (11\text{-}29)$$

In equations (11-21) through (11-29), all terms are at time level 2 except for the one portion of the skew Laplacian specified previously.

The Gauss-Seidel method is used to solve the resulting system of equations. In an effort to reduce the computer time necessary for the solution of the prediction equations, successive over-relaxation is utilized.

Initial Conditions

The following technique is used to specify every point in the H and B fields to start the model. At time zero, the upper layer is assumed to be in Laplacian flow and geostrophic balance. At the same time, the lower layer is assumed to be at rest (i.e., all B values are equal to zero). Values of the baroclinic volume transport stream function, Ψ_H, are specified around the perimeter of the model and an initial upper layer thickness, H_O, is chosen. Values of Ψ_H around the perimeter are converted to layer thickness, H, by utilizing the geostrophic assumption:

$$H = \left\{[2f/(g\epsilon)]\,\Psi_H + H_o^2\right\}^{1/2}$$

With H specified around the boundaries, the relaxation of $\nabla^2 H = 0$ is carried out to obtain the initial H field.

Boundary Conditions

Equations (11-15) and (11-16) are fourth order spatially in the prognostic variables H and B. This requires two boundary conditions around the entire perimeter of the model. Figure 11-1 shows a small grid with two ports which will be used to demonstrate the various boundary conditions applied to the model.

The values of H along the solid boundaries are specified from the initial conditions and are kept constant with time. Since the baroclinic transport through the southern port drives the model, the values of H across this port are specified at each time step. The eastern port is downstream in the flow, therefore the model is allowed to specify the baroclinic velocity distribution across this port, although the total transport through the port is fixed.

Considering the preceding conditions, H must be predicted for all grid points labeled X in Figure 11-1, including those across the eastern port. As the "computing star" approaches a solid boundary, for example (3, 4) in Figure 11-1, the vorticity is required on the associated boundary points, (2, 4). To provide this vorticity, a fictitious row of grid points is employed which surrounds the main grid. The inclusion of lateral friction in the model requires that the tangential flow vanish at the solid boundaries. To meet this condition, the derivative of H normal to the boundary must be equal to zero. Considering the centered finite difference form, the H values at the fictitious points are set equal to the values just inside the boundary [i.e., $H(1, 4) = H(3, 4)$]. This process is carried out around the perimeter, including the ports, which implies that the flow through the ports is perpendicular to the openings. Using these fictitious H's, $\nabla^2 H$ can be determined on the boundaries.

The only problem remaining is at the eastern port. In order to predict H at the grid points across the port, ζ must also be specified at points (11, 3) and (11, 4) on the demonstration grid. The philosophy was adopted that ζ should be determined upstream. That is, if the flow is directed outward, the model should specify ζ based on advection of potential vorticity. If the flow is inward, ζ is arbitrarily set outside the model. The following equation was used to determine ζ at these points where Z is potential vorticity, as previously defined:

$$Z^{(n+1)}(i, j) = \begin{cases} (1 - \gamma - \alpha)\, Z^{(n)}(i, j) + \gamma Z^{(n)}(i - 1, j) + \alpha Z^{(n)}(i - 1, j^*) & \text{for } \gamma > 0 \\ f/H & \text{for } \gamma \leqq 0 \end{cases}$$

where

$$j^* = \begin{cases} j - 1 & \text{for } \nu \geqq 0 \\ j + 1 & \text{for } \nu < 0 \end{cases}$$

and

$$\gamma = u(i, j)(\Delta t/\Delta S), \quad \alpha = |\, v(i, j)(\Delta t/\Delta S)\, |$$

This scheme is also used at the corner point (10, 5).

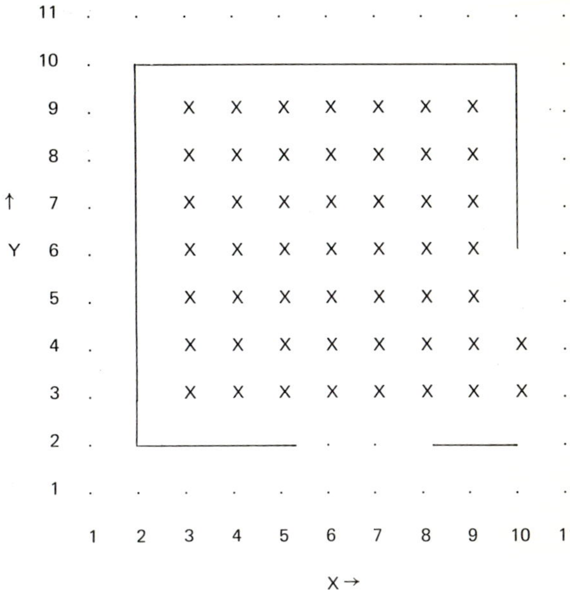

Figure 11-1. Demonstration grid for boundary conditions.

In the case of the barotropic prediction, the same approach is used. The only difference is that the eastern port is closed to barotropic flow and therefore only the southern port exists for the bottom layer. The barotropic flow through this port is handled similarly to the baroclinic flow through the eastern port.

Parameter Selection

The first group of parameters selected was the spatial parameters. A 20-km grid spacing was chosen as a trade-off between resolving major circulation features and computer core and time requirements. The model size of 1420 x 800 km (74 x 43 grid including the fictitious points) with the associated ports was then adopted and is superimposed over a chart of the Gulf of Mexico in Figure 11-2. The decision was made to use "block" topography so that a parametric study of the effect of topography could be made in the future. The block approximation to the shelf areas of the Gulf is included in Figure 11-2. A model depth of 3000 m was assumed with a shelf depth of 500 m to insure that assumpton 2 was satisfied.

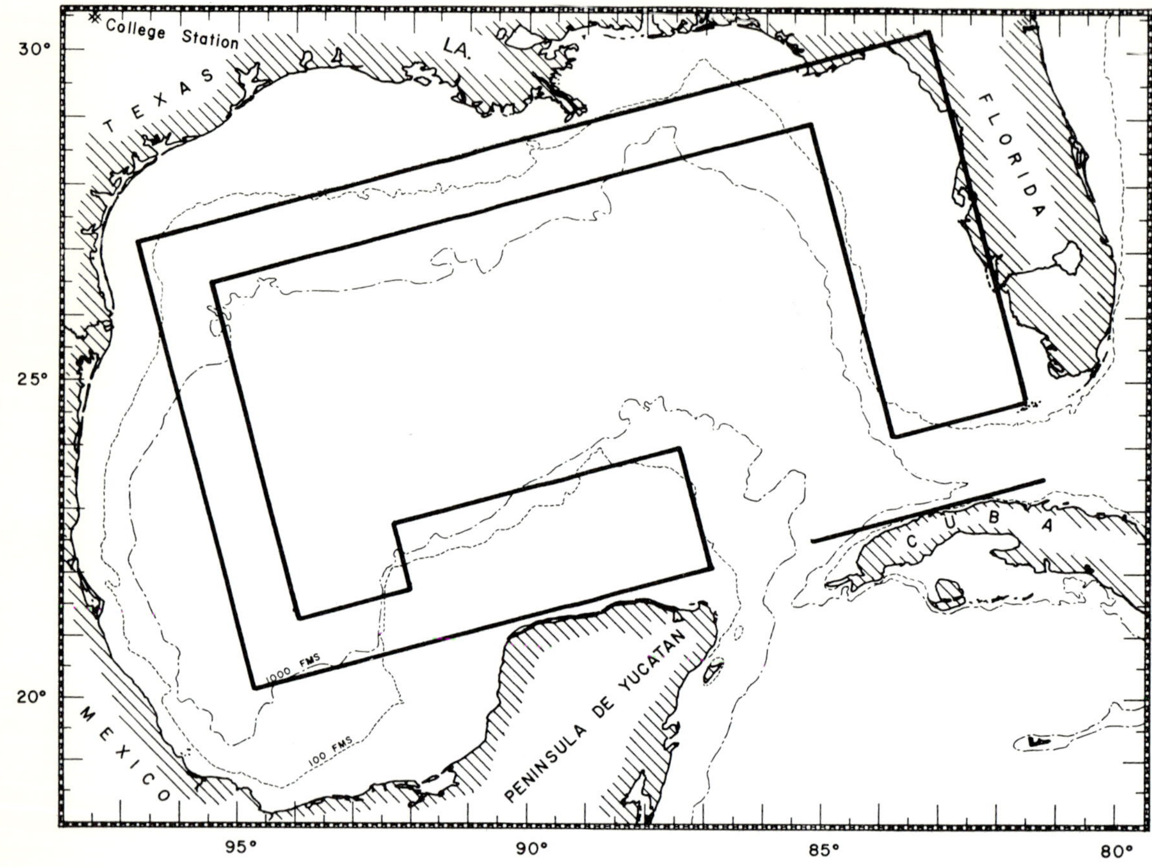

Figure 11-2. The outline of the model superimposed over a chart of the Gulf of Mexico.

In the preliminary tests, the model was unstable with the above topography. This instability was eliminated by smoothing the block topography with the following smoothing function:

$$D^{(n+1)}(i,j) = [D^{(n)}(i+1,j) + D^{(n)}(i-1,j) + D^{(n)}(i,j+1) + D^{(n)}(i,j-1) + 4D^{(n)}(i,j)] / 8$$

where the superscripts in parentheses refer to the smoothing pass. This smoothing function was applied eight times in order to remove the sharp gradients from the topography. Figure 11-3 shows the contours of the final smoothed topography along with an example of the grid in the upper right-hand corner and across the ports.

The friction coefficients play an important part in the model, and a parametric study would have been desirable but was not possible. The lateral friction coefficient, K, was chosen as 5×10^6 cm^2/sec. The selection of the vertical friction coefficients σ and σ' was based on considerable experimentation. The ratio of these coefficients appears to control the ratio of the barotropic high to the baroclinic high in the loop current, while the value of σ controls the high in the baroclinic

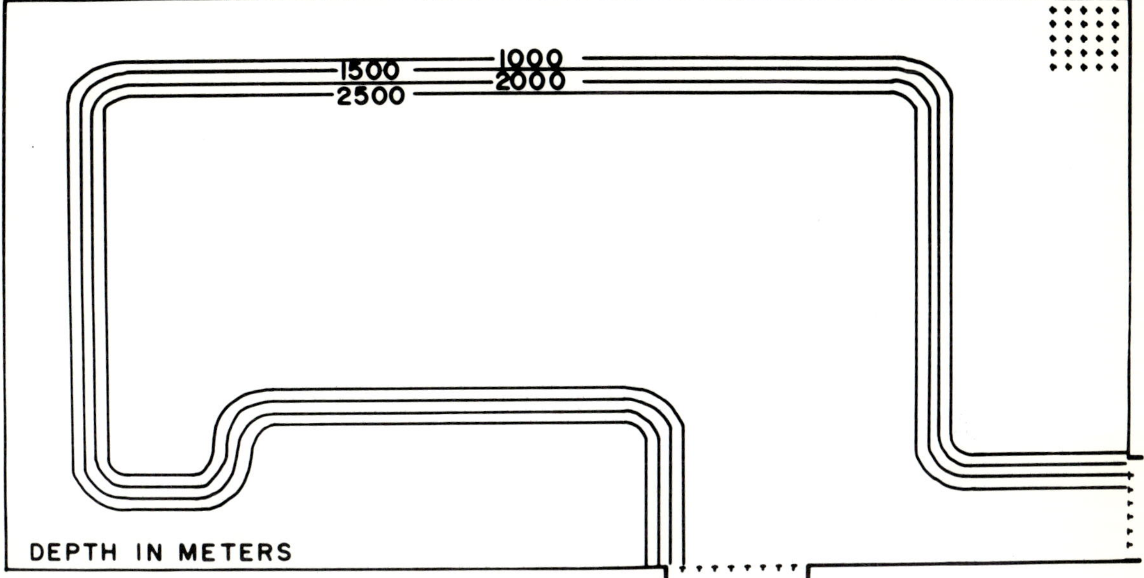

Figure 11-3. Contours of the bathymetry used in the model.

mode. The values of $\sigma = 1 \times 10^{-3}$ cm/sec and $\sigma' = 50 \times 10^{-3}$ cm/sec were selected based on comparison of quasi-steady state circulation patterns.

Beta = 2.06×10^{-13}/sec cm and $f_o = 5.35 \times 10^{-5}$ /sec were chosen for the beta-plane approximation appropriate for the Gulf of Mexico. It was felt that tilting the grid 15° would not have a major effect on the circulation. Therefore, the grid was assumed to be orientated N-S with f a function of y only.

The time step Δt, was selected for model stability. Physical limitations due to advectional velocity allow a maximum time step of 2 hours if a maximum velocity of 2 m/sec is assumed. The 2-hour step was initially used; however, after 50 days prediction, instability became evident. The time step was then reduced to 1.5 hours and the instability disappeared. The 1.5 hour time step was continued for the remainder of the prediction.

The initial top layer thickness, H_O, was chosen as 100 m to approximate the depth of the thermocline.

Figure 11-4 shows the variation of baroclinic volume transport stream function specification at 90-day intervals for the southern port. The distribution was varied sinusoidally, over a 1-year period, about the mean while maintaining a constant 30×10^6 m^3/sec flow. The effect of this variation was to intensify the flow to the west around the 90-day point and to minimize this intensification around the 270-day point.

The model started to exhibit instability after about 100 steps. This instability vanished upon reducing the solution to a single time step, rather than the normal double time step, for one step. As an extra safety margin, this process was carried out every 50 steps.

Results

One 360-day prediction was made using the described model. In order to start this prediction at the most reasonable point, the model was spun up in two stages using the mean flow as an input.

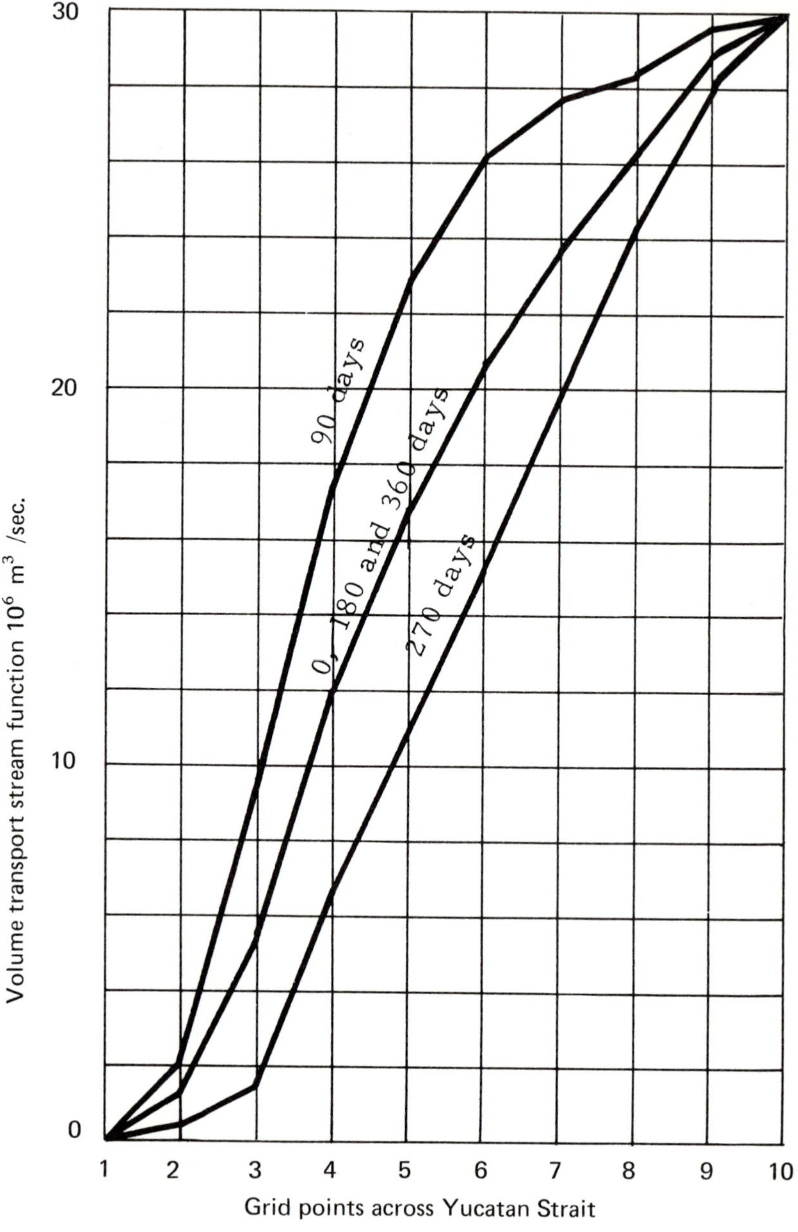

Figure 11-4. Baroclinic volume transport stream function across the Yucatan Strait at 90-day intervals.

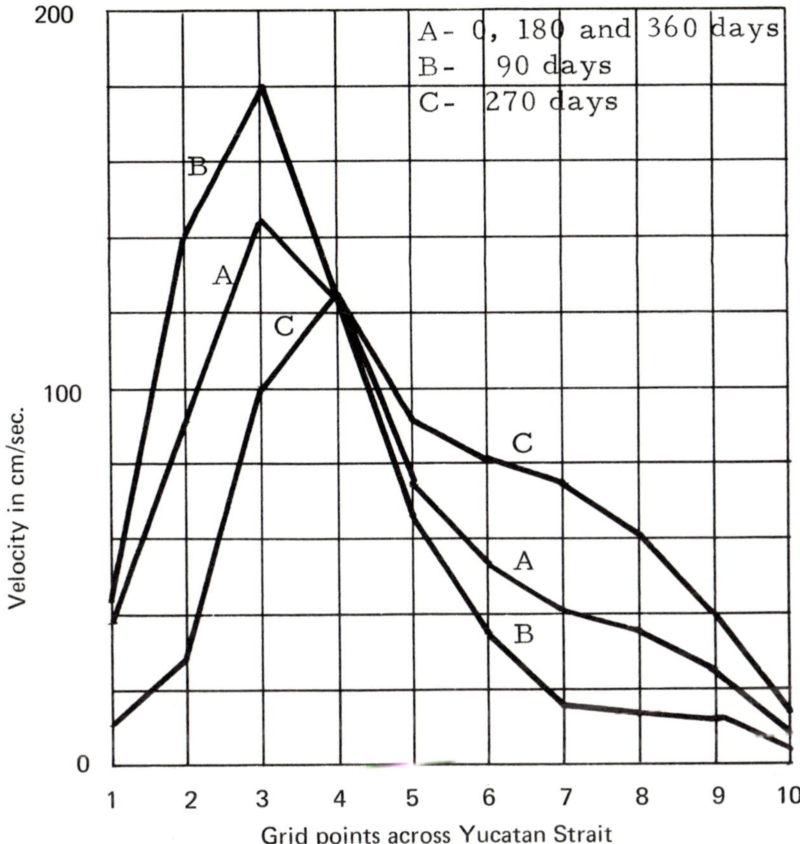

Figure 11-5. Baroclinic velocity profiles through the Yucatan Strait at 90-day intervals.

First, the baroclinic mode was spun up (700 steps at 2 hours/step) from the Laplacian flow specified by the initial conditions. During this period, the barotropic mode was kept dynamically inert (by holding all $B = 0$) to reduce the effects of initial transients. Then the barotropic mode was released and both layers were allowed to proceed for an additional 1000 steps. At this point, a quasi-steady state was reached which was used as the starting point for the main prediction run (i.e., day 0).

The main prediction run was made using the inflow specification detailed previously. At selected intervals, the H and B fields were converted to volume transport stream functions using the equations

$$\Psi_H = [g\epsilon/(2f)]H^2 - H_O^2]$$

and

$$\Psi_B = (g/f)\ D\,B.$$

At 60-day intervals, a total volume transport stream function, Ψ_T, was obtained by adding the baroclinic and barotropic volume transport stream functions.

Figures 11-6 through 11-25 are the outputs obtained at 15- to 20-day intervals. The change in

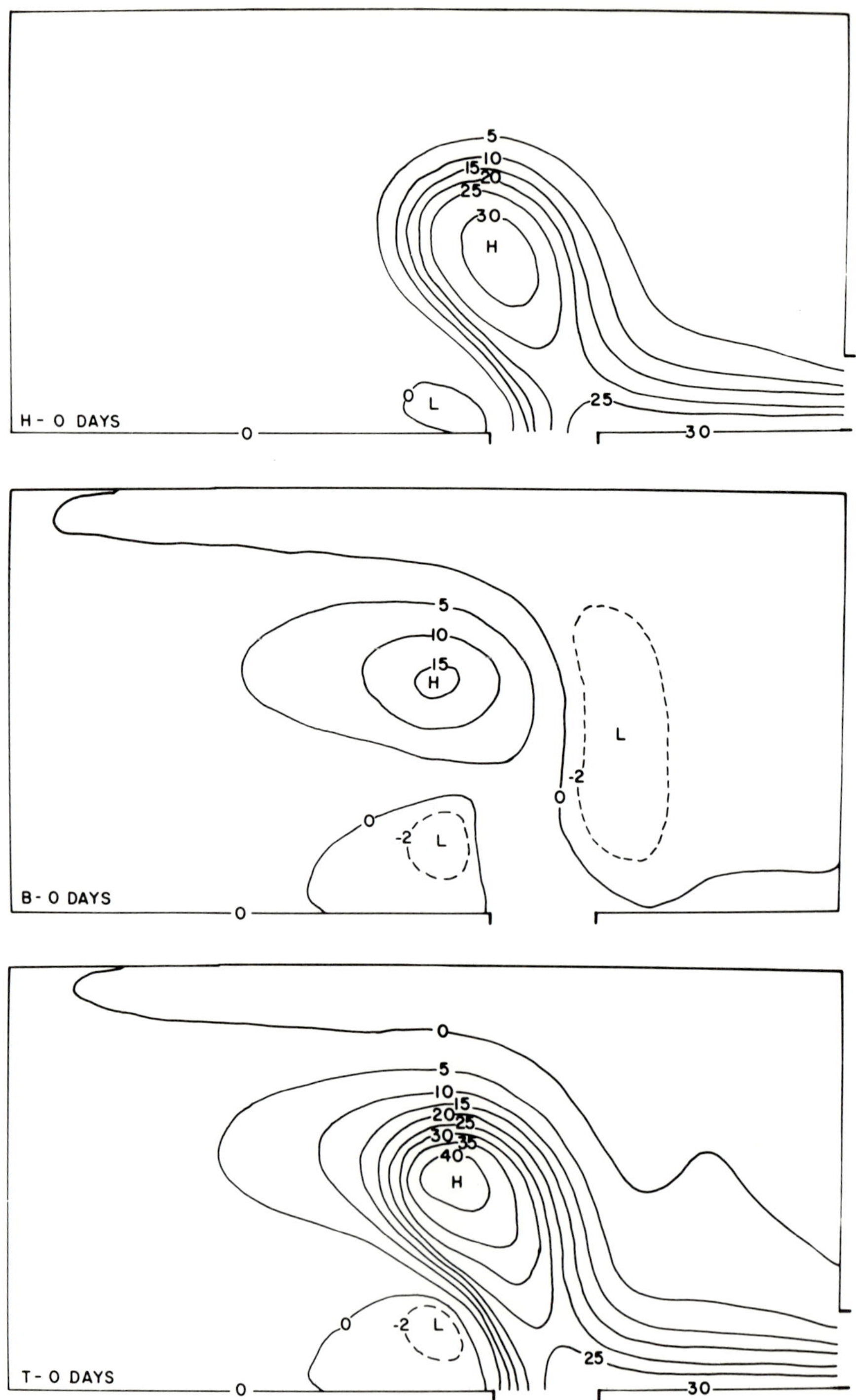

Figure 11-6. Initial contours of volume transport stream functions in $10^6 m^3/sec$.

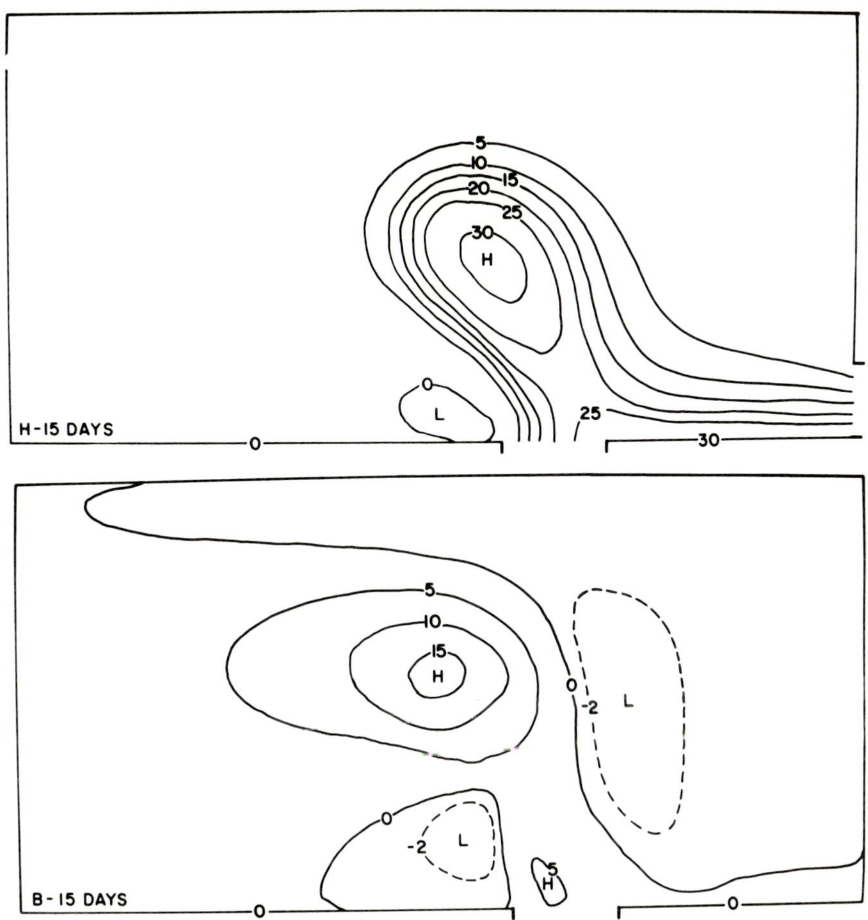

Figure 11-7. Contours of volume transport stream functions in $10^6\, m^3/sec$ after 15 days.

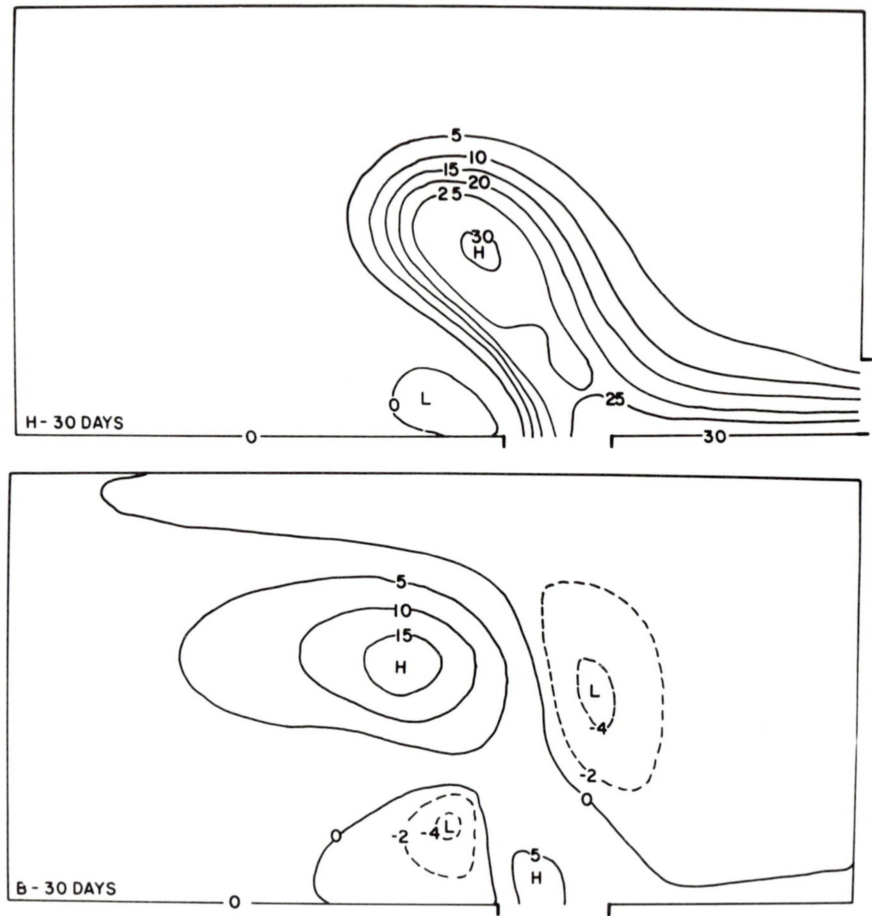

Figure 11-8. Contours of volume transport stream functions in $10^6\ m^3/sec$ after 30 days.

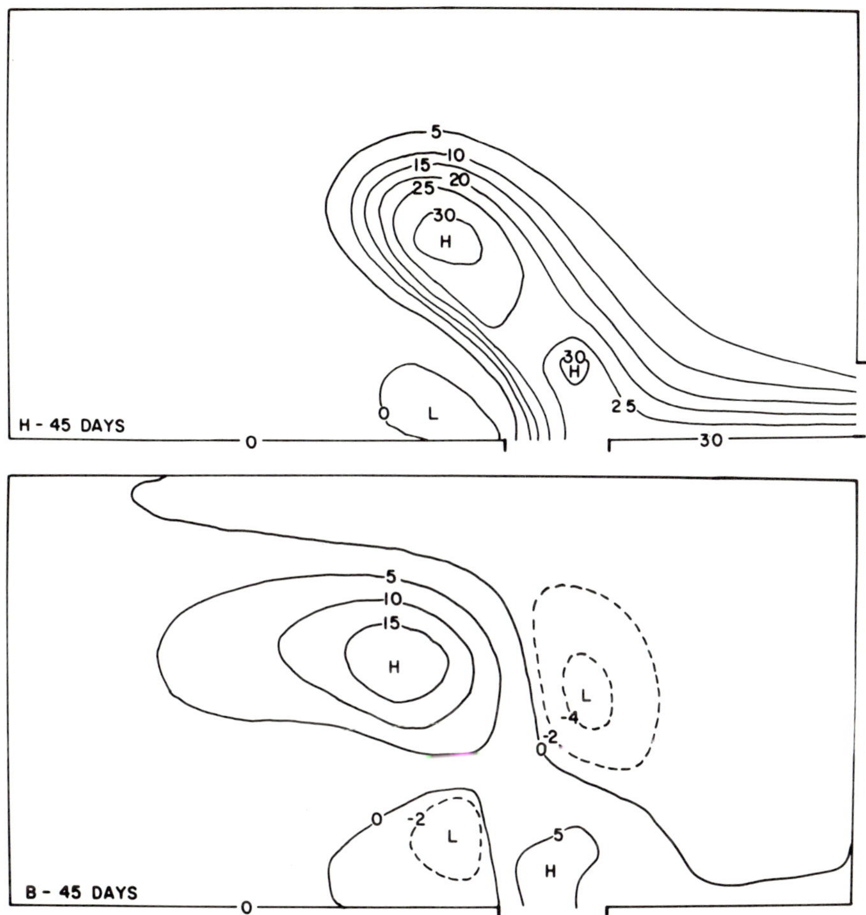

Figure 11-9. Contours of volume transport stream functions in $10^6 m^3/sec$ after 45 days.

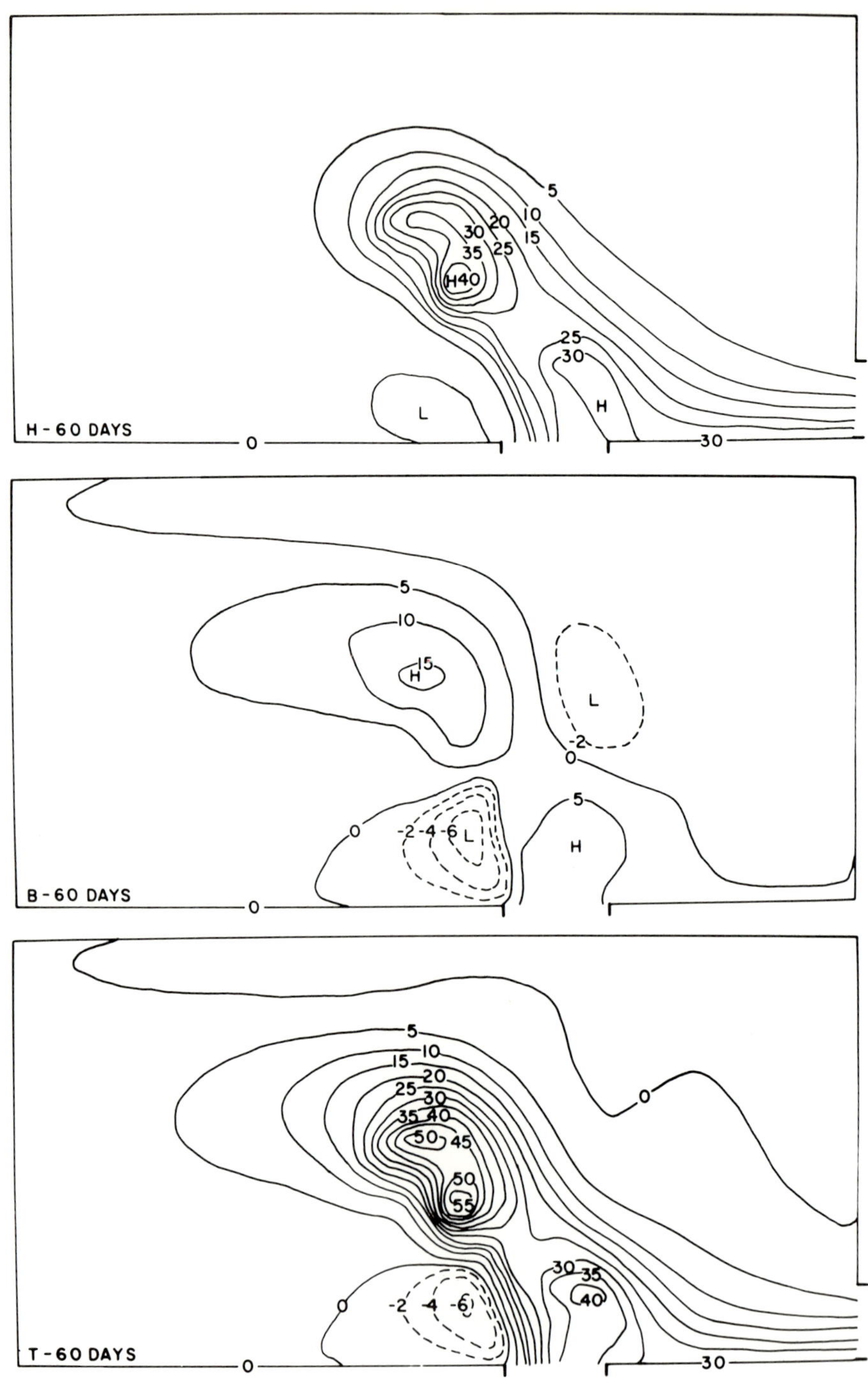

Figure 11-10. Contours of volume transport stream functions in $10^6 m^3/sec$ after 60 days.

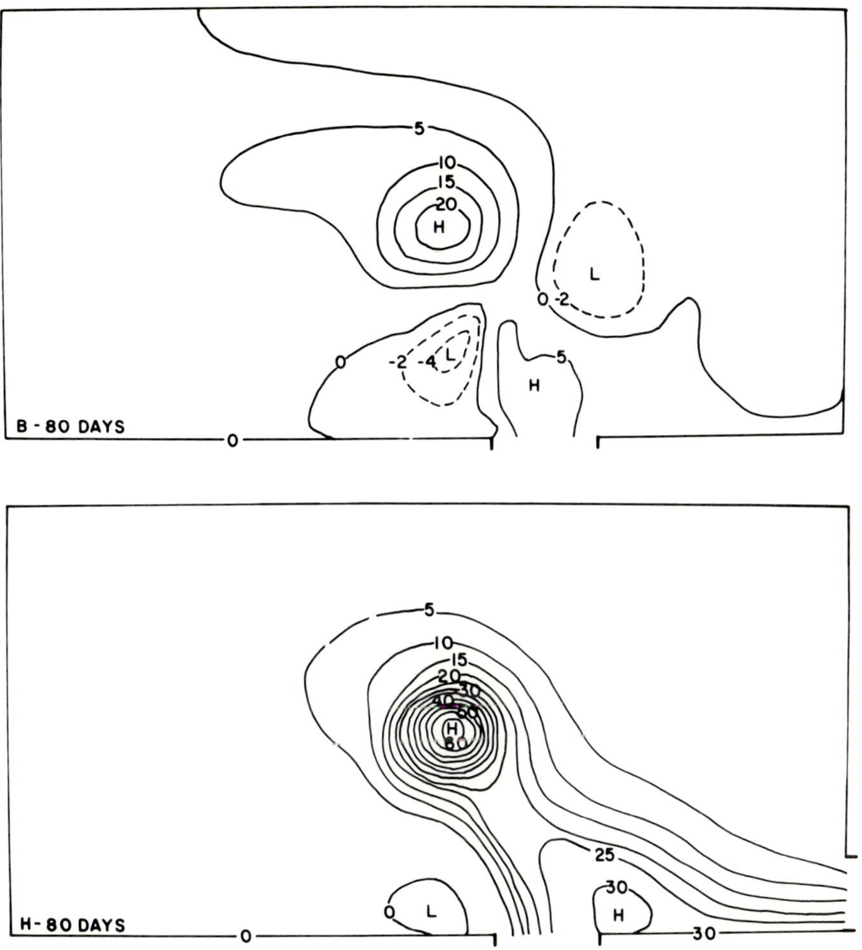

Figure 11-11. Contours of volume transport stream functions in $10^6 m^3/sec$ after 80 days.

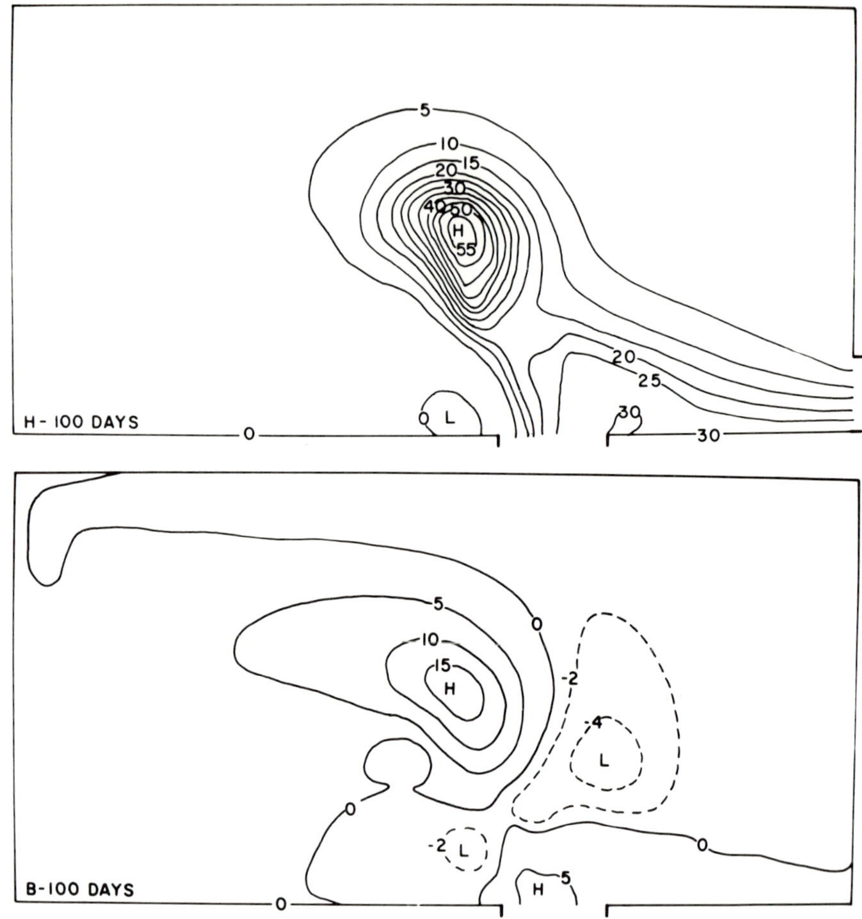

Figure 11-12. Contours of volume transport stream functions in $10^6 m^3/sec$ after 100 days.

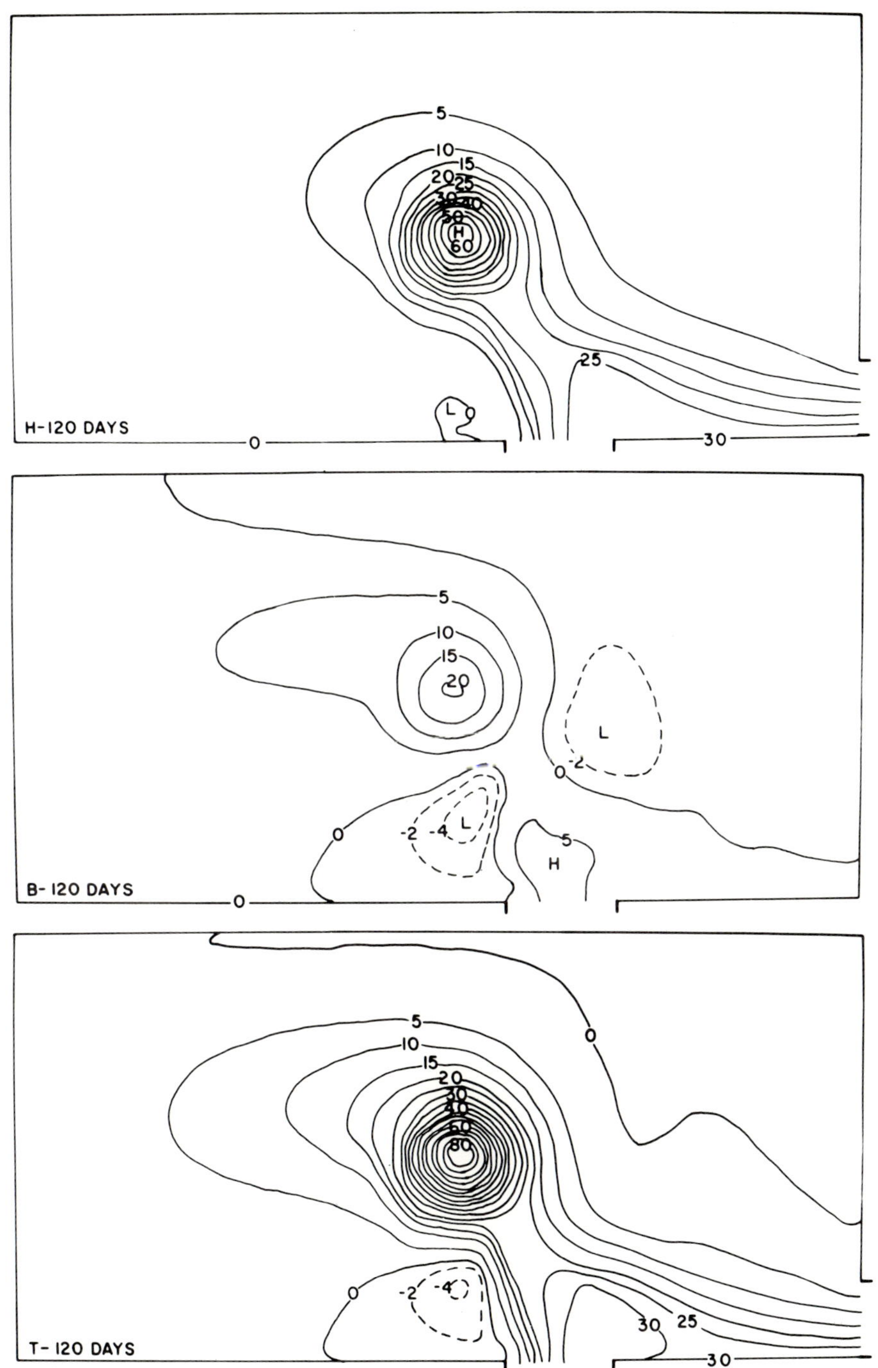

Figure 11-13. Contours of volume transport stream functions in $10^6 m^3/sec$ after 120 days.

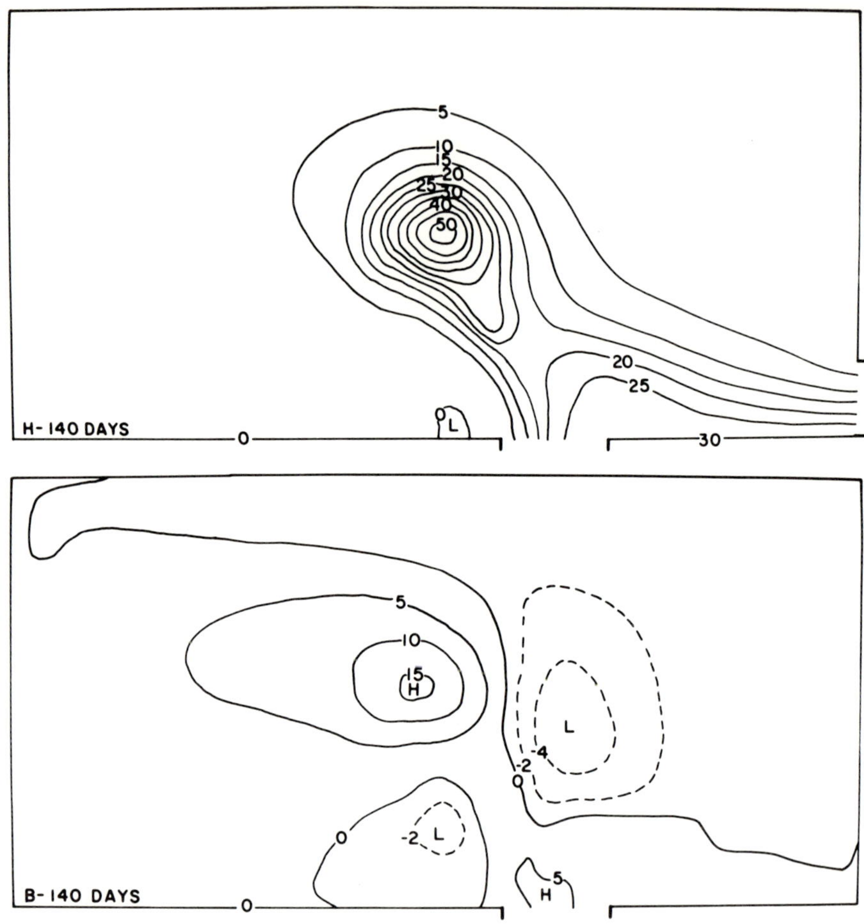

Figure 11-14. Contours of volume transport stream functions in $10^6 m^3/sec$ after 140 days.

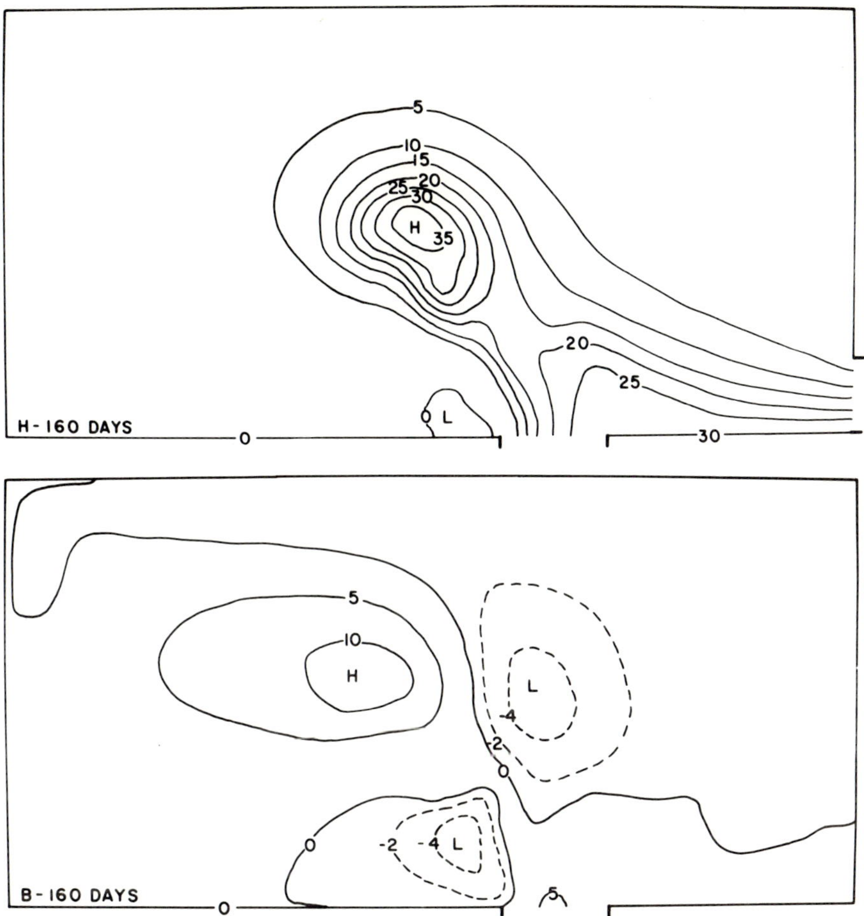

Figure 11-15. Contours of volume transport stream functions in $10^6\,m^3/sec$ after 160 days.

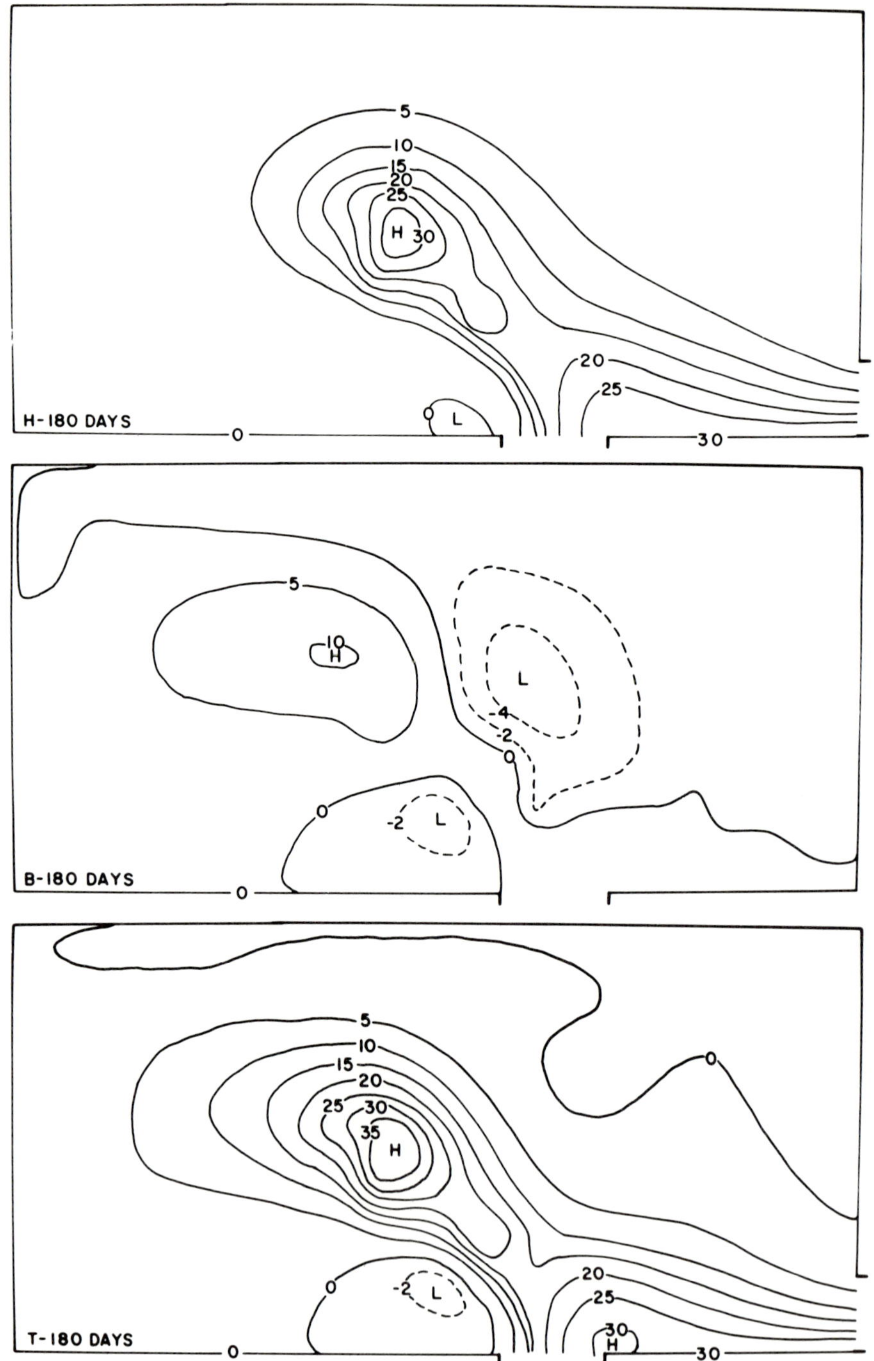

Figure 11-16. Contours of volume transport stream functions in $10^6 m^3/sec$ after 180 days.

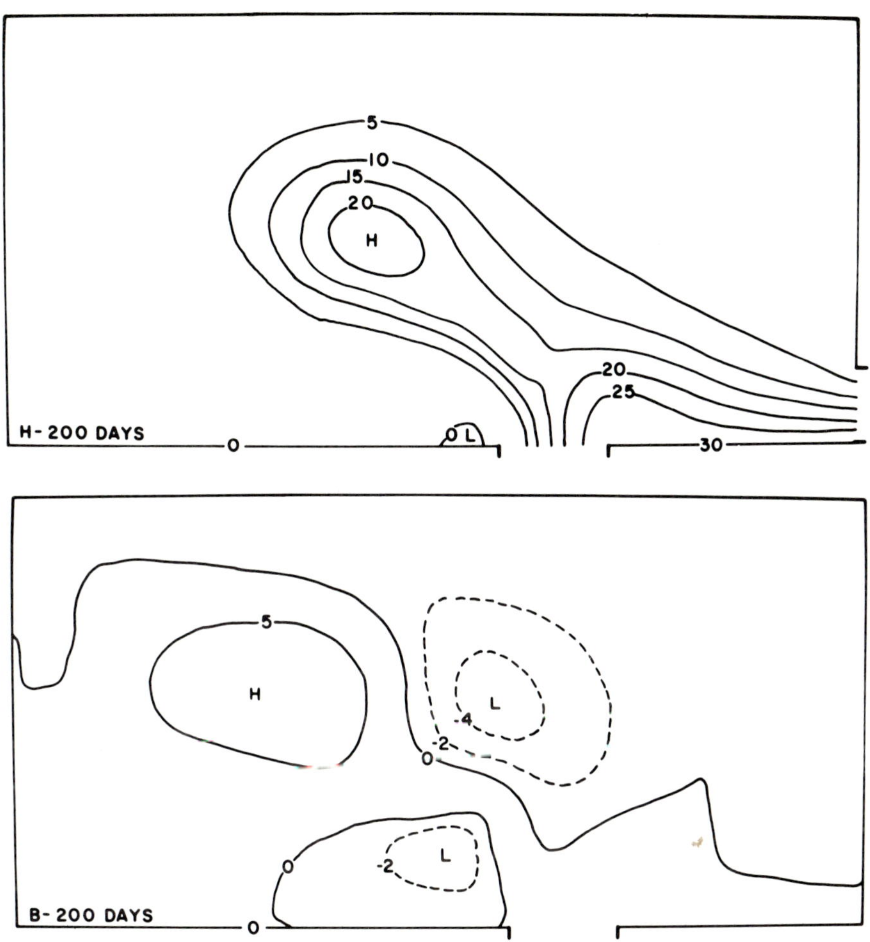

Figure 11-17. Contours of volume transport stream functions in $10^6 m^3/sec$ after 200 days.

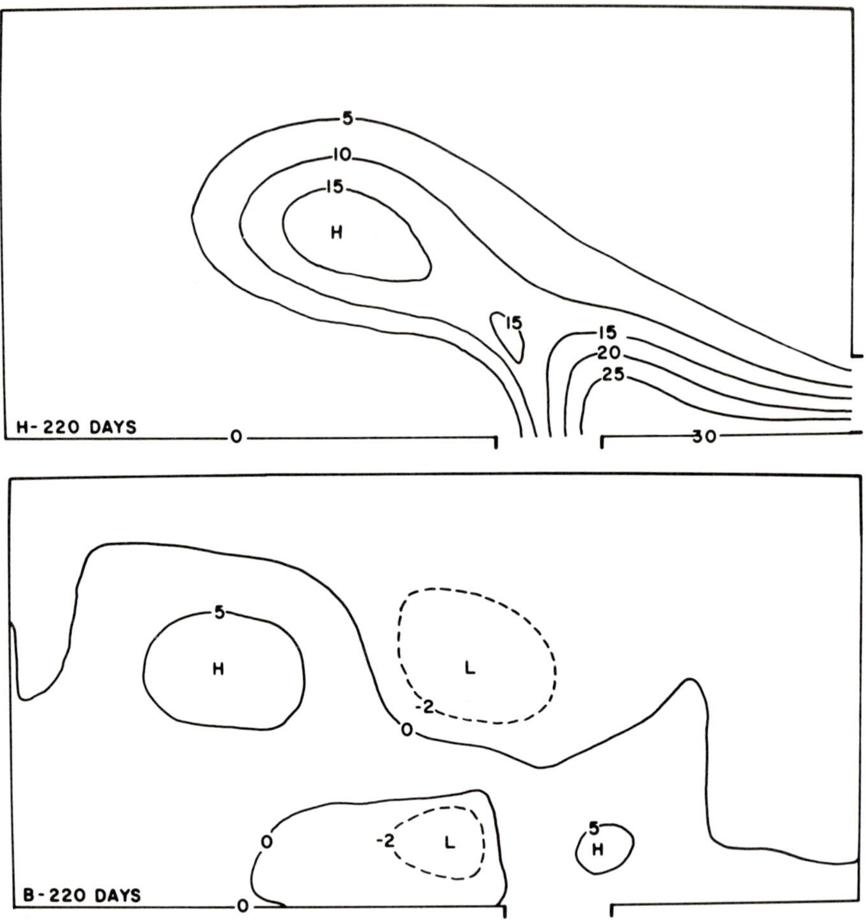

Figure 11-18. Contours of volume transport stream functions in $10^6 m^3/sec$ after 220 days.

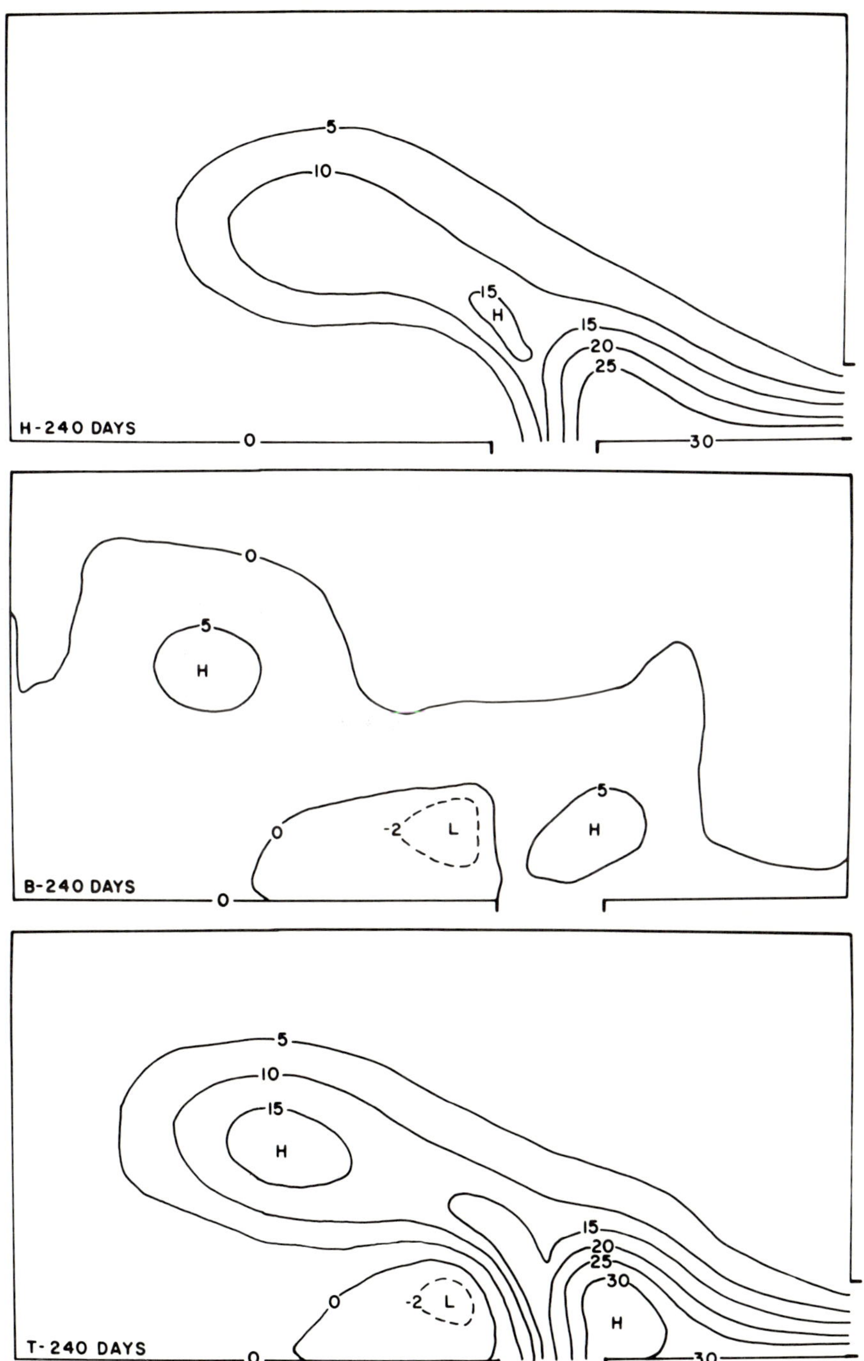

Figure 11-19. Contours of volume transport stream functions in $10^6\,m^3/sec$ after 240 days.

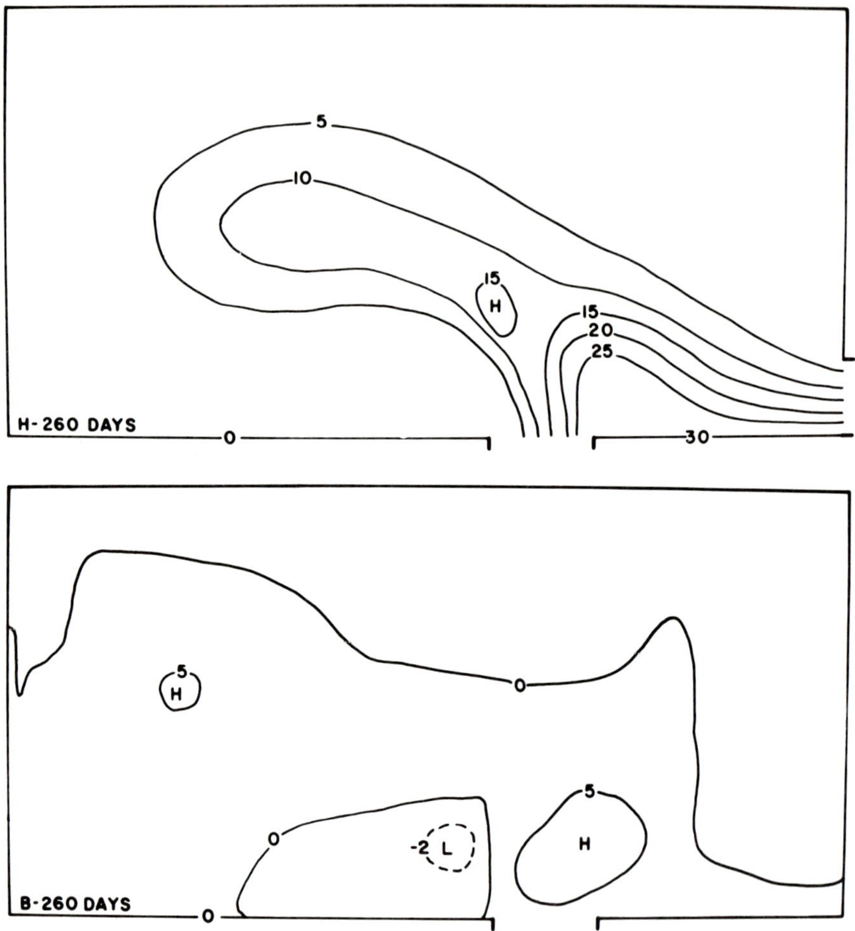

Figure 11-20. Contours of volume transport stream functions in $10^6 m^3/sec$ after 260 days.

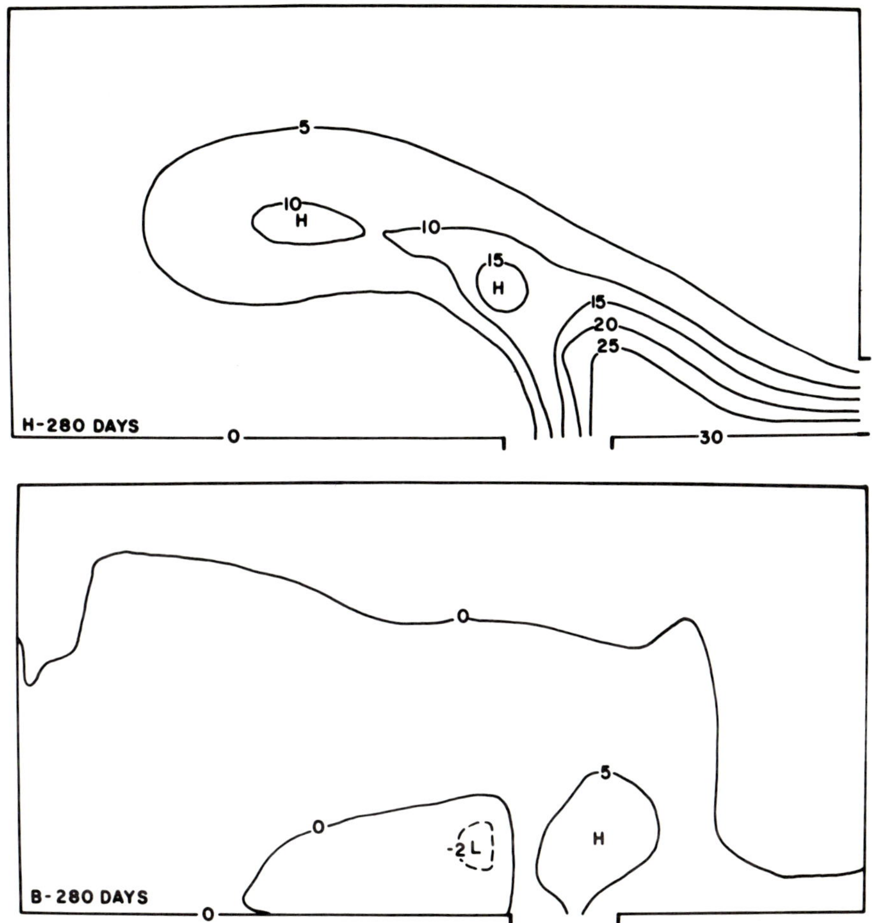

Figure 11-21. Contours of volume transport stream functions in $10^6\,m^3/sec$ after 280 days.

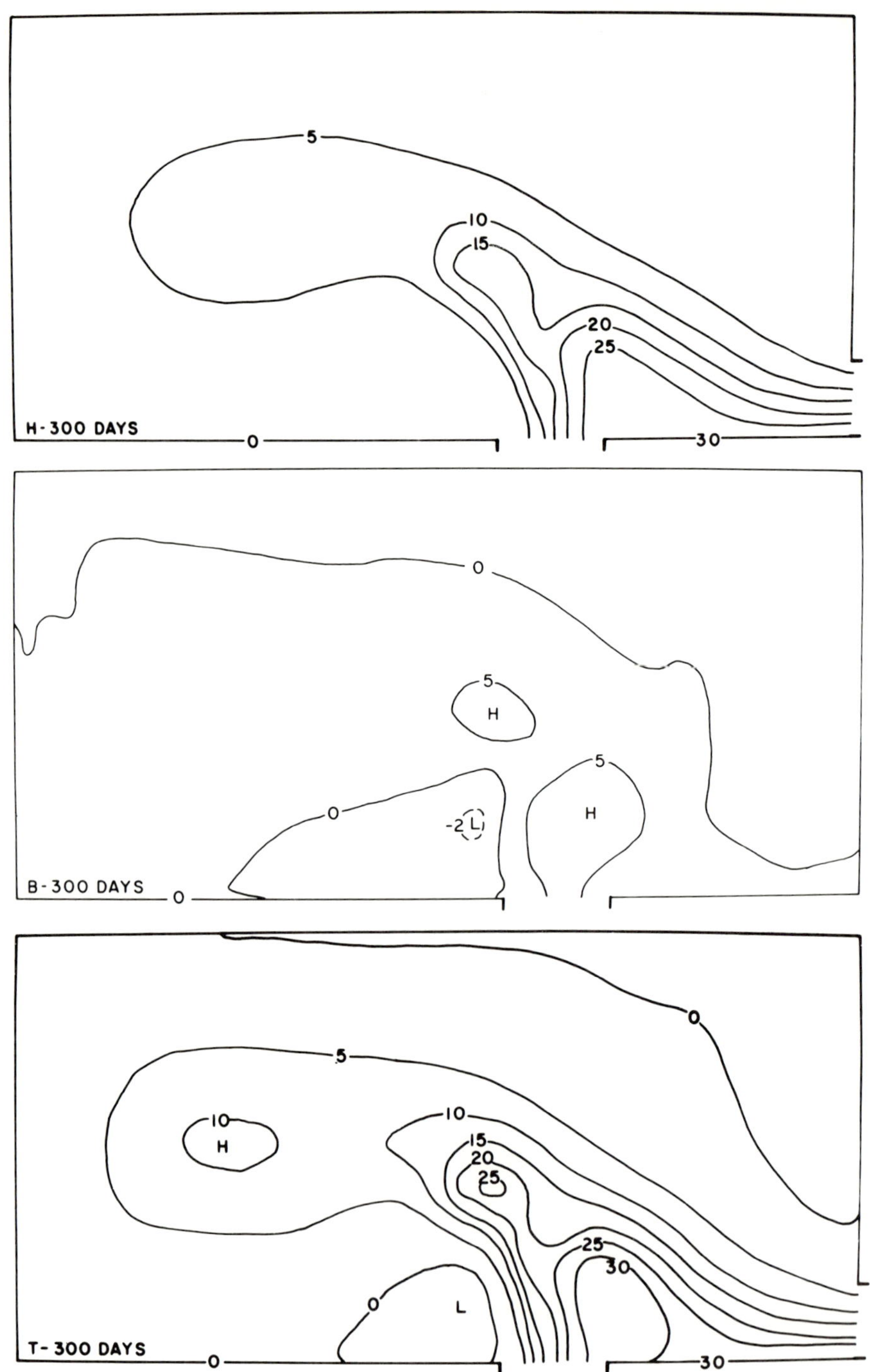

Figure 11-22. Contours of volume transport stream functions in $10^6 m^3/sec$ after 300 days.

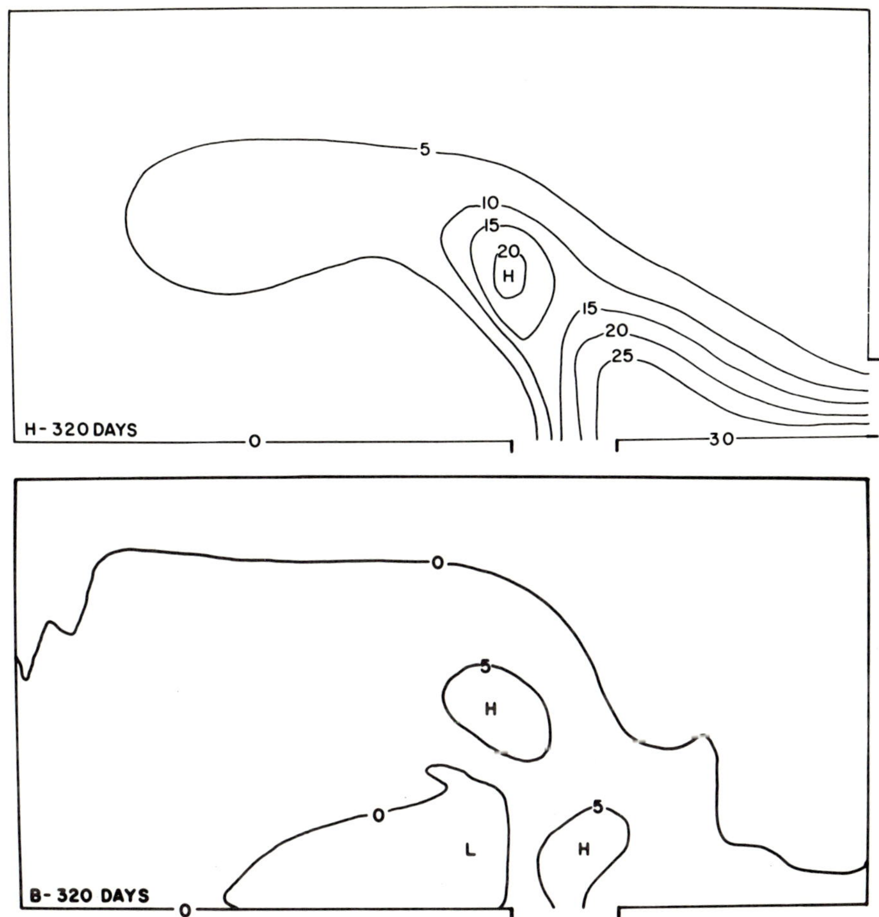

Figure 11-23. Contours of volume transport stream functions in $10^6 m^3/sec$ after 320 days.

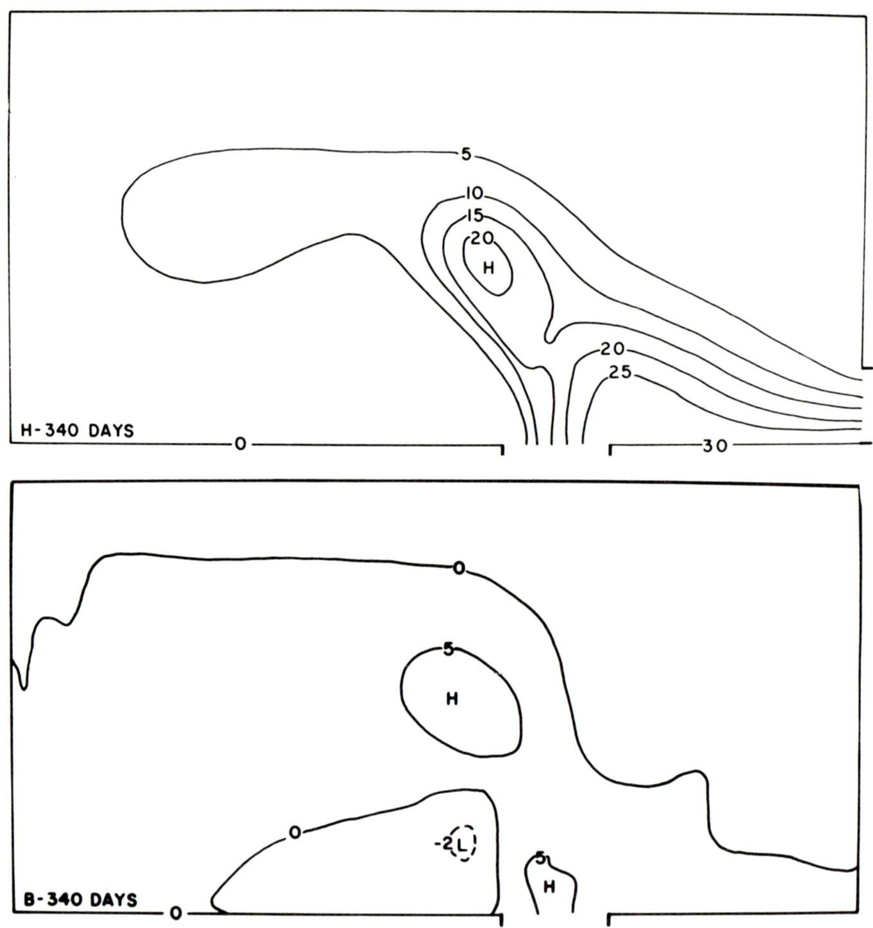

Figure 11-24. Contours of volume transport stream functions in $10^6 m^3/sec$ after 340 days.

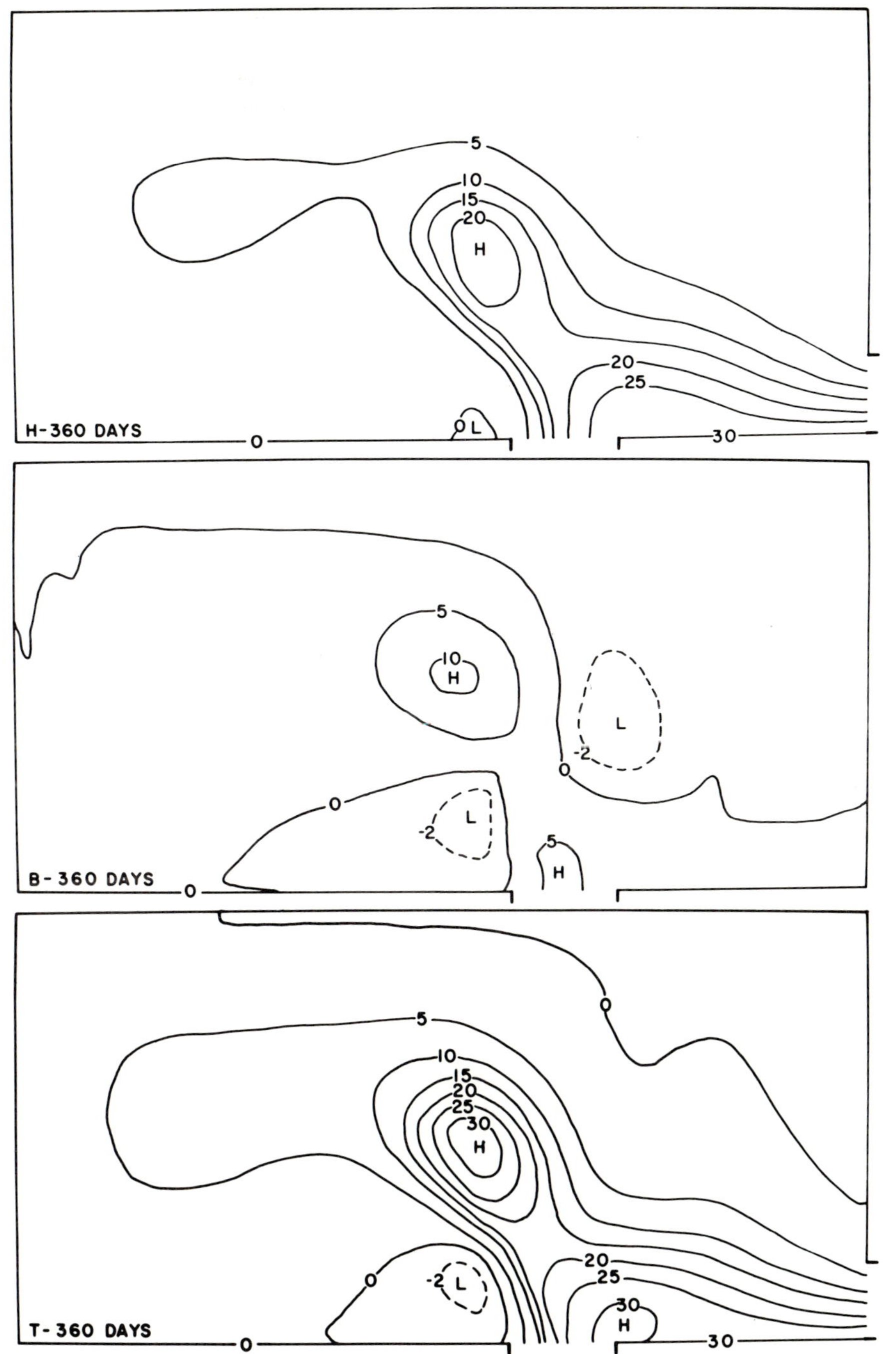

Figure 11-25. Contours of volume transport stream functions in $10^6\, m^3/sec$ after 360 days.

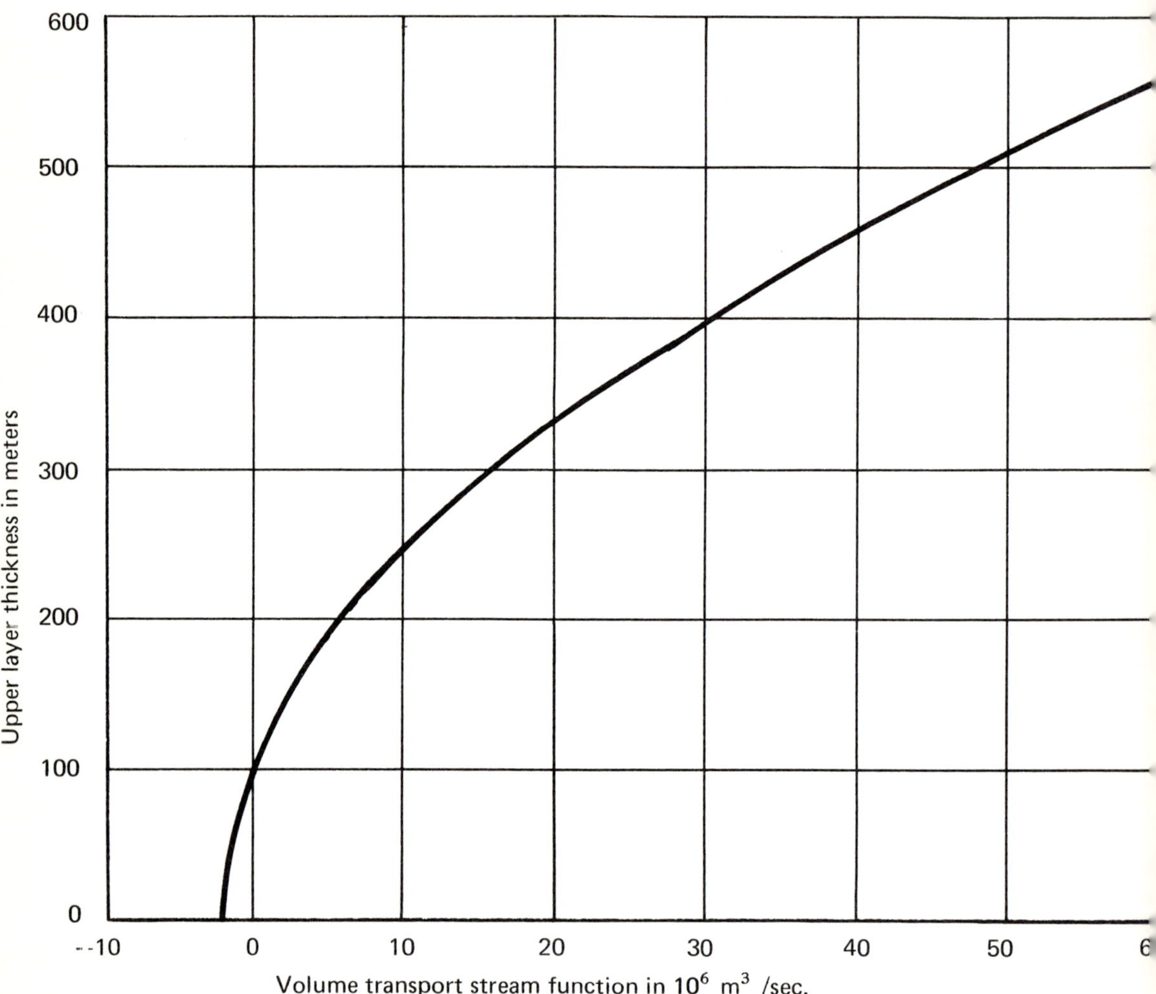

Figure 11-26. Nomograph for conversion of baroclinic volume transport stream functions to upper layer thickness.

output interval was caused by the change in time step which occurred at the 50-day point. The baroclinic volume transport stream function in these figures can be converted back to upper layer thickness through the use of the nomograph in Figure 11-26.

It is interesting to note that the spin up process took 3 hours of CDC-6500 CPU time while the 360-day prediction took 10 hours or an average of 50 minutes/month prediction. This explains why a parametric study was not pursued in this limited study.

Discussion

The circulation predicted in the baroclinic mode has the same basic features as the observed data in the Gulf of Mexico. The predicted results should more closely match the observed data if a 1-year prediction could be made using the 360-day results as a starting point. This would remove most of the transient effect of the spin up process.

Measurement of the barotropic mode of circulation requires deep, direct current measurements which are difficult and costly. Consequently, little or no data are available for comparison with the model.

Aside from the second-year prediction, it would be valuable to use this model for a parametric study of the effects of inflow, topography and friction coefficients.

The next logical step in model development would be to allow topography to intersect both the layer interface and the surface. This would allow full topographic effect on both the baroclinic and barotropic modes and also allow the actual outline of the Gulf of Mexico to be used rather than the present box shape.

Acknowledgments

The authors wish to express their appreciation to the personnel at Fleet Numerical Weather Central, Monterey, California, for their cooperation and assistance while using their computer facility.

This research was sponsored by the Office of Naval Research under contracts Nonr 2119 (04) and N00014-68-A-0308 (0002).

References

Baer, L., Adamo, L.C. and Adelfang, S. 1968. Experiments in oceanic forecasting for the advective region by numerical modeling, 2, Gulf of Mexico. *J. Geophys. Res.*, 73(16):5091-5104.

Hamm, D.P., and Lesser, R.M. 1968. Experiments in oceanic forecasting for the advective region in numerical modeling, 1, The Model. *J. Geophys. Res.*, 73(16):5081-5089.

Paskausky, D.F. 1969. *A barotropic prognostic numerical model of the circulation in the Gulf of Mexico.* Dissertation, Dept. of Oceanogr., Texas A&M Univ., College Station, Texas.

Richtmyer, R.D. and Morton, K.W. 1967. *Difference methods for initial-value problems.* New York: John Wiley & Sons.

Smith G.D. 1965. *Numerical solution of partial differntial equations.* New York: Oxford Univ. Press.

Thompson, Philip D. 1961 *Numercial weather analysis and prediction.* New York: MacMillan.

Veronis, George. 1963 *a*. On the approximation involved in transforming the equations of motion from a spherical surface to the β-plane. 1. Barotropic systems. *J. Marine Res.*, 21(3):110-124.

Veronis, George. 1963*b*. On the approximations involved in transforming the equations of motion from a spherical surface to the β-plane. 2. Baroclinic systems. *J. Marine Res.*, 21(3):199-204.

Section 4
Experimental Modeling

12

Experimental Circulation Modeling within the Gulf and the Caribbean

Takashi Ichiye

Abstract

Experiments were carried out with scale models of the Gulf of Mexico and Caribbean Sea in a circular plexiglass tank with a diameter of 120 cm mounted on a turntable rotated at speeds between 4 and 8 rpm. The reduction ratios of the model are 5.5 x 10^{-7} and 3.3 x 10^{-7} for the Gulf and Caribbean, respectively, in horizontal scale. Vertical exaggeration is 100 times. The driving force for the Gulf was the inflow and outflow system maintained with a reservoir or a circulating pump. By adjusting the rotation rate and the flow rate, the Rossby number of the model circulation was set between 0.001-0.02. The β-effect was simulated by tilting the model by an angle of about 2° in the meridional direction. For the ranges of Reynolds number from 2 x 10^2 to 5 x 10^3 and Rossby number from 10^{-3} to 4 x 10^{-3}, the flow patterns show similarity with those observed. The patterns for larger Rossby and Reynolds numbers indicate that the inflow reaches far westward and is more turbulent than those observed.

The driving force for the Caribbean model was wind. In the experiments a boundary current was developed along the southern and the western coast without the β-effect. The model flow pattern for the winter wind situation with a moderate wind speed at 4 rpm agrees well with the volume transport streamlines computed with the geostrophic method from the hydrographic data for winter seasons but not with the streamlines of the surface or intermediate water inferred by geostrophic and core methods. The bottom flow comes in the Caribbean mostly through the Windward Passage.

Introduction

The Gulf of Mexico and the Caribbean Sea form the "American Mediterranean." The Caribbean is divided into five major basins from east to west: Grenada, Venezuela, Colombia, Cayman and Yucatan. The circulation in the American Mediterranean constitutes the western segment of the southern leg and the sourthern part of the western boundary current of the anticyclonic circulation system of the North Atlantic Ocean. The semi-closed nature of the two seas affects hydrodynamic characteristics, particularly a vorticity balance, different from those of a source region of the western boundary current in the North Pacific or in theoretical models of the ocean circulation. The rather complicated shorelines and bottom topography of the two seas prohibit numerical modeling of the circulation except with oversimplified bottom and coastal boundaries.

One approach to treating the circulation in a limited area of the ocean is to use a laboratory scale model. Hydraulic models for estuaries and bays have long been used by engineers and scientists, but model experiments of the large-scale circulation in the ocean were only recently initiated (Von Arx, 1952). Since the large-scale circulation is dominated by geostrophic equilibrium, the Coriolis effect due to the earth's rotation must be incorporated when modeling such motion. For this purpose, the whole system in which the scale model is placed should be rotated with the proper rate. When simulating a large area of the ocean, the effect of latitudinal variation of the Coriolis parameter (β–effect) should be incorporated. Von Arx (1952) could successfully simulate the wind-driven circulation in the northern hemisphere oceans with a circular basin mounted on a rotating table. In his model the β–effect was simulated by the increase, from the center to the rim, of the equilibrium depth of the water in the basin caused by a centrifugal force.

For circulation simulation in the American Mediterranean, the Coriolis effect is essential, and thus the scale models of both seas were mounted on a rotating table. It was assumed that the β-effect was not important, though the bottom of the Gulf of Mexico model was tilted during some experiments to simulate the β-effect.

The main purpose of the experiments was to understand the effects of bottom and coastal configurations of the two seas on the geostrophic current of barotropic mode. The vertical structure of the flow and the details of the horizontal current patterns are not a subject of this study.

Experimental Procedure for the Gulf of Mexico

Since the major current system in the Gulf of Mexico is assumed to be quasi-geostrophic, the Coriolis force plays an essential part in the Gulf's circulation, and the rotation must be included in any realistic model experiment. To simulate the Gulf, its scale model of fiberglass was set in a circular basin of plexiglass with a diameter of 120 cm and a depth of 25 cm. The basin was mounted on a turntable which was rotated by a gear system with speeds ranging from 4 to 8 rpm. The ratio of model to prototype is 5.5×10^{-7} in horizontal scale and 5.5×10^{-5} in vertical scale. Therefore, the model is exaggerated vertically 100 times.

The main purpose of the experiment was to determine the circulation pattern due to the inflow through the Yucatan Channel. Two features of the prototype could not be simulated exactly by experiments: the effect of the shelf on the major circulation in the Gulf and the vorticity distribution in the flow through Yucatan Channel.

Even though the vertical to horizontal exaggeration is 100, the depth of the shelf in the model is less than 1 cm. Therefore, the surface roughness of the model shelf has stronger effects on the flow than in the prototype. However, in the model as in the prototype, the influence of the shelf on the major barotropic circulation in the Gulf is slight compared with the effect of the inflow through the Yucatan Channel, though the shelf may cause formation of distinct water types. Therefore, this feature might not be serious enough to cause major discrepancies between the model and the prototype.

The current through the Yucatan Channel was simulated by a flow through a glass nozzle connected either to a reservoir or to a circulating pump. When using the reservoir, the same volume of water was removed from outside Florida Straits with a siphon. Dye showed that the flow through the Yucatan Channel was vertically uniform due to the effect of rotation (Taylor, 1917), indicating the predominance of a barotropic flow. However, simulation of horizontal distribution of the current's vorticity was attempted but unsuccessful.

The flow rate can vary from 0.3-3 cm^3/sec with the reservoir and from 2-30 cm^3/sec with the circulatory pump. The velocity of the jet through the nozzle could be adjusted from 0.1-100 cm/sec by changing the shape and size of the nozzle. The flow from the Yucatan Channel could be adjusted in its depth, width, velocity and direction by changing nozzles. The flow pattern in the model Gulf was made visible by introducing ink or dye into the tank. Sixteen mm time-lapse motion pictures were taken with a speed of four frames per rotation.

Dynamic Similarity between the Model and the Prototype

The characteristic velocity of the Gulf of Mexico was taken as the mean velocity of the flow through the Yucatan Channel. Since total transport and the cross-sectional area of the Channel are 30 x 10^6 m^3/sec (Sverdrups) and 1.67 x 10^8 m^2, respectively, the mean velocity equals 18 cm/sec, which was rounded to 20 cm/sec. The characteristic distance in the horizontal direction was taken as the average north-south distance of the Gulf, equaling 910 km. Other characteristic values of the prototype and the model are listed in Table 12-1. The Rossby number and the Froude number can be almost equalized between the prototype and the model by adjusting the rotation speed and the inflow. The Reynolds number of the model was 10^3 with molecular viscosity. Since motion in the prototype is dissipated through eddy viscosity, instead of through molecular viscosity as in the model, the Reynolds number of 10^3 corresponds to horizontal and vertical eddy viscosity of 1.8 x 10^6 and 1.8 x 10^4 cm^2/sec, respectively. It is admitted that this value of vertical eddy viscosity is unrealistically large. However, the effect of viscosity is negligible in the internal flow of the model because the current in the model is predominantly horizontal and vertically constant except in the bottom boundary layer owing to the rotation (Taylor, 1917; Greenspan, 1968).

Table 12-1

Comparison of Dimensional and Nondimensional Characteristic Quantities for the Model and Prototype of the Gulf of Mexico

		Model	Prototype
Ω (sec^{-1})	Angular Velocity	0.63	3.1 x 10^{-5}
L (cm)	Horizontal Length	50	9.1 x 10^7
H (cm)	Vertical Length	11	2 x 10^5
U (cm/sec)	Characteristic Velocity	(0.2)	20
Q (g/sec)	Mass Transport	(10)	3 x 10^{13}
$L\Omega$ (cm/sec)		31	2.8 x 10^3
$R_o = U/\Omega L$	Rossby Number	0.006	0.007
$F = U/(gh)^{1/2}$	Froude Number	1.9 x 10^{-3}	1.4 x 10^{-3}
ν(cm^2/sec)	Molecular Vicsocity	10^{-2}	10^{-2}
ν_h (cm^2/sec)	Eddy Viscosity (Horizontal)		(1.8 x 10^6)
ν_v (cm^2/sec)	Eddy Viscosity (Vertical)		(1.8 x 10^4)
$R_e = UL/\nu$	Reynolds Number	10^3	
$R_e' = UL/\nu_h$	Eddy Reynolds Number		10^3

Table 12-2

Ranges of Critical Values
Used in the Gulf Experiments

	Maximum	Minimum
Ω (rad sec^{-1})	0.54	0.83
U (cm sec^{-1})	0.05	0.56
R_o	0.001	0.021
R_e	2.5 x 10^2	2.8 x 10^3

Although the Rossby number (R_o) of the prototype may vary only within a fraction of an order of magnitude, it can be easily changed by an order of magnitude in the model by changing the rotation rate (Ω) and the mean velocity (U) in the inflow. The Reynolds number (R_e) of the model also can be changed by changing U. The ranges of Ω, U, R_o and R_e realized in the experiments are listed in Table 12-2.

The planetary vorticity tendency term is represented as the β–effect in the vertically averaged vorticity equation. Formation of an anticyclonic loop current in the Gulf was explained by including the β-effect in the vorticity equation (Ichiye, 1962). Therefore, simulating the β-effect in the model appeared to be essential. It can be done by meridional change of depth. The depth necessary for the model can be determined by equating the planetary vorticity term of the prototype to that of the model as

$$T^{-2}\,(\beta v)_{Proto} = -\,(h^{-1}\,dh\,/\,dy\cdot 2\,\Omega v)_{Model},$$

in which T is a dimensionless reduction ratio in time, y is a coordinate increasing toward the north, v is the y-component of velocity, h is the depth and Ω is the angular rotation rate of the model. For the model of the Gulf this equation yields

$$(-h^{-1}\,dh\,/\,dy)_M = (\beta)_P\,(2\Omega)^{-1}_M\,(L'T)^{-1}$$
$$= 2.6 \times 10^{-3}\ (cm^{-1}), \quad (12\text{-}1)$$

in which L′ is a reduction ratio in horizontal distance and

$$(\beta)_P = 2.1 \times 10^{-13} (sec^{-1}\ cm^{-1}),$$

$$(\Omega)_M = 0.2\ (sec^{-1}),\ T = 1.16 \times 10^{-4}$$

Integration of equation (12-1) with y yields the depth of the model which exponentially decreases northward with y. However, the exponential coefficient is very small and a total length of y in the model is only 50 cm. Therefore, the northward decrement of h is practically linear with y. In order to test the β-effect, the scale model, which was built without the depth modification, was tilted by an angle of about 2° in the meridional direction. Comparing the flow patterns of the tilted and the horizontal models indicates that tilting caused no substantial difference. This suggests that the β-effect is at least not as important as other vorticity tendencies in the present experiments.

Results of the Experiments on the Gulf of Mexico

The circulation patterns produced by the experiments were governed by two parameters: the mean speed (U) of the inflow through the Yucatan Channel and the rotation speed (Ω). Variations in these two parameters can be expressed in terms of variations in the Rossby number R_o ($= U\,/\,\Omega L$) and the Reynolds number R_e ($= UL\,/\,\nu$), as indicated in Tables 12-1 and 12-2.

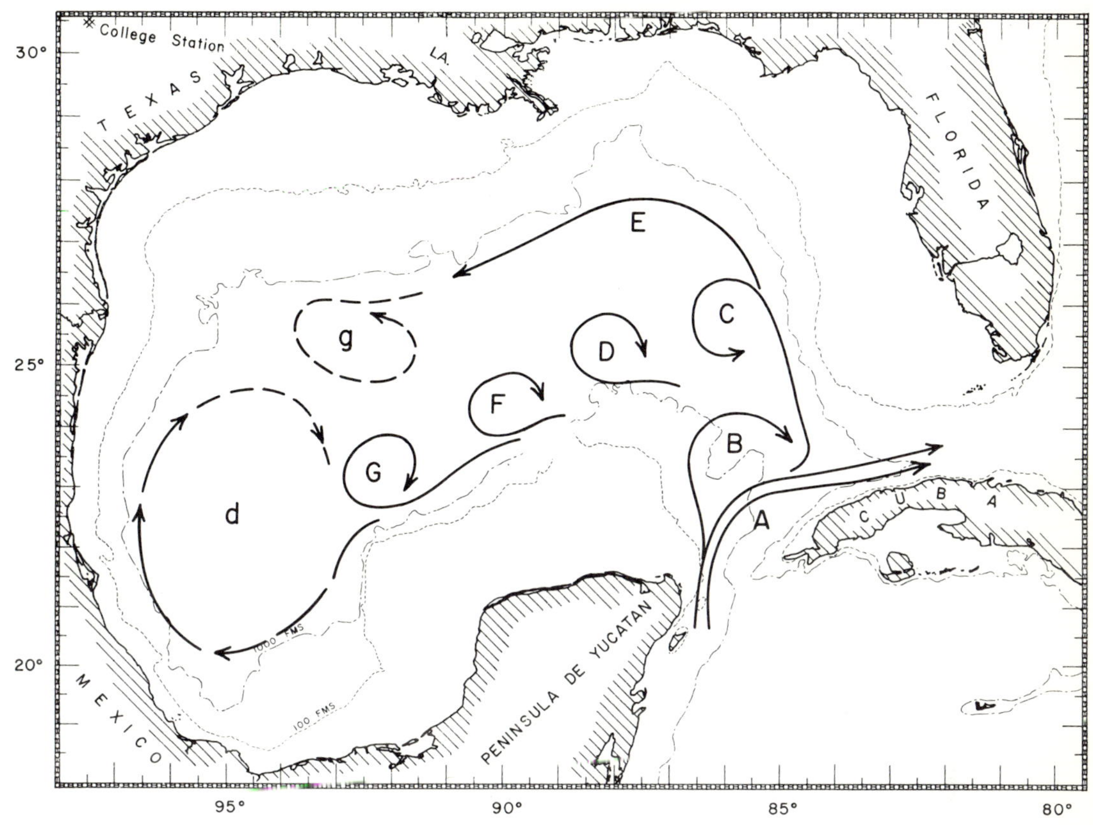

Figure 12-1. Flow pattern derived from dyed water with mean inflow U *= 0.15 cm/sec and rotation rate* Ω *= 0.6 rad/sec.*

The development of the flow pattern in one experiment for U = 0.15 cm/sec and Ω = 0.6 radian/sec is discussed now as an example. These values of parameters yield R_O = 0.005 and R_e = 750. These Rossby and Reynolds numbers are classified as medium in the ranges realized by the experiments. After about 300 revolutions (rev) of the table (40 minutes) from the start of the experiment, dye was injected in the nozzle. For the next 60 rev the current flowed along the Cuban coast as indicated by A in Figure 12-1. Then an anticyclonic loop B appeared and lasted for about 20 rev. Then a flow branched from the current A and moved northward along the eastern shelf, forming a cyclonic eddy C within less than 20 rev. This eddy C was rather intense and lasted almost 200 rev. After about 60 rev from formation of C, weak anticyclonic eddy D was formed west of C and lasted for about 30 rev. During that period, a branch of C extended westward, forming a current E. Also in the southwestern basin a steady anticyclonic gyre d was formed. (The lower case letters designate features of weak flow.) An anticyclonic eddy F was formed about 40 rev after the disappearance of D and lasted another 20 rev. Then a branch of F moved westward and after about 60 rev formed another anticyclonic eddy G which lasted for another 20 to 30 rev. Almost simultaneously with G, an end of E moved into the western basin and formed a very weak cyclonic eddy g. This example, as well as other sets, indicates that anticyclonic eddies were usually of short

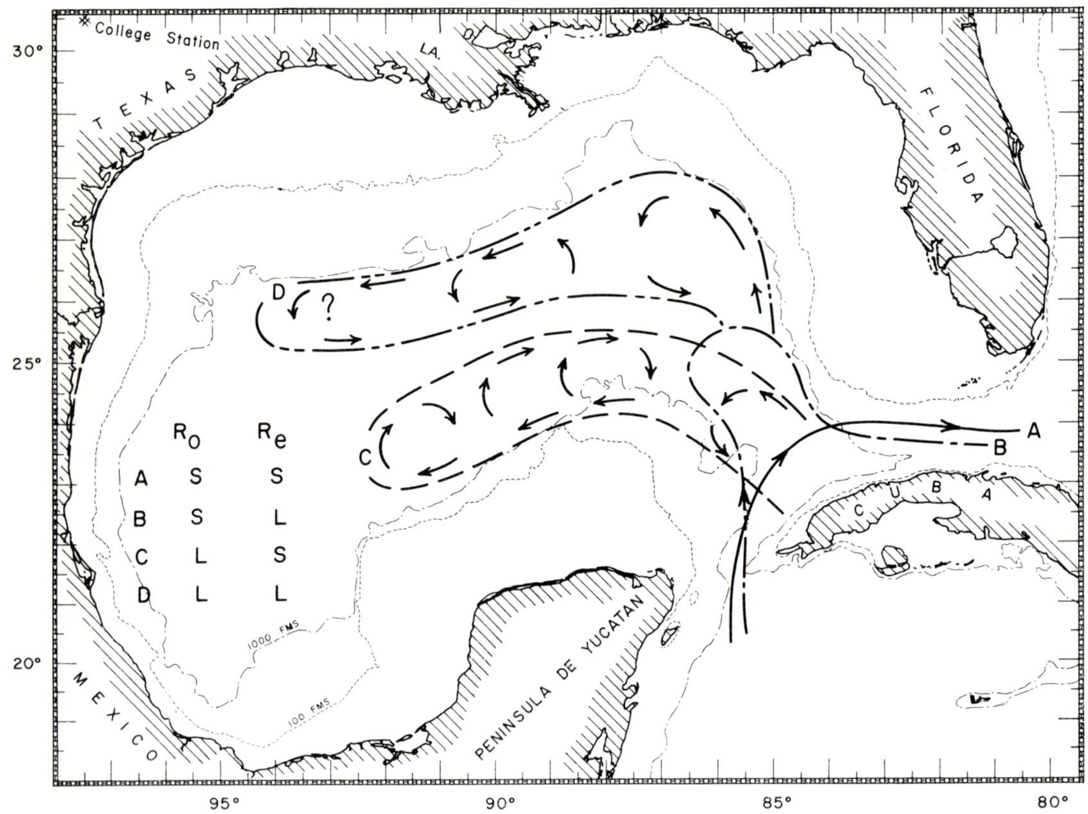

Figure 12-2. Areas of penetration for four combinations of large (L) or small (S) Rossby (R_o*) and Reynolds (*R_e*) numbers.*

duration and apparently moved westward, whereas cyclonic eddies lasted about 10 times longer and stayed north of the Yucatan Channel. The anticyclonic eddy *d* seemed to be due to topographic effect of the western basin rather than to the effect of the inflow through the Yucatan Channel.

A cause of eddy *B* seemed to be an excessive negative vorticity in the right-hand side of the stream axis in the Yucatan Channel. The persistent eddy *C* was due to formation of a cyclonic curvature by a strong positive shearing vorticity to the left-hand side of the flow axis. Eddies *D, F* and *G* were generated by friction of the coastal boundary. Formation of eddy *C* has never been observed in the prototype, because the initial positive relative (shearing) vorticity may decrease northward baroclinically in the prototype and also because the Reynolds number of the model is higher than is usually the case within the prototype.

In the experiment various combinations of inflow velocity and rotation speed (Rossby number and Reynolds number) were tried. Since it would take too much space to describe each individual case, only the characteristic features of four different combinations of Rossby and Reynolds numbers are discussed. Arbitrarily, Rossby number is classified large or small for a range of 0.01-0.02 or 0.001-0.004 and Reynolds number large or small for ranges of 10^3-3×10^3 or 2×10^2-5×10^2. To show flow pattern features, the area of influence is defined as an area penetrated by dyed water by the time of about 200 rev after the dye leaves the nozzle. In many cases the boundary of the area was demarcated by eddies which were generated on the sides or at the end of the continuous flows. Therefore, the boundaries do not necessarily correspond to the streamlines. In Figure 12-2 the area of influence is schematically shown in four combinations of small or large Rossby and Reynolds numbers.

For large Rossby and Reynolds numbers, the inflow from the Yucatan Channel reaches the northern coast (actually the edge of the shelf along the Gulf) and then flows along the coast almost to the western coast, generating one cyclonic eddy just north of the Channel and one or two smaller cyclonic or anticyclonic eddies northwest of it. The flow patterns in general showed rather irregular features having short-lived eddies with scales smaller than the major eddies. For large R_o and small R_e, the inflow turns to the west along the Campeche Bank but does not reach the western coast. In this case there is one almost stationary cyclonic eddy north of the Yucatan Channel and two or three slightly smaller anticyclonic eddies of much shorter duration. Eddies and the area of influence in this case correspond roughly to the flow pattern in the later stage of Figure 12-1. For small R_o and large R_e, the main part of the inflow reaches the eastern coast but does not turn to the west. An anticyclonic eddy is occasionally formed north of the Yucatan Channel and lasts between 10-30 rev. It is dissipated by turbulent motion rather than being detached and migrating from the main current as inferred by Leipper (1970) or as in the case shown in Figure 12-1. For small R_o and R_e, the water from the Yucatan Channel follows the northern coast of Cuba and flows out of the Gulf, without appreciably influencing the main body of the Gulf. The actual flow regime of the Gulf may be in a range of small Rossby numbers and small to medium Reynolds numbers, as evidenced by observed flow regime.

Figure 12-3 shows photographs enlarged from 16 mm moving picture frames for four typical cases of *A*, *B*, *C* and *D* denoted in Figure 12-2. (Case *D* of the picture represents an intermediate flow regime between case *C* and case *D* denoted in Figure 12-2). Each frame was taken more than 100 rev after injection of dye. In the photographs, dyed water is shown with white area, since the prints were made from positive color film.

Experimental Procedure for the Caribbean Sea

The scale model of the Caribbean Sea was built of plywood with a thickness of 1.3 cm. Each sheet was cut following 400-m depth contours and glued to form the bottom topography. The horizontal and vertical reduction ratios are $1/3 \times 10^{-6}$ and $1/3 \times 10^{-4}$, respectively. The characteristic angular velocity of the prototype is represented by the value at 15°N, equaling 1.86×10^{-5} rad/sec. The rotation speed of the model was variable from 0.42 rad/sec (4 rpm) to 0.84 rad/sec (8 rpm). The characteristic horizontal dimension *L* of the prototype is 1500 km, the average north-south distance of the basin. The characteristic velocity of the prototype is taken as 1 cm/sec, which corresponds to the mean zonal velocity in the basin. A more exact value of the latter is 0.6 cm/sec, which can be obtained by dividing the total transport of 30×10^6 m^3/sec (30 Sverdrup) at the Yucatan Channel (representing the average eastward net transport in the basin) by the mean meridional cross-section area of 5×10^9 m^2. Therefore, the Rossby number of the Caribbean is 4.3×10^{-4}. The mean zonal current in the model corresponding to this Rossby number is extremely small, equaling 1.35×10^{-3} cm/sec at the rotation speed of 6 rpm.

Since the primary driving force of the cirCualtion within the Caribbean is wind stress, small fans were mounted on the table rim. To each fan was attached a diffuser. Wind speeds from each fan could be varied by regulating the power supply.

As with the Gulf of Mexico experiments, 16 mm motion pictures were taken at a rate of four frames per rotation. Subsurface and bottom currents were indicated by dye injected at several locations, while the surface current was visible by the movements of small paper patches. It was usually observed that the subsurface current was slower by an order of magnitude than the surface

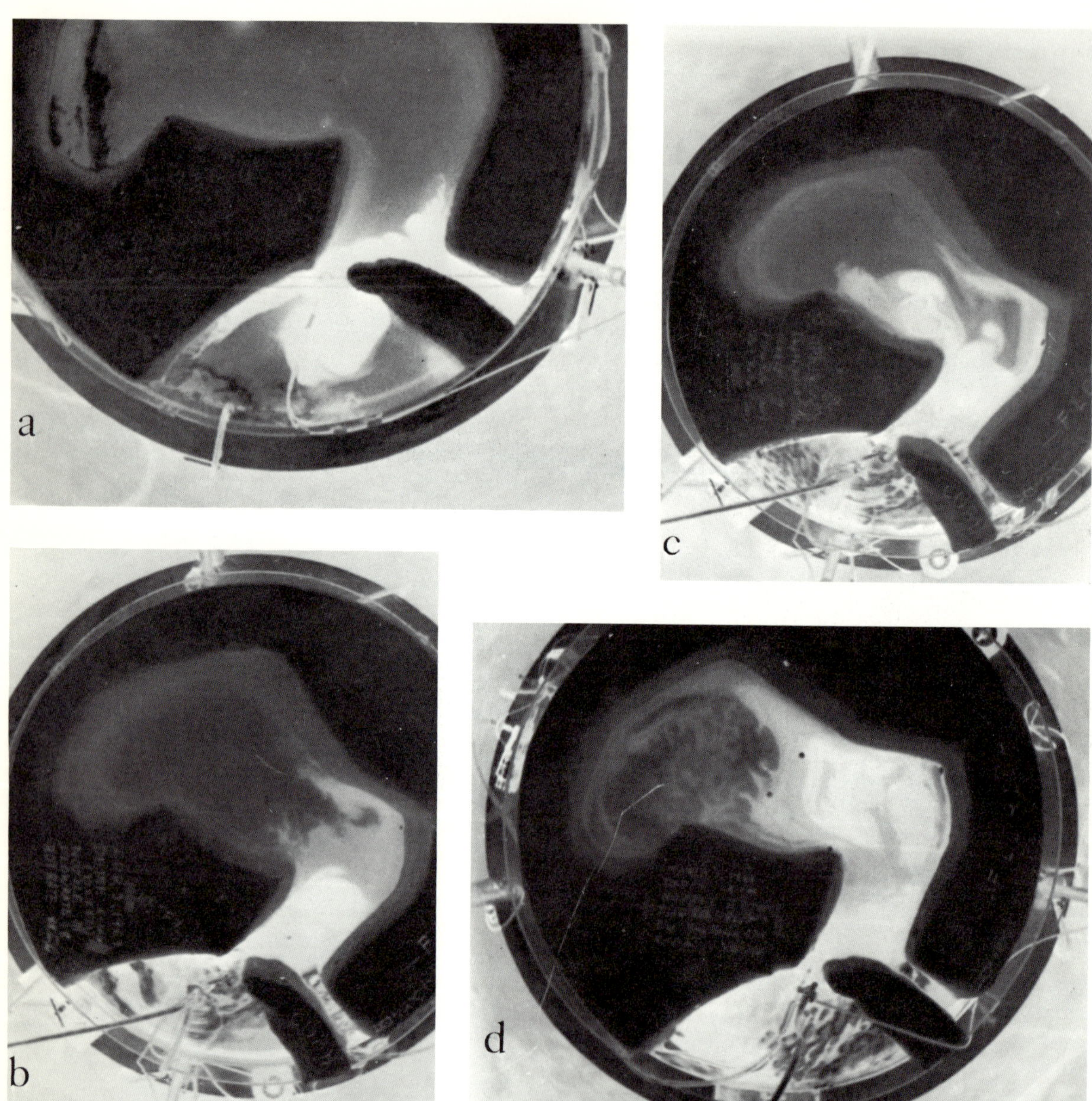

Figure 12-3. Photographs of the four types of flow patterns corresponding to cases A, B, C *and* D *defined in Figure 12-2.*

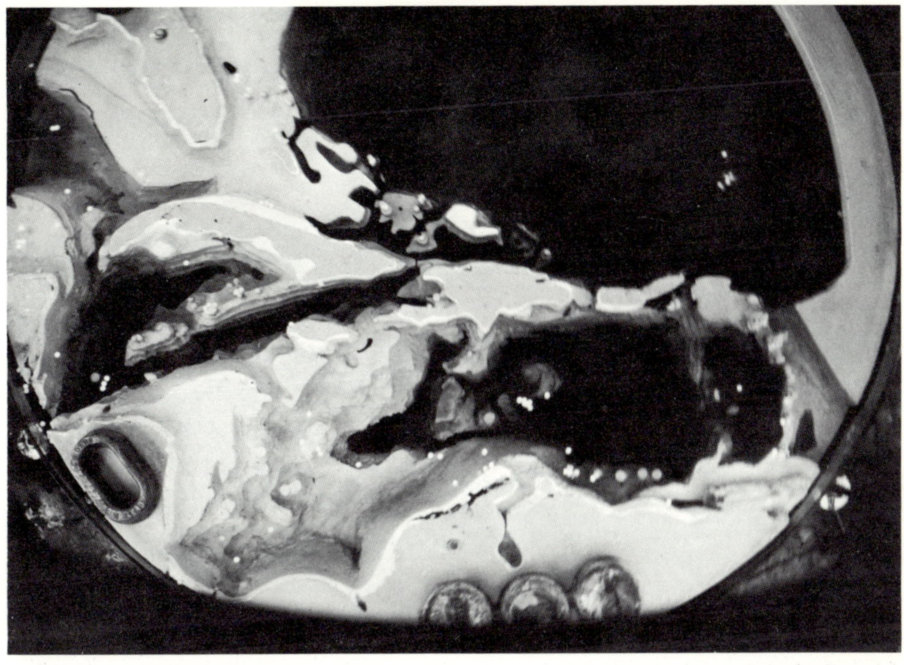

Figure 12-4. Photograph of the model of the Caribbean Sea.

current, which had speeds of 10^{-2} to 10^{-1} cm/sec. The surface current appears to be limited only within a thickness of a few sheets of paper. A photograph of the model is shown in Figure 12-4. The model included the eastern half of the Gulf and part of the Atlantic Ocean east of 50°W and south of 40°N, although the bottom topography of the Atlantic was made flat except for the continental shelves. In the picture the black area indicates that dye moved near the bottom filling the basins.

Owing to difficulties in measuring wind speed directly in the model, the wind strength was regulated only in three degrees (weak, moderate and strong), corresponding to skin speeds of the orders of 10^{-3} cm/sec, 10^{-2} cm/sec and 10^{-1} cm/sec, respectively. Two types of wind distributions were applied: one with only an easterly wind over the sea, representing the summer condition; and the other with an additional westerly wind over the northern part of the sea, representing the winter situation. Since the wind distributions and strengths have six combinations, the rotation rate was limited to only three values (4, 6 and 8 rpm) for each set of the wind system.

Results of the Experiments Modeling the Caribbean Sea

Discussion of details for each combination of different wind conditions and rotation rates would be too lengthy. It turned out that the strong wind made the circulation far more turbulent than might be expected in the prototype and that the weak wind failed to generate the observed concentrated flow through the Yucatan Channel. Since there are no comprehensive hydrographic data for the summer season to be compared with the experiments, only the results corresponding to the winter season are discussed. Figures 12-5 and 12-6 show the circulation patterns for moderate wind at rotation rates of 8 and 4 rpm, respectively. The streamlines were drawn by tracing dyed water at the subsurface levels in 16 mm film. The fast or slow rotation with the same wind strength corresponds to small or large Rossby number with the

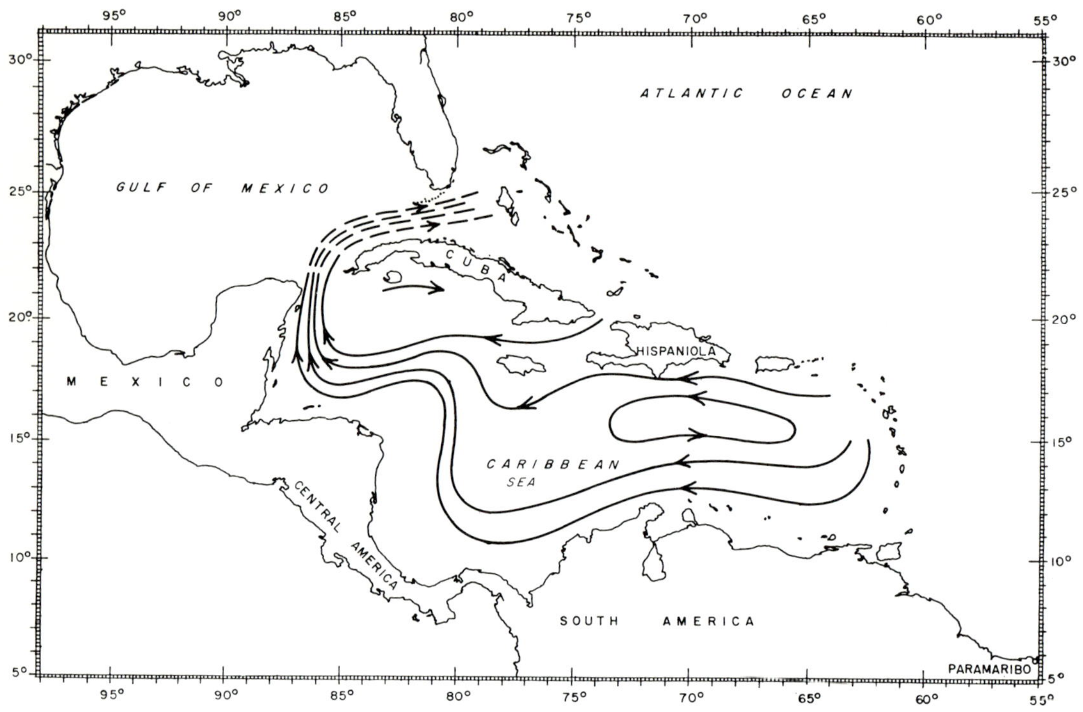

Figure 12-5. Experimental flow pattern of the Caribbean Sea for medium wind in the winter condition at rotation rate of 8 rpm.

same Reynolds number, though the model Rossby numbers in both cases are of the order of 10^{-3}.

A difference between the two cases can be recognized, though there is some similarity. In Figure 12-5 (small Rossby number), a well-defined current flowed along the southern coast and was intensified as it flowed northward along the western coast. A substantial inflow from Windward Passage joined the flow from the east in Cayman Basin. A weak counter-current was observed south of Cuba. A weak, zonally elongated, cyclonic eddy was recognized by thin vertical walls of dyed water in the Venezuela Basin. Figure 12-6 (larger Rossby number) indicates a more complicated system of eddies, although again the narrow current flowed along the southern coast, was joined by an inflow from Windward Passage and was intensified within the Cayman Basin. There were two cyclonic eddies in the Venezuela Basin. The flow along the southern coast of Hispaniola did not join directly the current through the southern sea but split into branches, forming small anticyclonic and cyclonic eddies in the Colombia Basin. For the larger Rossby number the counter-current south of Cuba was slightly stronger and an anticyclonic eddy was formed south of the Yucatan Channel.

Compared with those of the Gulf, hydrographic data from the Caribbean are scarce. Dynamic topographies relative to the 1200-db level (Gordon, 1967) and the flow pattern of the intermediate water derived by the core method (Wüst, 1963) do not agree well with a conspicuous feature common for Figures 12-5 and 12-6; that is, the well-defined current along the southern and western coasts adjoined by a substantial inflow through Windward Passage, although both studies were based on the data collected during the cold season. The main reason for such a discrepancy seems to be that the streamlines derived by the

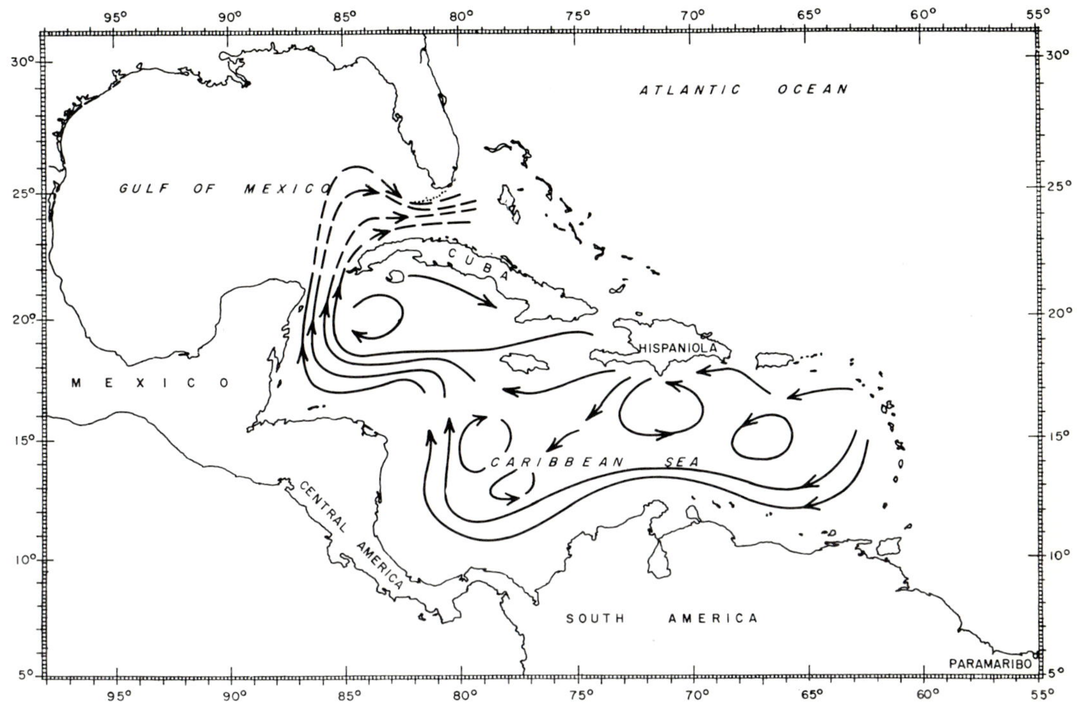

Figure 12-6. Experimental flow pattern with the same wind as in Figure 12-5 at a rotation rate of 4 rpm.

experiments represent the vertically average current or the volume transport instead of the current at different levels.

To compare the experimental results with observed data it was necessary to construct the transport streamlines instead of dynamic topographies. Transport streamlines are in general more difficult to construct because of uncertainty of the level of no motion for the dynamic calculation. However, the total transport across a section can be determined in the Caribbean because of the semiclosed nature of its western section. The transport between a pair of stations may be estimated by using this total transport as a reduction factor, as discussed below.

The mean transport through the Yucatan Channel is fairly accurately determined because of numerous hydrographic sections of high quality. The value is estimated as 30 Sverdrups (Nowlin and McLellan, 1967). The transport within the Caribbean can be calculated by using the five hydrographic sections analyzed by Gordon (1967). The sources of data at these sections are listed in Table 12-3 and the section locations are plotted in Figure 12-7. The net transport through Section 4 or 5 should be equal to the net transport through the Yucatan Channel because there is no lateral opening between these three sections. The volume transport through Sections 4 and 5, determined by using the newly calculated geostrophic velocities, yields the same value of 38.1 Sverdrups. This should be equal to 30 Sverdrups. It is assumed that the reduction ratio 30/38.1 = 0.787 might be applied to the transport computed from geostrophic velocities between any pair of stations on Section 1 to 3, though there is no theoretical justification. Actually, the reduction of the total transport across these sections does not change the pattern of the transport streamlines. In Table 12-3, the volume transports determined by the dynamic

Table 12.3.

Hydrographic Sections, Calculated and Standardized Volume Transport

Section	Longitude	Ship	Date	Volume Transport (from geostrophic calculations)	(Sverdrup) (standardized)
1	64°30	Crawford	Feb. 1958	27.9	22.0
2	68°30	Atlantis	Nov.-Dec. 1954	18.4	14.5
3	73°50	Crawford	Feb.-March 1958	26.6	20.9
4	79°40	Crawford	March 1958	38.1	30
5	84°30	Crawford	March 1958	38.1	30

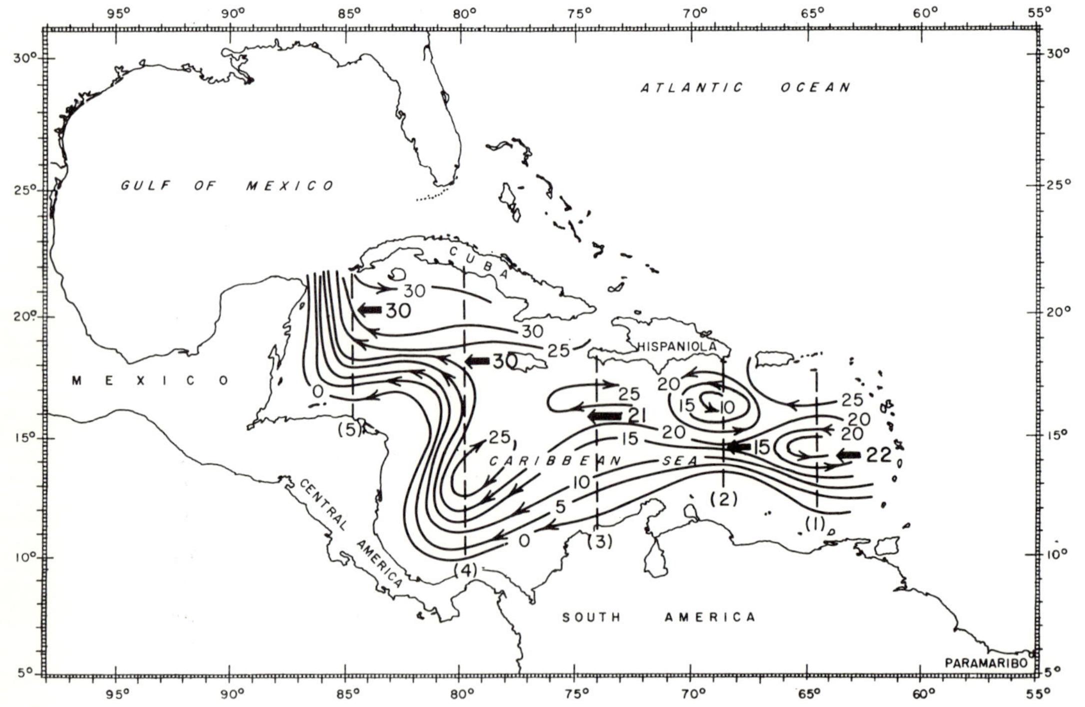

Figure 12-7. Streamlines of volume transport determined at five hydrographic sections standardized for the total transport through the Yucatan Channel.

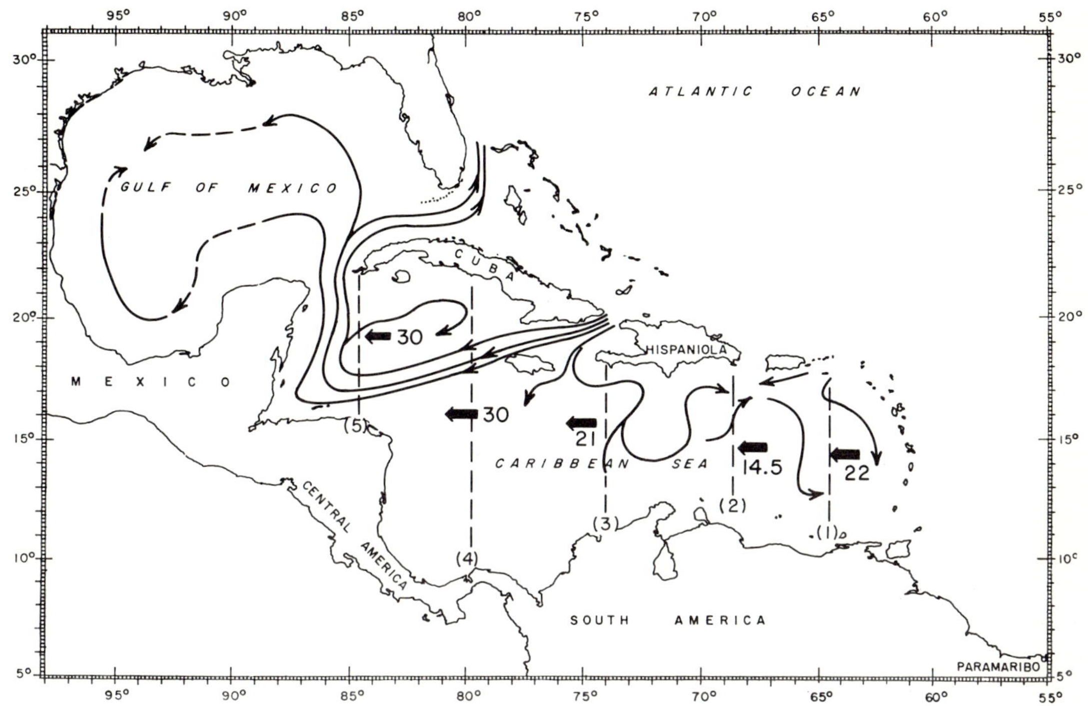

Figure 12-8. Composite flow paths of bottom water determined by various sets of experiments.

method with reference levels close to the bottom and those standardized for the Yucatan Channel transport are listed for each section.

The streamlines based on the standardized transport are shown in Figure 12-7, which clearly indicates concentration of the transport along the southern coast in agreement with the experimental data. The similarity between Figure 12-7 and Figure 12-6 is quite striking. Two distinct cyclonic eddies in Venezuela Basin are recognized in Figure 12-6. The transport through Windward Passage reaches 9 Sverdrups. Even a weak counter-current south of Cuba is seen in Figure 12-6. The only major difference is that two rather weak anticyclonic eddies are present in the Colombia Basin in Figure 12-7, instead of the streamlines divergence and the cyclonic and anticyclonic eddies of Figure 12-6.

Experiments on the difference of circulation due to seasonal variations of the wind system were made only qualitatively. To simulate the winter condition, the westerly wind was generated in the area north of about 30°N. Comparing the streamlines with or without the westerly wind indicates that the westerly wind intensified the concentration of the current along the southern and western coast and in the Yucatan Channel. Cochrane (1966) observed that intensification of the transport across the Yucatan Channel reaches maximum in spring and early summer and minimum in fall. The discrepancy may be due to the situation that the experimental streamlines were obtained after they reached a steady state condition.

Water motion near the bottom of the model as made visible by dense dye was different from and much slower than motion within the main body of the water. The pattern of bottom flow usually showed more similarity among different combinations of Rossby and Reynolds numbers and was consistent for each set of combinations.

This suggests that the bottom topography and the friction in the bottom boundary layer had dominant roles on the bottom current in the model over the driving force and the rate of rotation utilized.

Figure 12-8 shows the composite flow paths near the bottom as determined from various sets of Caribbean Sea experiments. The paths in the western Gulf of Mexico are derived from experiments of the Gulf.

Although similarity in bottom roughness between the prototype and the model may be far from perfect, the model bottom flow of the Caribbean has some similarity to the pattern derived by Sturges (1965) from the potential temperature-salinity diagrams. In the model, a large amount of the bottom flow enters the Cayman Trench through Windward Passage. A small part of the inflow crosses Jamaica Ridge, entering into the eastern basin. A greater part of the bottom water flowing out through the Yucatan Channel turns eastward through the Florida Strait and a smaller part flows into the western Gulf along the northern and southern coasts. In reality, a part of the bottom water returns into the Cayman Sea.

Concluding Remarks

The results of the experiments show rather striking similarities to the observed circulation within the prototypes, particularly in the Caribbean, in spite of the rather crude simulation. Larger discrepancies in the Gulf of Mexico model with the prototype than in the Caribbean model may be partly due to more pronounced effects of density stratification in the prototype Gulf and partly due to the condition that the model circulation was driven by the Yucatan Channel inflow, which is very hard to simulate with correct vorticity distribution across the flow.

Acknowledgments

The experiments on the Gulf of Mexico were carried out at the Oceanographic Institute of Florida State University from 1961 to 1963 under a grant from the National Science Foundation with assistance of Noel B. Plutchak. The experiments on the Caribbean Sea were carried out at Lamont Geological Observatory of Columbia University from 1966 to 1968 under sponsorship of the Atomic Energy Commission and the Office of Naval Research with assistance of Frank Husson and other summer students. The present analysis was prepared under support by a contract with the Office of Naval Research, N00014-68-A-0308 (0002), and under grants from the National Science Foundation, GA-13430 and GA-26498. The author is particularly grateful to Dr. Worth D. Nowlin, Jr., who read the manuscript critically and gave many important comments.

References

Cochrane, J.D. 1966. *The Yucatan Current.* Progress Report of Dept. of Oceanogr., Texas A&M Univ., Reference 66-23T.

Gordon A. 1967. Circulation of the Caribbean Sea. *J. Geophys. Res.,* 72(24):6207-6224.

Greenspan, H.P. 1968. *The theory of rotating fluids.* Cambridge Univ. Press, 30-38, 225-233.

Ichiye, T. 1962. Circulation and water mass distribution in the Gulf of Mexico *Geofisica Internacional* (Mexico City), 2(3):47-76.

Leipper, D.F. 1970. A sequence of current patterns in the Gulf of Mexico. *J. Geophys. Res.,* 75(3):635-657.

Nowlin, W.D., and McLellan, H.J. 1967. A characterization of the Gulf of Mexico waters in winter. *J. Marine Res.*, 25(1):29-59.

Sturges, W. 1965. Water characteristics of the Caribbean Sea. *J. Marine Res.*, 23:147-162.

Taylor, G.I. 1917. Motion of solids in fluids when the flow is not irrotational. *Proc. Roy. Soc.* (London) A93:99-113.

Von Arx, W.S. 1952. A laboratory study of the wind driven ocean circulation. *Tellus,* 6:116-123.

Wüst, G. 1963. Stratification and circulation in the Antillean-Caribbean basins. *Deep Sea Res.,* 10:165-187.

Section 5
Hurricane Effects

13

Circulation Changes Caused by Hurricanes

Takashi Ichiye

Abstract

Observations in the Gulf of Mexico of temperature and salinity changes due to passing hurricanes are reviewed. Observations for Hurricane Carla (1961) and for Hurricane Inez (1966) were made on the continental slope in the northwestern and the western Gulf, respectively. The data from the latter case indicate upward displacement and deepening of the thermocline near to and to the left hand side of the hurricane center, respectively. The data from Hurricane Hilda (1964) were obtained on several transects across the track in the central Gulf and are the most comprehensive. Comparison of hydrographic data with those of the undisturbed state indicate upward and downward displacement of the thermocline at and outside the track of the eye, respectively.

Theoretical models are also reviewed. These include steady state circulation induced in a homogeneous ocean by a circular storm, perturbation currents (both horizontal and vertical) in a homogeneous ocean due to a moving wind system of arbitrary space distributions, the response of a two-layer ocean with the motionless lower layer to a stationary circular storm starting suddenly or to a storm moving with a constant velocity and the effects of a shelf or a coast on the response of the two-layer ocean to a storm with arbitrary space and time distributions.

Introduction

Meteorologists have confirmed that the supply of energy from the sea to the atmosphere plays an important role in the formation and intensification of tropical cyclones. Particularly, sea-surface temperature (SST) data are readily available and, thus, several authors tried to correlate path or central pressure of hurricanes with the sea-surface temperature during their passage. It was recognized that a hurricane has a tendency to move toward the area of the warmest SST or the area of the highest energy transport from the sea to the atmosphere in a zone between the tropical easterly wind and the mid-latitude westerlies (Fisher, 1958; Perlroth, 1962 and 1967; Dunn and Miller, 1964).

Except for storm surges in the coastal areas, however, responses of the ocean to these storms were rather poorly understood because oceanographic observations in the open sea do not yet reach a stage of synoptic type. Particularly, it is almost impossible to obtain sequential synoptic patterns of currents or water properties by using surface research vessels alone during and after a storm passing over the ocean. These storms could conceivably cause drastic change in distributions of current and water properties because of their extremely large wind stresses over the sea surface.

In the western North Pacific Ocean, Ichiye (1955) reported four examples of surface isotherm changes south of Japan after passages of tropical cyclones. Since the undisturbed sea-surface isotherms in this area are almost parallel to the streamlines of the Kuroshio, the passages of cyclones caused meanders of the isotherm as well as decrease in temperature, particularly in the area where a cyclone path crossed the Kuroshio. Therefore, the change of the surface isotherms due to passage of a cyclone in this area can be attributed to four causes: (1) forced instability waves along the shear flow at the ocean current boundary; (2) upwelling caused by surface divergent wind transport; (3) mechanical mixing of surface waters with subsurface waters; (4) removal of warm offshore water by the wind drift. The Kuroshio area south of Japan is most frequently visited by tropical cyclones but is not ideal for studying the ocean's response to a storm because of strong and complicated currents and large horizontal temperature and salinity gradients, but it is an ideal place for studying ocean current meanders caused by atmospheric disturbances.

Jordan and Frank (1964) and Jordan (1965) used 15-day mean sea-surface temperature charts and two sets of individual observations of sea-surface temperature by merchant ships, respectively, before and after typhoons in the area south of 30°N and west of 140°E. These authors contend that vertical mixing by strong winds is the primary factor in sea-surface cooling in typhoons because the temperature decrease was larger on the right-hand side of the storm track. However, this is inconclusive, since the areas studied by them on the right-hand side on the tracks correspond to the area of shoaler thermocline.

Historical Hurricanes in the Gulf of Mexico

The Gulf of Mexico is a sea area where tropical cyclones pass most frequently in the North Atlantic Ocean. Among 94 memorable U.S. hurricanes from 1873 to 1967 (Sugg and Carrodus, 1969), 61 passed on the Gulf. Also, among 251 hurricanes in the whole North Atlantic Ocean from 1901 to 1957, 50 reached hurricane wind force from tropical cyclones in the Gulf (Dunn and Miller, 1964). However, the effect of hurricanes in the Gulf was studied only recently. Fisher (1958) constructed daily SST charts during 11 hurricanes from 1953 to 1954 to determine the correlation between SST and hurricane tracks. Most of his SST charts were off the east coast of the United States and several were for the western Caribbean. There was no case for the Gulf of Mexico. His aim was to establish the relationships between the present SST and future hurricane track. He could not find definite evidence of the hurricane's effect on the SST change.

Change of Water Characteristics after Carla (1961)

Stevenson and Armstrong (1965) studied the effect of Hurricane Carla, which entered the Gulf through the Yucatan Straits on September 7,

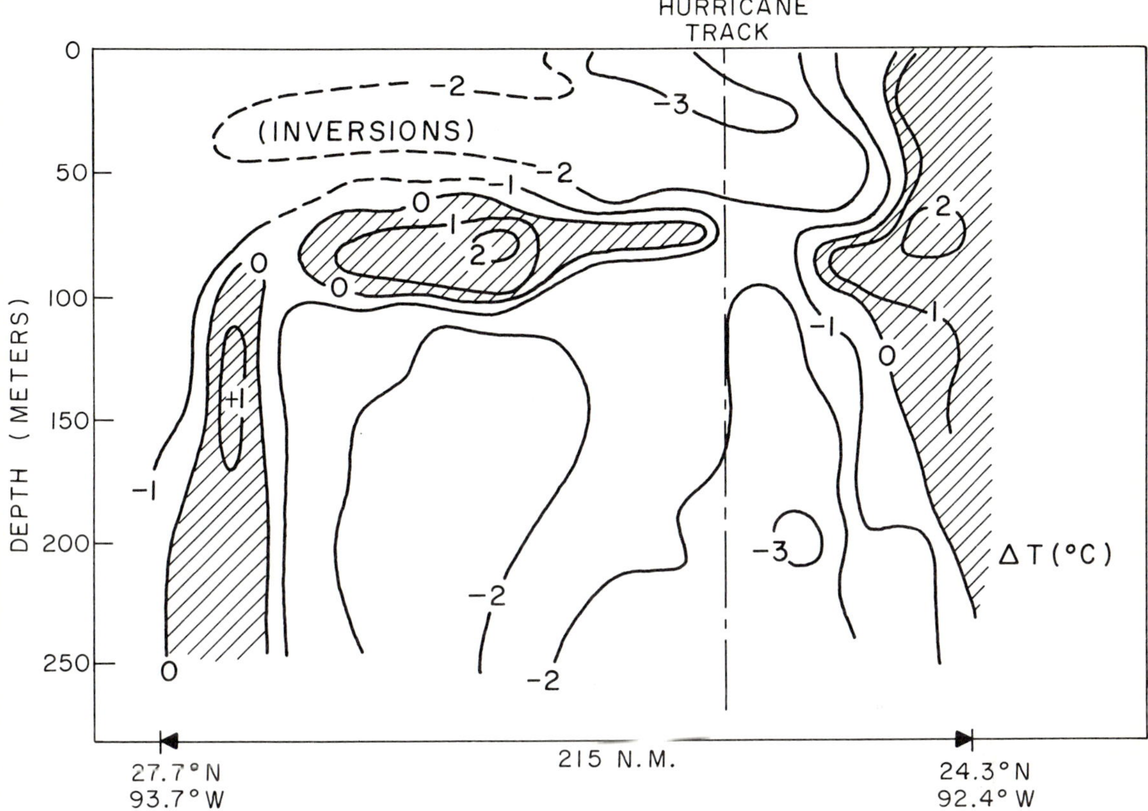

Figure 13-1. Temperature changes (°C) from August 23 to September 15, 1961, determined from data with a thermistor chain on board the R/V Hidalgo.

1961, and landed about 120 km northeast of Corpus Christi in the afternoon of September 11, reading lowest central pressure of 930.0 mb. The upper layer temperature profiles crossing the hurricane track were obtained on August 23 and September 15 from 22°38′N, 91°04′W to 27°39′N, 93°40′W and from the latter station to 24°18′N, 92°24′W, respectively, with a thermistor chain towed by the R/V *Hidalgo*. The pre-hurricane profile indicates an anticyclonic eddy of about 130 nautical miles diameter between 27°20′N and 25°20′N. This might be a cutoff eddy migrating from the Loop Current. The post-hurricane profile still shows this eddy, but it also indicates intensification of a cyclonic eddy south of the anticyclonic one. The temperature difference between the pre- and post-hurricane profiles shows overall decrease for the post-hurricane profile except near the thermocline. The decrease is largest in the surface layer near the hurricane center. At sub-thermocline depths, the decrease is large both near the center of the hurricane and the anticyclonic eddy. It is also reported that the BT traces showed many inversions (temperature increase with depth) in the upper layer above 70 m in the northern part of the post-hurricane profile (Figure 13-1).

A month after the hurricane, between October 4-9, the *Hidalgo* made a hydrographic survey of the northwestern part of the Gulf. The winds over the northwest Gulf in the weeks following the hurricane were generally from the southeast and low, and there was not much meteorological disturbance to cause drastic change of oceanographic conditions. Therefore, it was claimed that the condition a month after the hurricane was still representative of the condition immediately following it. However, this contention is ambiguous, because the temperature distribution at a depth of 50 m on October 4-9 shows low-temperature regions centered at about 27°N, 94.5°W, but the thermistor chain profile on September 15 does not show a low-temperature dome expected by this cold water region.

The October 4-9 hydrographic data are difficult to interpret for studying Carla's effects for three reasons:

1. There are no data before the hurricane;
2. The data were taken too late; and
3. The area covered includes the continental shelf and slope where the water properties are complicated and change greatly with time.

Even with such shortcomings, however, conspicuous features of the temperature and salinity distributions can be distinguished. One is wide occurrences of the temperature inversions in the upper 40-50 m along the hurricane track. Another is temperature decrease up to 4°C over the area in the upper layer at least down to 250 m.

The temperature inversions after the storm appear to result mainly from surface layer cooling and not much from upwelling. The salinity of the area was 30-31 per mil through depths of 40 m and 36 per mil or more below 100 m. Therefore, when the surface layer temperature drops by more than 5°C, density stratification is still stable and there is no overturning of the water. In the deep-sea area (Figure 13-1), the inversions are abundant on the right-hand side of the hurricane track, suggesting stronger wind stresses. Lack of inversions on the shelf and slope areas southeast of Galveston Bay may result because the salinity of the upper layer is higher in this area than east of Corpus Christi, and temperature decrease of the upper layer thus causes mixing with the lower layer of high salinity. The authors estimate the total heat loss from the water as more than 2.2×10^8 cal/day.

On the other hand, a general decrease of temperature at least more than 250 m may not be explained solely as a result of cooling from the surface, since the temperature profiles indicate presence of the thermocline. This suggests upwelling in a wake of the hurricane. The wider upwelling area on the right-hand side of the hurricane track may be due to the effect of the continental shelf. The water transport on the shelf is almost parallel to the wind stress, whereas the one in the deep water is to the right of the wind, causing divergence on the right-hand side and convergence on the left-hand side of the track at the surface layer between the shelf and the deep sea, as a hurricane moves normally toward the coast.

Hydrographic Conditions at the Continental Slope Before and After Hurricane Inez (1966)

A hurricane's effects on the distributions of water properties and movement at the continental slope are more clearly seen in the study made by Franceschini and El-Sayed (1968). Their data consist of the profiles along the continental slope of the western Gulf before and after Hurricane Inez in 1966.

The hurricane entered the Gulf in the early morning of October 5 through the Florida Straits. During October 6, the maximum wind zone with surface winds of 100-120 mph was 12-15 miles from the center. The hurricane reached its maximum intensity late on October 8, with a central pressure of 948 mb and with maximum surface winds of about 135 mph. Hurricane force winds extended 60 miles from the center, which was about 200 miles from the coast of Mexico, moving at 8-10 mph and gale force winds extended out 150-200 miles.

The hurricane crossed the track of the *Alaminos* at about 23°N, 97°W at 10 GMT on October 10 with westward speed of 6 mph and estimated maximum surface wind of 130 mph. The storm landed about 40 miles north of Tampico, Mexico, causing storm tides of 10-12 feet north of the eye and precipitation exceeding 10 inches in 24 hours.

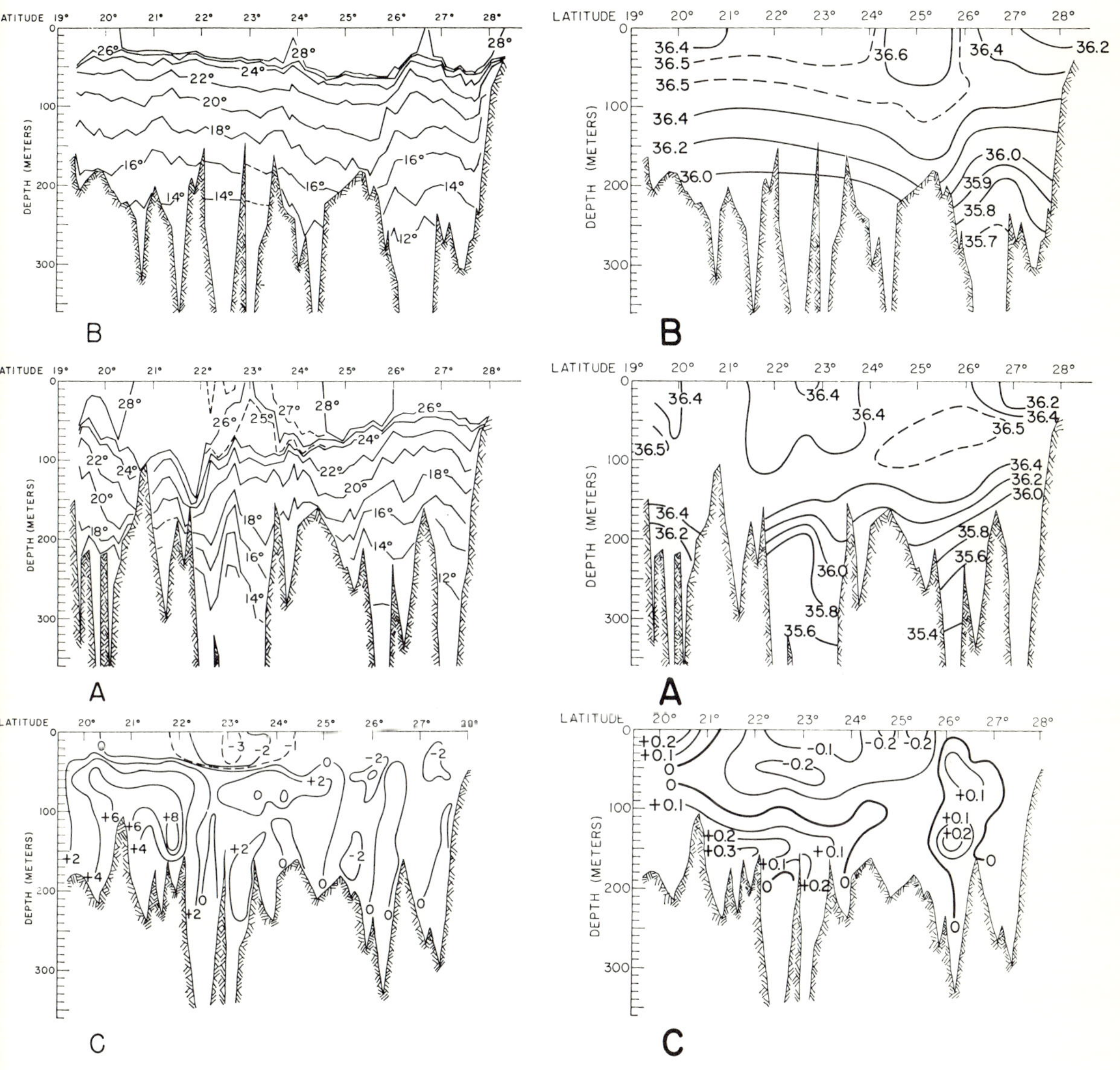

Figure 13-2. Vertical sections of temperature (°C) and its change associated with Hurricane Inez, 4-14 October 1966 from Alaminos Cruise 66-A-14 (B - before, A - after, C - change). From Franceschini and El-Sayed, 1968.

Figure 13-3. Vertical sections of salinity (per mil) and its change associated with Hurricane Inez, 4-14 October 1966 from Alaminos Cruise 66-A-14 (B - before, A - after, C - change). From Franceschini and El-Sayed, 1968.

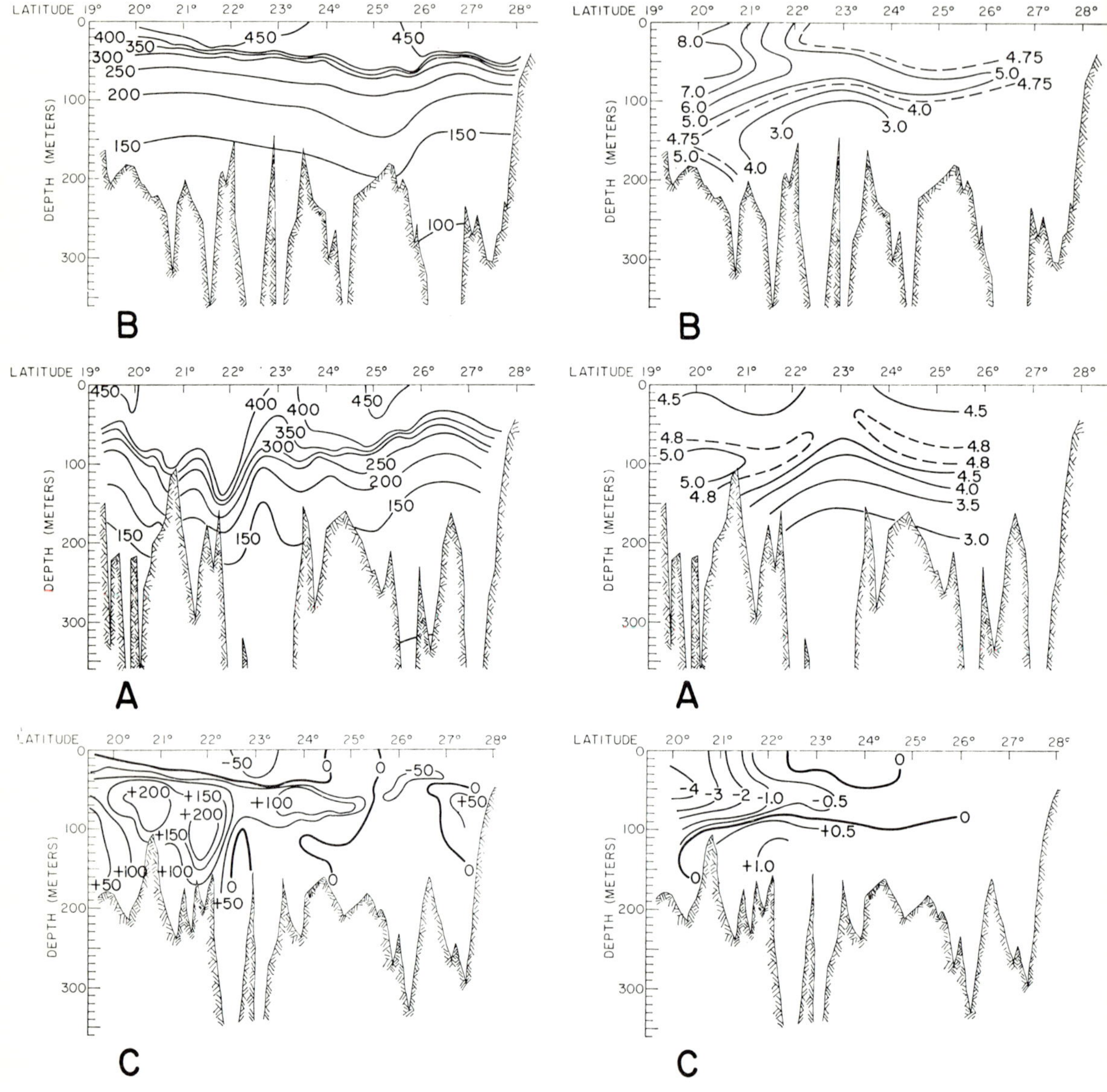

Figure 13-4. Vertical sections of thermosteric anomaly (cl per ton) and its change associated with Hurricane Inez, 4-14 October 1966 from Alaminos *Cruise 66-A-14 (*B *- before,* A *- after,* C *- change). From Franceschini and El-Sayed, 1968.*

Figure 13-5. Vertical sections of dissolved oxygen (ml per liter) and its change associated with Hurricane Inez, 4-14 October 1966 from Alaminos *Cruise 66-A-14 (*B *- before,* A *- after,* C *- change). From Franceschini and El-Sayed, 1968.*

The ship departed from Galveston October 3 and made a section of 54 BT stations and 10 hydrocasts almost parallel to the 200-m depth contour south of 28°N. On the October 11 return trip, 64 BT stations and 13 hydrocasts were taken. The location at which the storm crossed the first leg was reached about 40 hours after the eye of the hurricane had passed.

The vertical cross sections of temperature, salinity, thermosteric anomaly and oxygen before and after the hurricane and the change are plotted in Figures 13-2 to 13-5. The temperature section before the hurricane shows that the surface mixed layer bounded by the 28°C isotherm decreases almost uniformly in depth southward from about 60 m at 26°N to 30 m at 20°N. The temperature pattern changed drastically after the hurricane. The average temperature of the surface mixed layer dropped by about 2°C, and its thickness increased overall but conspicuously by more than 80 m south of the hurricane center, which was located near 23°N.

The double thermocline and the maximum temperature decrease of 3.6°C near 23°N indicate upwelling around the hurricane center. Judging from the nearly vertical intersect of the 26°C isotherm at the surface after the storm, the upwelling seems to be limited only to 10-20 nautical miles from the center. This is reasonable because the maximum wind speed radius was estimated to be about 15 miles and, thus, the surface divergence should occur only between the center and the radius of the maximum wind stress. Therefore, the temperature decrease in the mixed layer beyond the upwelling area was caused mainly by outward horizontal advection of the upwelling cold water near the center rather than by vertical mixing.

Beyond the upwelling area, the sinking of the upper water occurred because the tangential wind stress seemed to decrease from its maximum value faster than the inverse of the radius. This is verified by evidences that the vertical temperature profiles by BT before and after Hurricane Hilda (Leipper, 1967) also indicate the temperature increase from below the mixed layer throughout the depth of 100 m or more. The sinking outside the eye of the hurricane appears to be more intense to the south than to the north of the center from the change in depth of the mixed layer and from the temperature increase after the hurricane. This appears to be due to the effect of the shelf topography, which produces the convergence and divergence of the transport of the upper layer respectively on the left-and right-hand side of a hurricane track crossing normally to the bottom contours.

The salinity section before the hurricane indicates the presence of the maximum salinity layer with salinity above 36.5 per mil near the thermocline south of 26°N. The horizontal gradient of isohalines are almost similar to the isotherms. The salinity section after the hurricane indicates the centrifugal movement of the salinity maximum layer from the area of the hurricane center. Sinking motion is again indicated by a salinity increase below the maximum salinity layer after the hurricane.

Four main features of circulation due to a passing hurricane are most clearly demonstrated by vertical sections of thermosteric anomaly. These are (1) strong upwelling in the narrow area around 23°N indicated by sharp upward trend of the isosteric lines; (2) broad sinking outside the eye indicated by downward trends of pycnoclines and isosteric lines; (3) stronger sinking to the left than to the right of the track indicated by the trends of isosteric lines and change in thermosteric anomaly values; and (4) centrifugal advective motion from the hurricane center in the upper mixed layer.

Oxygen distributions are rather incomplete because they were determined at only three stations–22°N, 23°N and 26°N (or 25°N on the northbound leg) of each section. Further, the oxygen values at 20°N on the southbound leg (before the hurricane) were higher than usually observed there. However, the section after the hurricane indicates intense upwelling at 23°N and centrifugal advection of the oxygen maximum layer from the center.

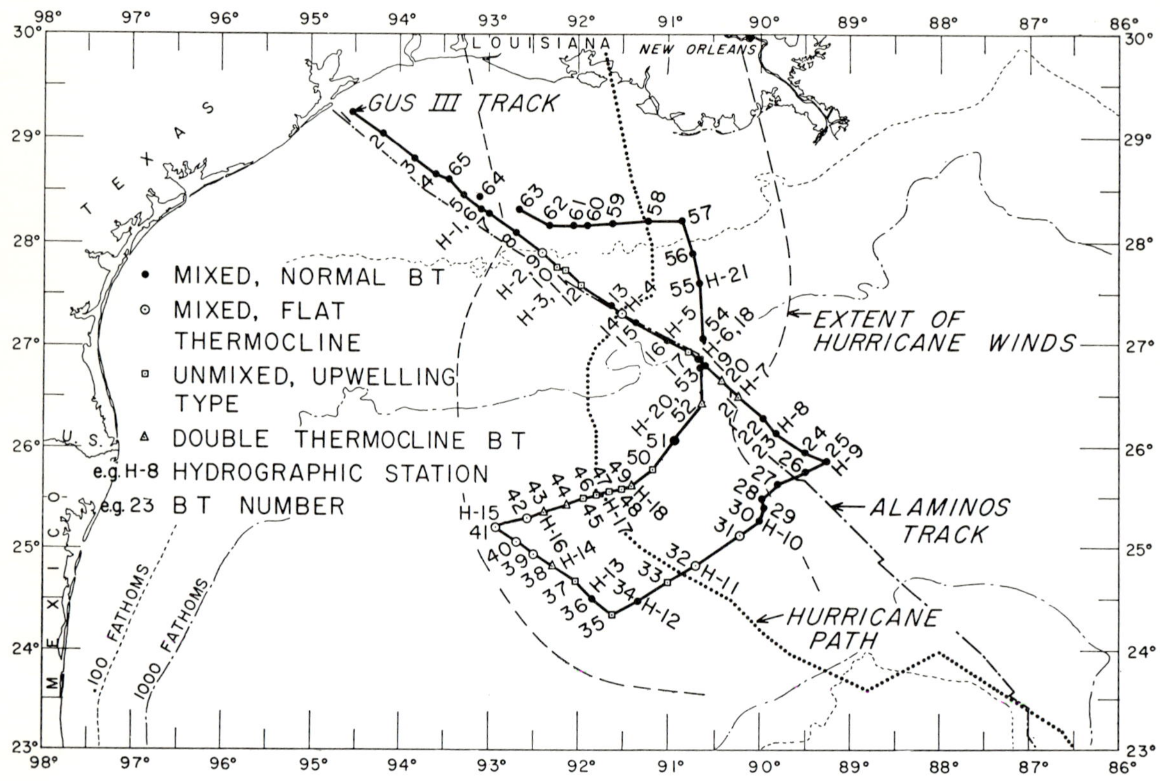

Figure 13-6. Location chart showing part of Hurricane Hilda, extent of hurricane winds, tracks of Alaminos *before Hilda and of* Gus III *afterward and locations of* Gus III *BT stations and hydrocasts. From Leipper, 1967.*

Ocean Conditions After Hurricane Hilda of 1964

The data obtained by Leipper (1967) on board the *Gus III* after Hurricane Hilda were more systematic than the previous two sets of data, since the cruise tracks included two deep-sea sections normal to the hurricane track as well as other sections in the slope and shelf areas (Figure 13-6). However, the data before the hurricane were only three BT stations at 26-27°N and 89-91°W. Therefore, Leipper had to use the hydrographic data taken at the same season but in 1965 as representative of the pre-hurricane condition.

Hurricane Hilda entered the Gulf on 30 September 1964 with wind less than 80 mph and intensified to the 150-mph stage when its center was about 250 miles offshore. The average moving speed was 6-8 kt and the width of the eye was about 35 miles in the northern Gulf. By October 4, the storm moved inland with winds diminishing less than 120 kt.

Directly observed changes in the Gulf before and after the hurricane are the data determined with the *Nomad* buoy and the sea-surface temperature data. The buoy was located at 25°N, 90°W and some 48 nautical miles from the hurricane path. The buoy records indicate that the winds reached 57 kt in the afternoon of October 2, but the sea-surface temperature began to drop early October 1 from 27.7 to 26.2°C in the morning of October 3 and then recovered to 27.8°C on October 4.

The sea-surface temperature data before Hilda were based on merchant ship measurements collected by the U.S. Weather Bureau between 24-30 September. The data after Hilda were based on the measurements collected by *Gus III* and other merchant ships and seem to be more accurate than the previous data. The sea-surface temperature difference before and after Hilda is shown in Figure 13-7. This indicates the general decrease of the order of 1°C in less affected parts of the Gulf but also a decrease of over 6°C in an area just to the left of the path. The area of greatest surface cooling occurred at the latitude where Hilda's average winds reached maximum but does not coincide with the minimum mixed layer. Therefore, the cold surface may be due to abvection of the water brought upwards by upwelling near the center but not due to the upwelling directly.

Three BT stations besides the three *Alaminos* stations in the central Gulf were obtained before Hilda in water of about 40 fathoms near the Gulf coast. Figure 13-8 compares these six BT data with the BT data taken near each station by the *Gus III* cruise after Hilda. The shallow water BT traces indicate a temperature decrease of 3-4°C in the surface mixed layer, an increase in mixed layer depth at two outward stations and a second mixed layer at St. 56, which is closer to the track of Hilda. The three deep BT stations before Hilda were made to the right of the storm track. The change of temperature profiles after Hilda shows common features with the shallow water BT data, that is, a decrease of the surface mixed layer by 2-3°C and an increase in mixed layer depth at outward stations. Only St. 16, which is closest to the track, shows a shoaler mixed layer after Hilda, suggesting direct upwelling. Comparing BT data before and after Hilda supports the concept of upwelling only near the center of the hurricane, sinking beyond that domain due to convergence in the wind-driven transport, and centrifigal transport and cooling and mixing of the upper water.

To compare the data after Hilda with an undisturbed state of the Gulf, another cruise, 65-A-11, was conducted with the *Alaminos* on August 10-24, 1965, since there was only a limited amount of data available before Hilda. It is assumed that the *Alaminos* data of 1965 represent

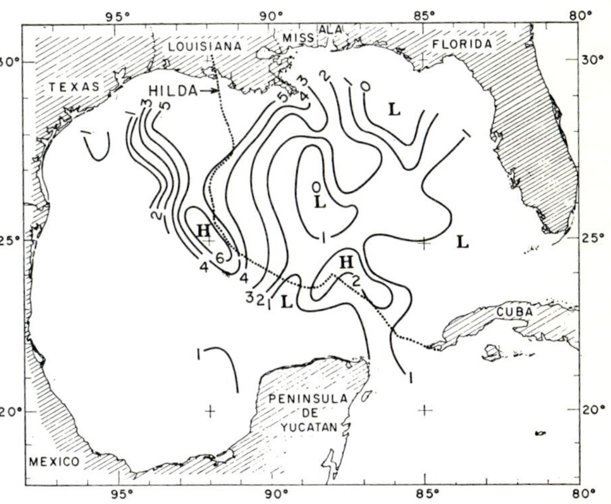

Figure 13-7. Sea-surface temperature decrease (°C) from 24-30 September to 1-13 October. From Leipper, 1967.

a typical late summer pattern of an undisturbed condition, since there had been no hurricanes or other marked weather phenomena prior to that cruise. The surface mixed layer in the undisturbed sections have almost uniform temperature of about 29°C to depths of 30-40 m, whereas in the sections after Hilda a surface mixed layer exists which is deeper on the average by 20-40 m, and the temperature of the mixed layer changes by more than 4°C from below 24°C to above 28°C in a horizontal direction. The coldest portion of each section after Hilda is to the west of the storm track. This does not agree with a concept held by some meteorologists (Jordan and Frank, 1964) that the surface cooling of the sea after a tropical cyclone is caused by stronger winds on the right-hand side of the track.

Density sections at *B*, *C*, and *D* after Hilda and at *C* and *D* in the undisturbed state in 1965 are shown in Figure 13-9. Since salinity was determined only at discrete depth intervals, the surface mixed layer thickness has smaller resolution than by BT profiles. However, the rise to the surface of isopycnals of sigma-t 24.5 clearly indicates the upwelling near the track of the hurricane. The horizontal extent of the mixed layer of sigma-t above 24.5 was larger at section *C*, where the winds were

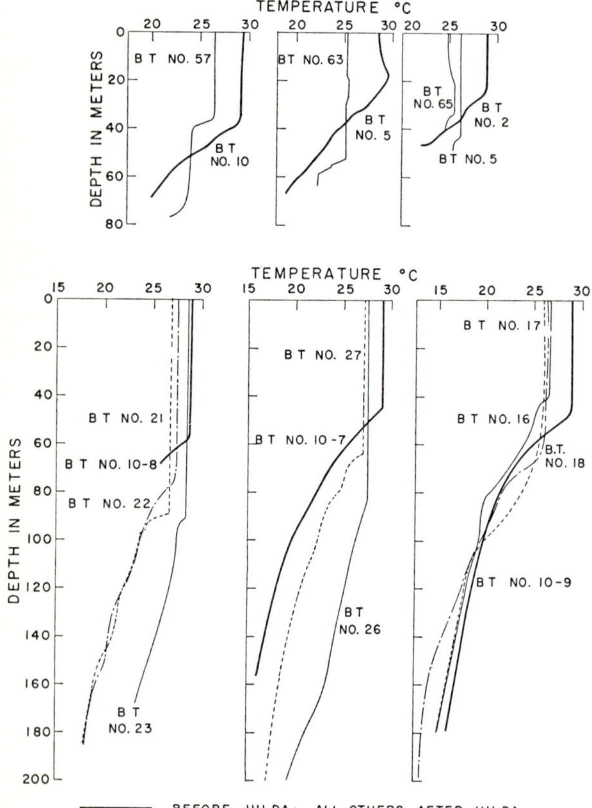

Figure 13-8. Comparisons of BTs obtained before and after Hilda at identical locations which may be seen in Figure 13-6. From Leipper, 1967.

at a maximum, than at sections *B* and *D* after Hilda. Comparison of the isopycnals after Hilda with those in the undisturbed state outside of the upwell region indicates that the sinking occurred after Hilda by 70-90 m below 50 m from the surface at about 100 nautical miles from the track. The centrifugal advection in the upper layer is inferred by pressence of lighter water with sigma-t less than 23.5 in the mixed layer more than 120 miles to the right of Hilda's track.

The heat lost from the ocean to Hilda could not be determined directly since there are few temperature data in the upper layer before the hurricane. Therefore, the estimation of heat loss is calculated on the basis of three assumptions: (1) there was no heat lost to the atmosphere from water colder than 25°C, since there is no significant surface mixed layer at these temperatures; (2) cooling to 28°C occurred with less than hurricane force winds and occurred over the entire Gulf; (3) all heat lost to the atmosphere during the hurricane took place from the water which had temperatures between 25°C and 28°C after the storm. Thus, the total heat loss can be computed from the difference of heat content of the volume of the 25-28°C water after the storm and heat content of 28°C of the same volume. The mean depth and mean temperature of this water volume are determined from the vertical temperature sections and the area is determined from the sea-surface temperature chart after Hilda. The computation yields separate estimates for the western and eastern parts of the track as 1.4×10^{18} cal and 9.4×10^{18} cal, respectively. The heat lost per cm^2 is about 4500 cal per 26 hours. This may be compared with the value of 3000 cal cm^{-2} day^{-1} for various hurricane situations calculated by Malkus (1962) based on the exchange formulas.

Theories on Circulation in a Homogeneous Ocean Induced by a Tropical Storm

Mathematical models of circulation induced by a tropical storm were developed by Hidaka and Akiba (1955) and by Ichiye (1955) for a homogeneous ocean. The former study was also based on a steady-state condition which is not realistic. The latter study includes nonstationary terms but is based on a system of linearized equations of motion.

Since Hidaka and Akiba assumed a steady state and linearized equations of motion, they obtained the solution in a closed form, and their results seem to be interesting compared with the more realistic later models. In their theory, circumsymmetrical wind stress and velocity distributions are assumed. The linearized equations of motion include the Coriolis terms, the pressure gradient in the radial direction and eddy viscosity

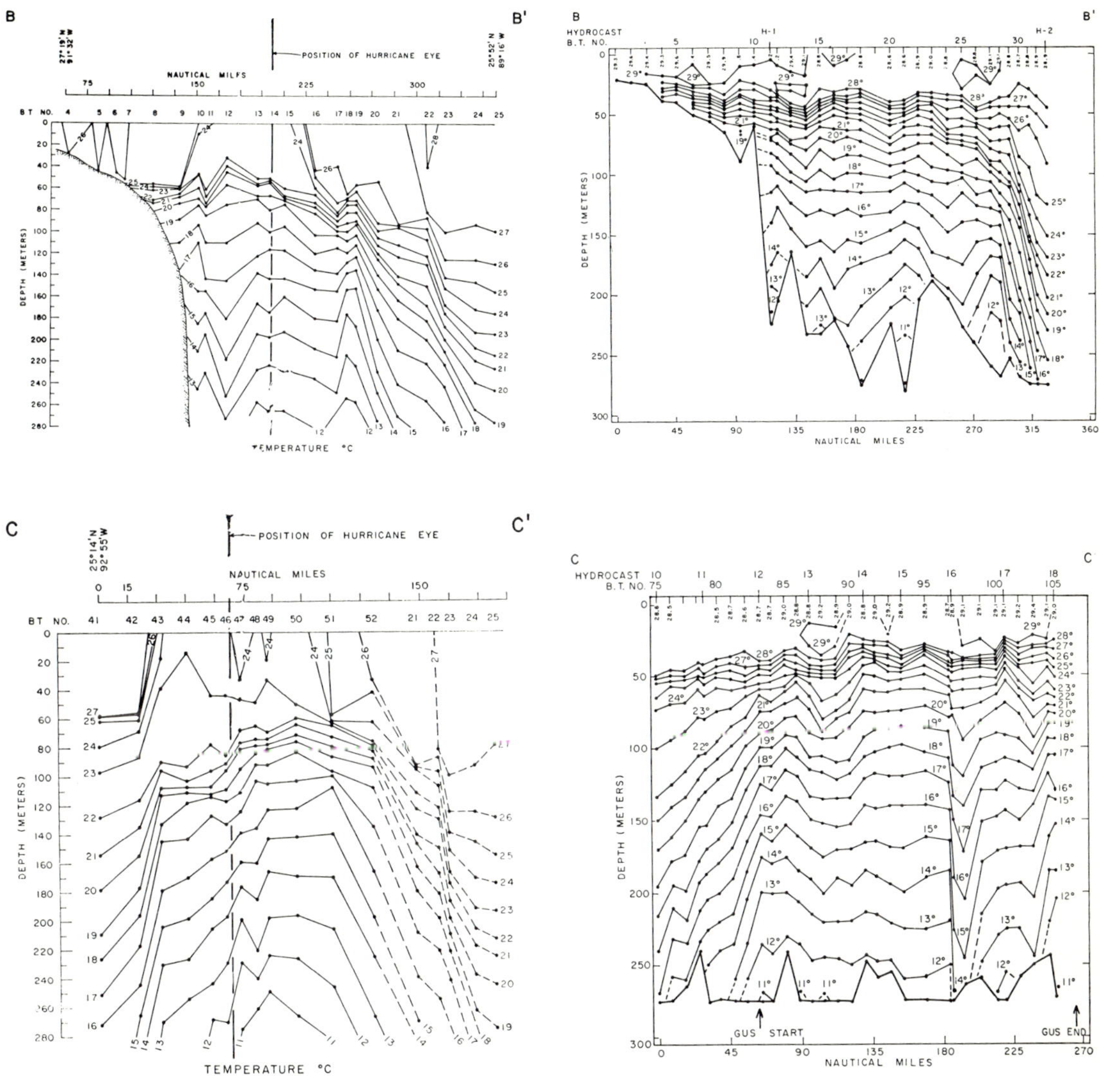

Figure 13-9. Depth of isotherms on sections crossing the path of Hilda, with the right and left figures representing the sections of the undisturbed state in August, 1965, and of the disturbed state after Hilda, respectively. From Leipper, 1967.

terms with different vertical and horizontal viscosity values. The surface and bottom boundary conditions are that the surface shear stresses equal the wind stresses and that there is no motion at the bottom, respectively.

The solution of the equations of motion satisfying the boundary conditions is expressed by the Fourier-Bessel integral as

$$u + iv = \int_0^\infty \lambda F(z, \lambda) J_1(r\lambda)\, d\lambda,$$

where u and v are velocity components in the radial and tangential directions, respectively, r and z are radial and vertical coordinates, J_1 is the Bessel function of the first order and λ is the integration variable. The vertical coordinate z is taken from the free surface and z and r are scaled, respectively, by characteristic lengths

$$D_v = \pi(2A_v/f)^{1/2} \text{ and } D_h = \pi(2A_h/f)^{1/2}$$

where f is the Coriolis coefficent, and A_h and A_v are horizontal and vertical eddy viscosity, respectively.

The function $F(z,\lambda)$ is defined by

$$F(z, \lambda) = \frac{T(\lambda) D_v}{\kappa A_v} \frac{\sinh \kappa (h - z)}{\cosh \kappa h}$$

$$+ \frac{2\pi^2 g G(\lambda)}{\kappa^2 f D_h} \left(1 - \frac{\cosh \kappa z}{\cosh \kappa h}\right)$$

where $\kappa^2 = \lambda^2 + 2\pi^2 i$ and h is a nondimensional depth.

The function $T(\lambda)$ and $G(\lambda)$ is the Fourier-Bessel transform of the wind stress $\tau(r)$ and the radial slope of the sea level $\zeta_r(r)$, defined, respectively, by

$$T(\lambda) = \int_0^\infty r\tau(r) J_1(r\lambda)\, dr$$

and

$$G(\lambda) = \int_0^\infty r\zeta_r(r) J_1(r\lambda)\, dr$$

where the wind stress $\tau(r)$ is a complex function $\tau(r) = \tau_r + i\tau_\theta$ consisting of the radial and tangential components τ_r and τ_θ.

The wind stress $\tau(r)$ is given as a known function, but the sea level slope $\zeta_r(r)$ or its Fourier-Bessel transform $G(\lambda)$ must be determined in terms of $T(r)$ by use of the vertically integrated equation of continuity

$$\int_0^h u dz = 0.$$

When the depth h is much larger than unity, $G(\lambda)$ is practically zero, indicating that the effect of the sea-surface slope is negligible. The numerical examples of the pure tangential wind and the wind blowing inward with an angle of 45° were worked out for the cyclonic wind stresses with $D_v = 100$ m, $D_h = 100$ km and the amplitude of the wind stress of 1 cm^2/sec^2. The wind stress increases with r to the distance r_o and decreases as r^{-1} beyond r_o.

The upwelling occurs mainly within a distance r_o from the center, though the authors considered this distance as D_h, since they used the same numerical value of r_o as D_h. The sinking occurs gradually beyond $r = 2D_h$, although the radial transport due to the assumed wind stress is non-divergent for $r \geqslant r_o$ in case of no horizontal eddy viscosity. The vertical current reaches maximum value at about $z = 1.6\, D_v$, whereas the horizontal current almost vanishes at $z = 0.5 D_v$. The upwelling domain reaches farther, but the maximum vertical velocity is smaller for the inward blowing wind than for the parallel wind. The maximum upward velocity is of the order of 1 to 2 x 10^{-3} cm/sec.

Ichiye (1955) determined the horizontal transport by transient wind stresses for the homogeneous ocean by including the nonstationary terms. He also obtained the vertical velocity by including the nonlinear inertia terms as a perturbation.

The horizontal transport vector **U** is given by

$$\mathbf{U} = f^{-1} \int_{-\infty}^{t} (\partial\tau/\partial t + \tau \times f\mathbf{k}) \sin f(t-\lambda) d\lambda \qquad (13\text{-}1)$$

where τ is the wind stress vector and **k** is the vertical unit vector. The time argument in τ is expressed by λ. Then the vertical velocity at the deep layer w is given by

$$(f + D_v^{-1}\phi)w = f \int_{-\infty}^{t} \Big\{ \mathrm{Curl}_z\tau \sin f(t-\lambda) + \mathrm{Div}\,\tau \cos f(t-\lambda) \Big\} d\lambda - D_v^{-1}\,\mathbf{U}\cdot \mathrm{Grad}\,\phi \qquad (13\text{-}2)$$

where ϕ is defined by

$$\phi = \int_{-\infty}^{t} \Big\{ \mathrm{Curl}_z\tau \cos f(t-\lambda) - \mathrm{Div}\,\tau \sin f(t-\lambda) \Big\} d\lambda, \qquad (13\text{-}3)$$

and $Curl_z$, *Div* and *Grad* indicate the z component of curl, divergence and gradient of a vector, respectively.

The effects of nonlinear inertia terms on the vertical velocity are represente by the terms with D_V^{-1} on both sides of equation (13-2). The corrections due to the nonlinear terms are expressed by the quantity

$$T_o\,(Lf^2D_v)^{-1} \qquad (13\text{-}4)$$

where T_o is the maximum wind stress and L is a horizontal dimension of the storm. The expression (13-4) can be obtained by comparing the first with the second term on both sides of equation (13-2) with approximations of *Curl, Div* and *Grad* by L^{-1} and of integration of the sine and cosine terms with time by f^{-1} in equations (13-1) and (13-3). When the characteristic values of T_o, L, D_V and f are taken as 10^2 cm^2/sec^2, 20 km, 50 m and 10^{-4} sec^{-1}, the ratio (13-4) becomes 1/10. Therefore, the effects of the nonlinear terms on the vertical velocity seem to be small even for an extremely intense storm. For working on a numerical example of equation (13-2), the wind stress magnitude which varies with a distance r from the center as $|\tau| = 2T_o(r/r_o)\{1 + (r/r_o)^2\}^{-1}$. This distribution indicates that the wind stress increases approximately in proportion to the distance r for $r<r_o$ and then it decreases as r^{-1} for $r \gg r_o$. The wind stress vector is assumed to make an inward angle with concentric circles. Also, it is assumed that the wind system moves with a constant speed c to a positive x-direction. Then the vertical velocity from equation (13-2) without the nonlinear correction terms is expressed by

$$w = \int_{-\infty}^{t} A(r) \sin\Big\{ f(t-\lambda) + \psi(r) \Big\} d\lambda \qquad (13\text{-}5)$$

where $A(r)$ and $\psi(r)$ are the amplitude and phase depending on $\mathrm{Curl}_z\tau$ and Div τ and r is expressed in terms of the Cartesian coordinates x and y as $r^2 = (x\text{-}c\lambda)^2 + y^2$. The right-hand side of equation (13-5) indicates that the vertical velocity changes with an inertial period after passage of the main part of the storm.

Change of the vertical velocity with time is computed from equation (13-2) at a point passed by a cyclonic wind center moving at a speed of 40 km/hour for an example of $T_o = 10^2$ cm^2/sec^2 and r_o = 40 km. When the storm approaches, a slight sinking motion starts 1 hour before the center, reaching maximum value of 0.03 cm/sec. It changes to upwelling about 1.7 hours after the center, with the upward motion reaching maximum of 0.2 cm/sec 8 hours after the center. About 13 hours after the center, the downward motion begins, reaching 10^{-3} cm/sec about 18 hours after the center. These upward and downward motions with an inertia period continue with diminishing amplitudes. When the vertical eddy viscosity is taken as 10^3 cm^2/sec, corrections on the vertical velocity due to the inertia terms are only about 10-15% of the linear solution.

Response of a Stratified Ocean to a Stationary Tropical Storm

The two previous models deviate from the real situation, since the surface slope caused by the wind stress is ignored under an assumption that the ocean depth is large compared with the Ekman layer thickness. In these models, the Ekman transport and the thickness of the Ekman layer are variable only with the wind stresses. However, in the real ocean the downward transfer of the momentum due to wind stresses may be partially inhibited by stratification of the ocean, and the thickness of the Ekman layer and the Ekman transport are dependent on the vertical density gradient as well as the wind stresses. The divergence of the Ekman transport is compensated by change of the surface elevation with time as well as vertical currents at the bottom of the Ekman layer. Therefore, it is necessary to include the effect of stratification to understand processes in the real ocean responsing to the storm.

The vertical transfer of the momentum in the ocean due to a storm can be determined by solving a system of equations of motion including inertia terms with vertical velocity and equations of conservation of mass, temperature and salinity. The system is almost hopelessly complicated. Therefore, O'Brien and Reid (1967) adopted an approximation of a two-layer ocean of uniform densities where the lower layer is motionless and there is no transfer of heat and salt between the two layers. In the subsequent model of O'Brien (1967), density of the upper layer is radially variable and transfer of heat and salt between the two layers is considered, although again there is no motion in the lower layer .

In the model of O'Brien and Reid (1967), a storm is assumed to have axially symmetric pressure and wind stress distributions which are variable with time, though the location of the storm is stationary. The assumptions of vertically uniform horizontal velocity in the upper layer, hydrostatic pressure and no motion in the lower layer lead to the pressure gradient in the upper layer given by $g\epsilon\ \nabla h^1$ where

$$\epsilon = (\rho_2 - \rho_1)/\rho_2, \tag{13-6}$$

ρ is density, g is the gravity constant, h_1 is the thickness of the upper layer, and indices 1 and 2 refer to the upper and lower layer, respectively. The Coriolis coefficient is assumed to be constant. The nonlinear inertia terms are included. A frictional force in the upper layer consists of the interior stress which is proportional to the square of the mean velocity and parallel to the velocity. Then equations of motion in the radial and tangential directions and of continuity in the upper layer are given by

$$\partial u/\partial t + u\partial u/\partial r - (f + v/r)v + g\epsilon\partial h/\partial r$$
$$= (\tau_r - \rho Kqu)(\rho h)^{-1} \tag{13-7}$$

$$\partial v/\partial t + u\partial v/\partial r + (f + v/r)u$$
$$= (\tau_\theta - \rho Kqv)(\rho h)^{-1} \tag{13-8}$$

and

$$\partial h/\partial t + r^{-1}\ \partial(uhr)/\partial r = 0 \tag{13-9}$$

where the index 1 is dropped, τ_r and τ_θ are the radial and tangential wind stress, respectively, q is the magnitude of velocity and K is the frictional coefficient.

The distribution of the wind stress magnitude is assumed to be proportional to

$$|\tau| \sim (1 + 10r^{-2})^{-1} \tag{13-10}$$

where r has the dimension miles. The vertical velocity w at the interface is given by

$$w = -(\partial h/\partial t + u\partial h/\partial r). \tag{13-11}$$

The solution of equations (13-7) and (13-8) is obtained by numerical integration of the characteristic difference equations derived from the system. The wind stress is assumed to start suddenly and remain constant. The numerical computation was carried out from the initial time up to 48 hours with time step of 18 minutes and radial grid spacing of one mile.

As numerical examples, ϵ = 2 x 10^{-3}, the radius of the maximum wind stress equaling 43 dynes/cm2 is 20 miles and the inflow angle is taken as 35 degrees. For numerical calculation, the wind stress follows equation (13-10) up to $r = 100$ miles and then decreases linearly to zero at $r = 200$ miles. In the region $r > 200$ miles, there is no wind stress and the disturbance propagates outward with an internal wave speed $(\epsilon gh)^{1/2}$.

The major features of the solution are (1) the radial velocity u is negative (inward) for all r up to 3 hours, begins to change sign near the center at about 3 hours and becomes positive for all r after about 6 hours, with the radius of maximum value shifting outward from about 40 km at 7.5 hours to 70 km at 26 hours to 120 km at 46.5 hours; (2) the tangential velocity is positive (cyclonic) for all r all the time but increases rapdily during the first 12 hours and becomes rather stationary after 24 hours, with the radius of maximum value being about 30 km; (3) a series of radial dispersive waves form near 30 km and propagate outward; (4) the thickness of the upper layer h increases (downwelling) near the center until about 6 hours but then decreases (upwelling) rapidly within $r < 80$ km from the initial value of 100 m to the minimum of about 25 m during the next 18 hours. The minimum thickness occurs not at the center but about 10-15 km, with maximum upwelling at about 30 km. Beyond 180 km, the thickness increases by about 10-15 m after 24 hours (Figure 13-10).

In the second model by O'Brien (1967), the radial change of density in the upper layer is considered. Therefore, the pressure gradient term on the left-hand side of equation (13-7) includes the term $\frac{1}{2}gh\partial\epsilon/\partial r$. The exchange of heat and salt at the interface changes the equation of continuity (13-9) into

$$\partial h/\partial t + r^{-1}\ \partial(hur)/\partial r = -K_s q\epsilon \qquad (13\text{-}12)$$

and requires the equation of heat transfer as

$$\partial\epsilon/\partial t + u\partial\epsilon/\partial r = -K_s q\epsilon/h \qquad (13\text{-}13)$$

where K_S is a constant exchange coefficient.

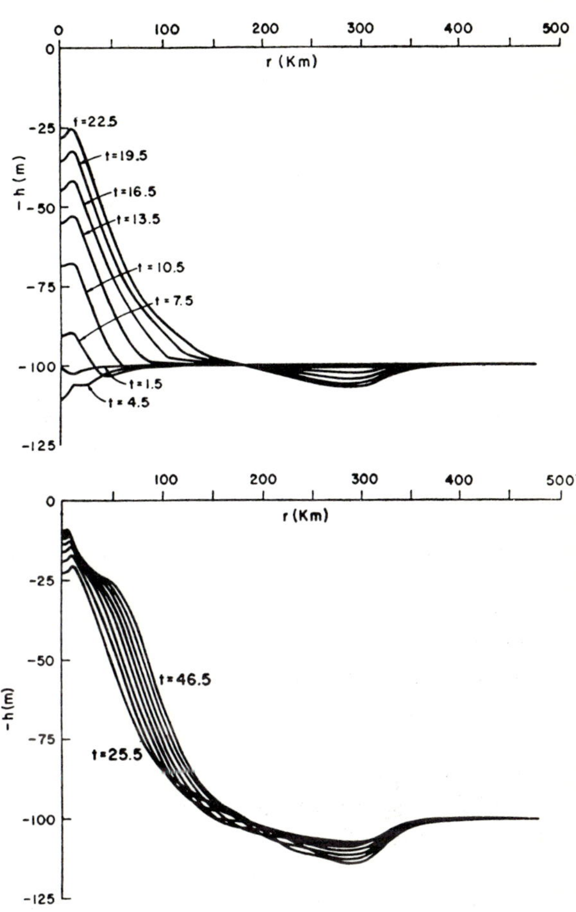

Figure 13-10. Depth of the upper layer h *as a function of* r *and* t *(in hours), with the top and bottom diagrams representing 0-24 hours and 24-48 hours, respectively. From O'Brien and Reid, 1967.*

The numerical integration of the system of equations (13-7), (13-8), (13-12) and (13-13) was carried out for similar physical constants by taking $K_S = K$. The results indicate that the horizontal velocity and the thickness of the upper layer are almost the same as the first model. Therefore, the effect of density change of the upper layer on the response of a two-layer ocean is negligible. The boundary conditions for ϵ are that ϵ is constant at the center and for $r > 200$ miles. The values of ϵ decrease rapidly from the center to the minimum at about $r = 30$ km and increase beyond that. The decrease of ϵ is rapid during the first 24 hours. The vertical section of computed density after 24 hours is similar to a vertical density section across Hurricane Hilda observed by Leipper (1967), although the computed density in the upper layer is uniform with depth. Since the velocity distributions are not substantially influenced by the computed horizontal density gradient, the actual density profile can be determined from the equation of mass conservation in the upper layer after the velocity is determined from the first model.

Reid and Gilbert (1970) generalized two models of O'Brien and Reid (1967) and O'Brien (1967) by assuming a continuously stratified sea, which is more realistic. In this model, the fluid velocity and density in the upper region depends on depth (z) as well as time and horizontal coordinates. Some additional assumptions and conditions besides those for the previous models are adopted: (1) the hydrostatic equilibrium for pressure is assumed; (2) turbulent exchange of momentum, heat and salt are included in gradient form with constant but horizontally and vertically different turbulent exchange coefficients which are regarded as identical for momentum, heat and salt; (3) at the base of the upper region, the pressure and density are constant and the horizontal component of velocity vanishes; (4) at a suitably large radial distance (R_∞), the horizontal velocity and the radial gradient of density vanish; (5) the vertical velocity vanishes at the sea surface; (6) at the surface, the density is prescribed and the shear stress is equal to the wind stress.

Then equations of motion in radial and tangential directions are expressed with a matrix form as

$$\frac{\partial}{\partial t}\begin{pmatrix}u\\v\end{pmatrix} = -\Gamma\begin{pmatrix}u\\v\end{pmatrix} + \left(f+\frac{v}{r}\right)\begin{pmatrix}v\\-u\end{pmatrix} - \begin{pmatrix}\partial\pi/\partial r\\o\end{pmatrix} + \Lambda\begin{pmatrix}u\\v\end{pmatrix} - \frac{A_h}{r^2}\begin{pmatrix}u\\v\end{pmatrix} \qquad (13\text{-}14)$$

instead of equations (13-7) and (13-8). The equation of heat transfer is given by

$$\partial\epsilon/\partial t = -\Gamma(\epsilon) + \Lambda(\epsilon) \qquad (13\text{-}15)$$

instead of equation (13-13). In these equations, Γ and Λ represent advection and diffusion operators and are defined by

$$\Gamma(\phi) = (1/r)(\partial/\partial r)(ru\phi) + (\partial/\partial z)(w\phi)$$

and

$$\Lambda(\phi) = A_h(1/r)(\partial/\partial r)[r(\partial\phi/\partial r)] + A_v(\partial^2\phi/\partial z^2)$$

The pressure π and vertical velocity w incorporating hydrostatic relation and (13-11), respectively, are expressed by

$$\pi = g\int_o^z \epsilon\, dz \qquad (13\text{-}16)$$

and

$$w = (1/r)(\partial/\partial r)\left\{r\int_z^h u\,dz\right\} \qquad (13\text{-}17)$$

where $z = 0$ or h corresponds to the base of the upper region or the sea surface, respectively.

Equations (13-14) and (13-15) are prognostic equations. The boundary conditions are represented by assumptions 3 to 6 with an additional condition similar to assumption 5 at $r = 0$. Equations (13-16) and (13-17) are diagnostic equations.

Numerical computation is carried out by using a staggered array of variables at each time level with "leap-frog" type operation in time. A grid point is designated with space (r and z) indices i and j and with time index l. The prognostic variables (u, v, ϵ) or the diagnostic variables (w, π) are evaluated at grid points of odd or even values of ($i + j + l$), respectively. The grid selected for computation employs $\Delta z = 3$ m and $\Delta r = 4$ km, yielding an associated value of $\Delta t = 10$ minutes for stable computations.

In a numerical example, the wind stress field similar to the previous two models was suddenly applied and sustained for 36 hours. Other constants include h = 120 m, R_∞ = 440 km, A_h = 1.33 x 10^8 cm^{2}sec^{-1}, A_v = 74 cm^{2}sec^{-1} and f = 6.2 x 10^{-5}sec^{-1}.

The main features of the results are (1) the radial flow is initially inward but reverses after an hour or two, producing upwelling in the core of the vortex; (2) at 18 hours after onset of the wind stress, the upwelling is confined to the inner 230 km, with two intense zones (one near the core and one centered at 200 km) reaching 8 x 10^6 m^3/sec. At 30 hours, two convection cells are developed with upwelling near the core and near $r = R_\infty$ and downwelling near r = 220 km; (3) at and near the surface, the tangential velocity is cyclonic with a maximum surface speed exceeding 3.5 m/sec; (4) in a subsurface region, anticyclonic flow occurs in contrast with the previous two models. At 18 hours, the maximum tangential velocity (anticyclonic) exceeding 0.6 m/sec is located at about r = 300 km and z = 20 m. At 30 hours, the maximum flow is located at r = 140 km and z = 30 m and is weakened to the order of 0.1 m/sec; (5) a maximum vertical-radial transport of about 10.8 x 10^6m^3/sec occurs at 14 hours. A maximum cyclonic tangential transport of 8.6 x 10^6 m^3/sec occurs at about 6 hours. A maximum anticyclonic tangential transport of 4 x 10^6m^3/sec is developed at 22 hours.

Relations for the rates of change of the kinetic and potential energy are derived from (13-14), (13-15), (13-16) and (13-17). The integrated values of the kinetic energy E_u and E_v, respectively, are due to the radial and tangential velocities. The kinetic energy and potential energy relative to the initial state E_p are defined by

$$E_u = \left\{ \tfrac{1}{2} u^2 \right\} ; E_v = \left\{ \tfrac{1}{2} v^2 \right\}, \qquad (13\text{-}18)$$

and

$$E_p = \left\{ gz \left[\epsilon_o(z) - \epsilon \right] \right\} \qquad (13\text{-}19)$$

where $\epsilon_o(z)$ is the initial distribution of ϵ and depends on z only. The braces in (13-18) and (13-19) imply an integral over the whole upper region from z = 0 to h and from r = 0 to R_∞.

The energy calculations associated with the 36-hour run indicate (1) strong oscillations of the inertial period occur for both E_u and E_v with maximum and minimum of the total kinetic energy of 8.6 and 4 x 10^{22} ergs at 11 and 26 hours, respectively; (2) the peak values of E_u at 14 hours and E_v at 6 hours are about 6 and 5 x 10^{22} ergs, respectively, and the ratio of these peak values is comparable to the ratio of the maximum convective transport to the maximum total cyclonic transport; (3) the value of E_p increases steadily to more than 3 x 10^{22} ergs at 36 hours, always remaining smaller than the total kinetic energy.

Response of a Two-Layer Ocean to a Moving Storm

The works of O'Brien and Reid (1967) and O'Brien (1967) are unrealistic as solutions of the problem of the ocean's response to a hurricane because their model hurricane stays at one place. If their results are valid for a case that a hurricane moves out of a region considered, the response of the ocean should almost vanish about 24 hours after the storm passes, with the depth of the mixed layer being restored to the pre-hurricane undisturbed level. However, the observations made by Leipper (1967) and Franceschini and El-Sayed (1968) still indicate the disturbed state in the depth of the mixed layer more than two days after passage of the hurricanes. In order to explain these facts, it was necessary to apply more realistic models.

The quasi-steady state which can be reached after 24 hours in the theories of O'Brien and Reid (1967) and O'Brien (1967) represents fairly well the observed distribution of the disturbed mixed layer depth. This suggests that their two-layer ocean model with the motionless lower layer simulates the actual situation during a hurricane passage except for the temporal sequences. It is necessary to include the translational movement of the hurricane system in order to improve their models.

The vectorial form and Cartesian coordinates are more convenient than the polar coordinates used in equations (13-7) to (13-9) to treat the effects of the propagation of a hurricane. The vectorial form of the equation of motion in the upper layer is given by

$$\partial \mathbf{u}/\partial t + \mathbf{u}\cdot\nabla\mathbf{u} + \mathbf{u}\times f\mathbf{k} = -g\epsilon\nabla h + \tau/h \tag{13-20}$$

where **u** and τ are the vectors of the horizontal velocity and wind stress, respectively, and h, **k** and ϵ are already defined in the two preceding sections. Equation of continuity in the vectorial form is given by

$$\partial h/\partial t + \nabla\cdot(h\mathbf{u}) = 0. \tag{13-21}$$

As before, it is assumed that the storm system moves with a constant speed c in the positive x-direction. When the moving coordinate $X = \mathfrak{X}-ct$ is introduced into equations (13-20) and (13-21) the time derivatives of these equations are changed into the X-derivatives and the system of the equations represents a steady state problem.

Equations (13-20) and (13-21) are scaled by use of characteristic horizontal distance L, thickness of the upper layer h_o and windstress T_o. The nondimensional equations of motion and continuity are

$$u_o\,\partial u_o/\partial X + v\partial u_o/\partial y - R_o^{-1}v = -\partial h/\partial X + m\tau_{\mathfrak{X}}, \tag{13-22}$$

$$u_o\partial v/\partial X + v\partial v/\partial y + R_o^{-1}u = -\partial h/\partial y + m\tau_y \tag{13-23}$$

and

$$\partial(uh)/\partial X + \partial(vh)/\partial y = 0. \tag{13-24}$$

The velocities are scaled by $(g\epsilon h_o)^{1/2}$ The velocity component u_o is defined by $u_o = u - \gamma$ where $\gamma = c(g\epsilon h_o)^{-1/2}$ is the ratio of propagation speed of the storm to the velocity of the internal gravity wave and may be called hydraulic Mach number as discussed below. The constants R_o and m are defined, respectively, by

$$R_o = (g\epsilon h_o)^{1/2}/(fL) \tag{13-25}$$

and

$$m = T_oL(g\epsilon h_o^2)^{-1} \tag{13-26}$$

The system of equations (13-22) to (13-24) are quasi-linear partial differential equations of the first order in the unknown functions u_o, v and h for two independent variables X and y. The system can be transformed into three ordinary differential equations along three different curves (characteristic curves) according to a general theory of partial differential equations (Courant, 1962).

The directions of the characteristic curves for the system of equations (13-22) to (13-24) can be obtained as roots of the secular equation derived from the matrices which are formed by representing equations (13-22) to (13-24) with the vectorial form. Since equations (13-22) to (13-24) are linear about derivatives of u_o, v and h about X and y, these can be expressed by

$$P\,\partial\alpha/\partial X + Q\partial\alpha/\partial y = \beta \tag{13-27}$$

where α and β are vectors defined by $\alpha = (u, v, h)$ and $\beta = (R_o^{-1}v + m\tau_x, -R_o^{-1}u + m\tau_y, 0)$ and P and Q are matrices defined by

$$P = \begin{pmatrix} u_o & o & 1 \\ o & u_o & o \\ h & o & u_o \end{pmatrix} \quad Q = \begin{pmatrix} v & o & o \\ o & v & 1 \\ o & h & v \end{pmatrix}$$

The directions dX/dy of the characteristic curves are determined from roots of the secular equation

$$|P - (dX/dy)Q| = 0. \tag{13-28}$$

One root of equation (13-28) represents the streamline as a characteristic curve. The other two roots are roots of the equation

$$(v^2 - h)(dX/dy)^2 - 2u_O v(dX/dy) + u_O^2 - h = 0 \qquad (13\text{-}29)$$

and represent the conjugate characteristic curves. The system of partial differential equation (13-27) becomes hyperbolic or elliptic according to real or complex roots of equation (13-29), respectively. When roots of equation (13-28) are real, the conjugate characteristic curves can be determined on the real X-y plane. Then solutions of equation (13-27) can be obtained as solutions of equation dinary differential equations on each characteristic curve when the initial values of the unknown functions are given on any curve transecting the families of the conjugate characteristic curves.

$$u_O^2 + v^2 > h. \qquad (13\text{-}30)$$

The following numerical values are used as the representative values for a hurricane and a two-layer ocean

$L = 2 \times 10^7\,(cm)$,

$T_O = 40(cm^2/\sec^2)$,

$\epsilon = 2 \times 10^{-3}$,

$h_O = 2 \times 10^4\,(\text{cm})$.

Then the scale velocity $(g\epsilon h_O)^{1/2}$ becomes 2 x 10^2 cm/sec and the nondimensional numbers R_O and m become 10^{-1} and 1, respectively.

If the wind stresses almost vanish beyond the distance L from the center, the magnitude of the velocity is negligibly small compared with unity for $r>1$. Therefore, the condition for hyperbolicity can be simplified as $\gamma^2>h$. Since h is nearly unity for $r>1$, this condition may be satisfied for a storm moving with a speed several times the internal wave speed. The number γ almost corresponds to Mach number of the compressible fluid (Courant, 1962) and justifies a name "hydraulic Mach number." On the other hand, even if γ is less than unity, the system of equations (13-22) to (13-24) may not always be elliptic but of mixed type because condition (13-30) can be satisfied in a part of the X-y plane, particularly when the wind stresses do not vanish completely beyond some radius from the center. This is because the equations for the storm response problem contain the forcing terms due to the wind stresses in contrast with those for the compressible fluid.

Solutions of Linearized Equations for Response of a Two-Layer Ocean to a Moving Storm

Behaviors of solutions of equation (13-27) become different outside the storm wind domain according to the characteristic type of the equation. The wind stresses in a hurricane become almost negligible compared with their maximum value beyond a certain distance from the center. Therefore, the equations of motions can be linearized outside a domain of the hurricane force wind. There are three major reasons why these linearized equations are useful for the response problem: first, linearization becomes valid outside the storm force wind where the characteristics type is most crucial for the features of the response; second, the observed data to be checked for the theoretical models were collected long after the passage of the storms; and third, solutions of these linearized equations demonstrate clearly differences between solutions of equations of the elliptic and hyperbolic types.

Since the average speed of a hurricane in the western North Atlantic Ocean is usually above 7 knots (Dunn and Miller, 1964), the values of γ are larger than unity when the depth of the mixed layer is about 200 m. Therefore, equation (13-27) is usually hyperbolic for the region outside the storm wind area. On the other hand, for the barotropic mode in which the ocean density is considered as vertically uniform to the bottom, the speed of long waves in the homogeneous ocean should be used for calculating γ in place of that of the internal waves. Therefore, equation (13-27) always becomes elliptic for the barotropic mode, since the speed of long waves reaches 400 knots for a depth of 4000 m, yielding γ far less than unity. Kajiura (1956) discussed the sea level elevation and the current for the barotropic mode caused by a moving storm by using elliptic type equations.

In case that. . .$\gamma \gg |u|$ and $|v|$, equations (13-22) and (13-23) become

$$\gamma \partial \mathbf{u}/\partial X + R_o^{-1}\ \mathbf{u} \times \mathbf{k} = -\nabla h + m\tau, \qquad (13\text{-}31)$$

where the x-component of the *del* operator is understood as the derivative about X. Equation of continuity (13-24) becomes

$$-\gamma \partial h/\partial X + h\nabla \cdot \mathbf{u} = 0. \qquad (13\text{-}32)$$

Equation (13-31) is linear in the unknown functions u, v and h, but equation (13-32) is not. To linearize (13-32), h of the second term may be replaced with its undisturbed value of unity.

When curl and divergence operations are applied to equation (13-31), two linear equations about the vorticity (vertical component) and divergence of **u** are obtained. Eliminating the vorticity from these equations and substituting the divergence for (13-32), we can eliminate u and v and obtain the equation for h. This equation can be integrated with X from positive infinity, where the disturbance of thermocline depth should be nonexisting. The equation for h is thus expressed by

$$-(\gamma^2 \partial^2/\partial X^2 + R_o^{-2}) \ln h + \nabla^2 h = m \left\{ \nabla\tau + (\gamma R_o^{-1}) \int_X^\infty curl\tau dX \right\} \qquad (13\text{-}33)$$

where ln indicates the natural logarithm. When h is close to unity, lnh approximately equals $h-1$ or H, deviation of the thickness of the upper layer from the undisturbed value. Then the left-hand side of equation (13-33) becomes

$$(1-\gamma^2)\, \partial^2 H/\partial X^2 + \partial^2 H/\partial y^2 - R_o^{-2} H. \qquad (13\text{-}34)$$

The linearized equation of (13-33) which is obtained by replacing the left-hand side with expression (13-34) can be solved analytically by using Green functions (Ichiye, 1970). For clarifying the physical meaning, it is assumed that the wind stresses vanish for $r > 1$. When $\gamma > 1$, the equation for H becomes hyperbolic. The solution indicates that the disturbance of the thermocline is limited to a wedge behind the storm area bounded by two straight lines which are tangent to the storm area, making angles $\pm\tan^{-1} (\gamma^2-1)^{-1/2}$ with the X-axis. Therefore, for $\gamma > 1$, the disturbance of the thermocline depth remains non zero after the passage of the storm. The wedge area affected by the storm becomes narrower as the value γ increases. When $\gamma < 1$, the linearized equation of (13-33) becomes elliptic. The Green function is expressed by the modified Bessel function of the zeroth order. The solution indicates that the disturbance of the thermocline decreases from the storm area exponentially.

The solution at (X, y) of the linearized equation of (13-33) can be expressed explicitly by

$$H(X, y) = \iint R(\xi,\eta)\ Gr(X,y;\ \xi,\ \eta)\ d\xi\, d\eta$$

where

$Gr(X,y\ ;\ \xi,\eta)$ is a Green function, and ξ and η are integration variables. The function R (ξ, η) is a forcing function which is obtained by dividing the right-hand side of equation (13-33) by $(\gamma^2 - 1)$.

For $\gamma > 1$, the Green function is given by

$$Gr\,(X,Y;\xi,\eta) = 0 \quad \text{for } |Y-\eta| > \xi - X > 0$$

$$= \tfrac{1}{4} J_o\, [n\ (\xi - X)^2 - (Y - \eta)^2]^{1/2}$$

$$\text{for } \xi - X > |Y-\eta| > 0 \qquad (13\text{-}35)$$

where $Y = |\gamma^2 - 1|^{1/2} y$, $n^2 = R_o^{-2} |\gamma^2 - 1|^{-1}$ and J_o is the Bessel function of the zeroth order.

The Green function (13-35) indicates that the forcing function of an infinite value at a point (ξ,η) causes the disturbance of the thermocline (and the currents in the upper layer) only in a wedge which is formed between two straight lines

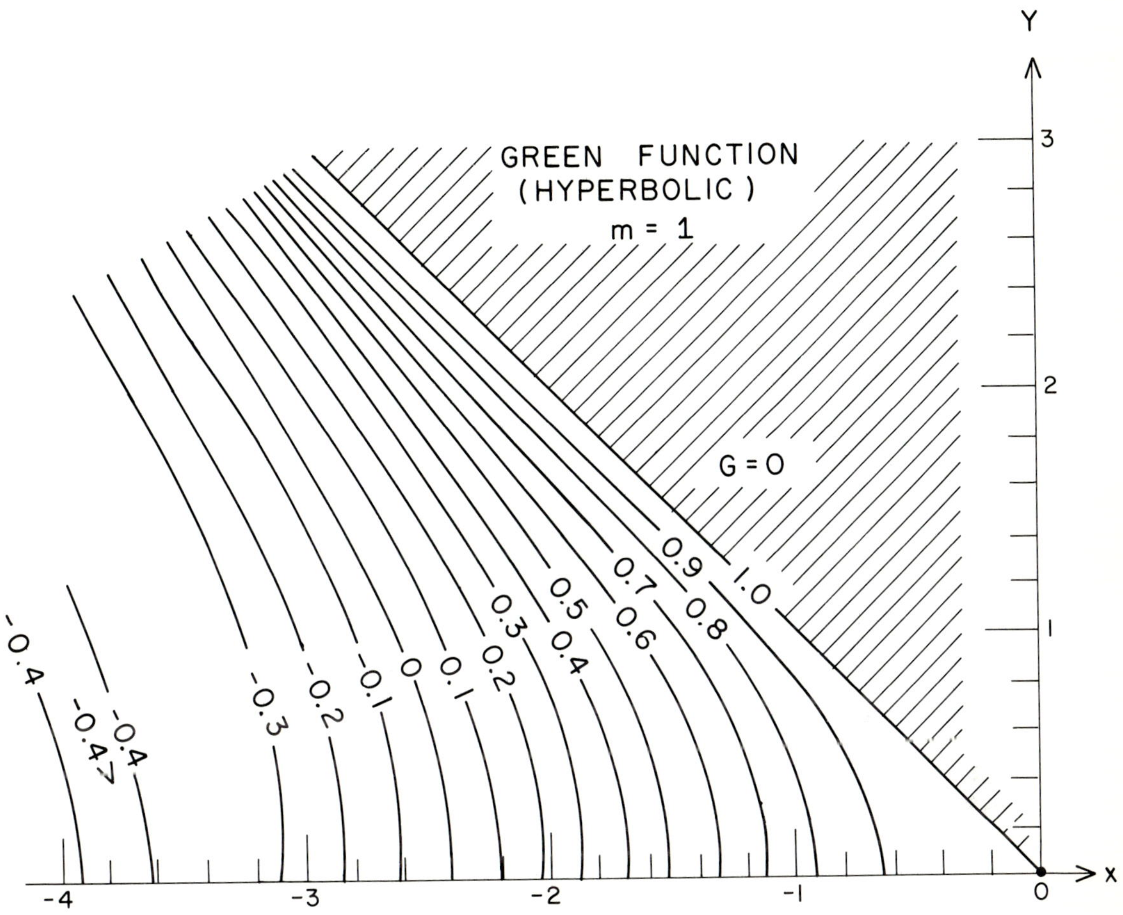

Figure 13-11. Green function for the hyperbolic case with n = *1.*

defined by X − ξ = ± (Y − η) on the *X-Y*, plane, extending in the direction of decreasing *X*.

When $\gamma < 1$, the Green function is given by

$$Gr\,(X, Y; \xi, \eta) = \tfrac{1}{2} K_O \left[n \left\{ (\xi - X)^2 + (\eta - Y)^2 \right\}^{1/2} \right] \qquad (13\text{-}36)$$

where K_O is the modified Bessel function of the zeroth order. This expression indicates that the disturbance due to an infinite forcing function at the point (ξ, η) decreases exponentially with a distance from the point. The Green function, expressed by (13-35) and (13-36), is computed for *n* = 1 and plotted in Figures 13-11 and 13-12, respectively.

When γ is substantially larger than unity, the term $\partial^2 H / \partial y^2$ in expression (13-34) can be neglected compared with other terms. Then the solution of the linearized equation of (13-33) can be approximately given by

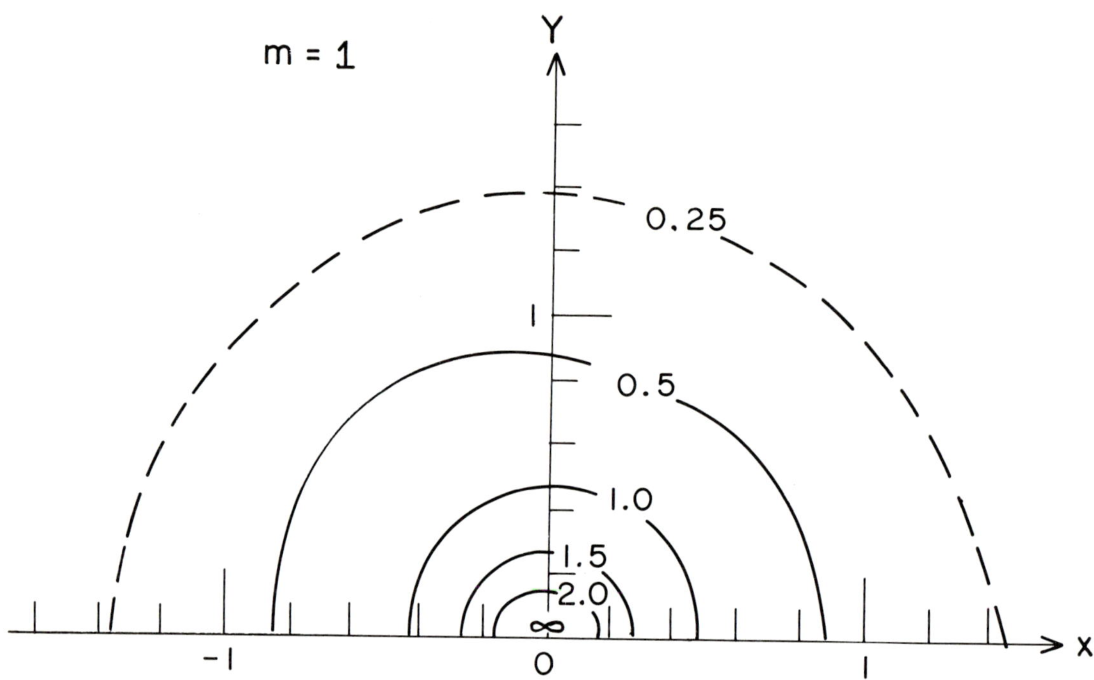

Figure 13-12. Green function for the elliptic case with n = *1.*

$$H = n^{-1} \left\{ \cos nX \int_X^\infty R(\lambda, y) \sin n\lambda d\lambda \right.$$

$$\left. - \sin n X \int_X^\infty R(\lambda, y) \cos n\lambda d\lambda \right\} \quad (13\text{-}37)$$

where λ is the integration variable corresponding to X in the forcing function R (X, y).

When h is substantially different from unity and γ is larger than unity, the Laplacian term of the left-hand side of equation (13-33)can be neglected against the *ln h* terms. Then the approximate solution of (13-33) can be obtained by replacing H with *ln h* and n with $(R_0\gamma)^{-1}$ on the left-hand and right-hand sides of equation (13-37), respectively.

To apply solution (13-37) or its modified form for *ln h* to the observations in the Gulf of Mexico, the wind stress magnitude is expressed by $|\tau| = T_o\ (r/a) \exp \left\{ - ½(r/a)^2 + ½ \right\}$. This expression indicates that the wind stresses become maximum at $r = a$. The forcing function R(X, y) due to these wind stresses is expressed by

$$R(X, y) = e^{½} m\ (\gamma^2 - 1)^{-1}\ T_o \Big[2 \sin\theta\ (1 - r^2/2a^2) \exp(-r^2/2a^2) - (\gamma R_o)^{-1} \cos\theta \Big\{ \sqrt{2}\, a\ Erfc\ (X/\sqrt{2}\, a) \times (3 - y^2/a) \exp(-y^2/a^2) + X \exp(-r^2/2a^2) \Big\} \Big] \quad (13\text{-}38)$$

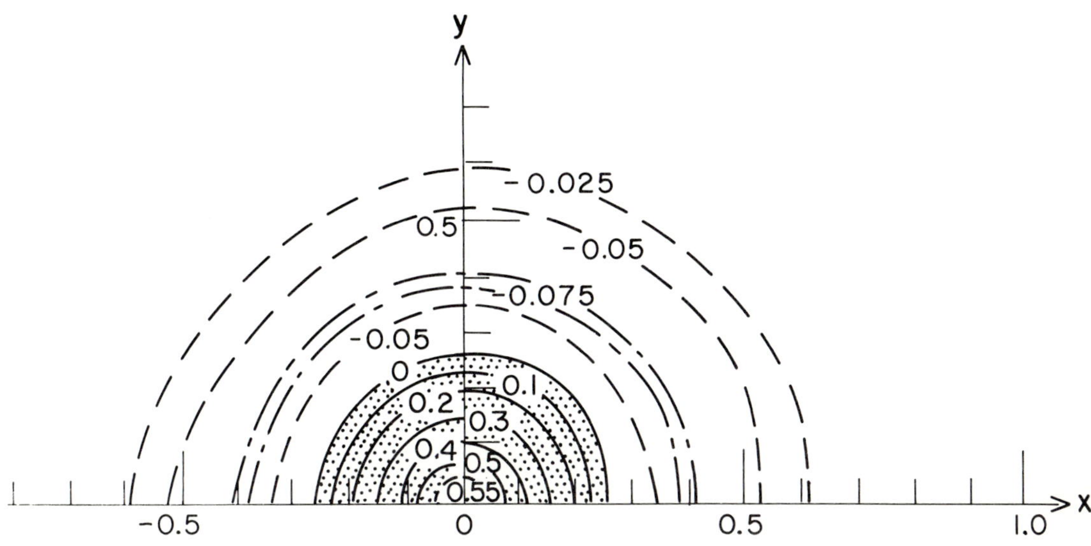

Figure 13-13. Forcing function for the divergence of the wind stresses with a = *0.2,* R_o = *1 and* tan o = *0.75.*

where θ is the inward angle of the wind stress vector and *Erfc* is the error function defined by

$$Erfc\,(x) = \int_x^\infty e^{-\lambda^2} d\lambda$$

As seen in the right-hand side of equation (13-33), the forcing function consists of two parts due to the divergence and curl of the wind stress vector. The part due to the divergence is circular symmetric around the moving center of the storm. The part due to the curl is asymmetric about X, and its magnitude is zero away in front of the storm but remains finite in the wake.

The divergence and curl terms of the forcing function of (13-38) are plotted in Figures 13-13 and 13-14 for a = 0.2, R_o = 1 and tan o = 0.75. The solution (13-37) or its modified form for *ln h* is calculated for h_o = 100 m and c = 4 m/sec, corresponding to the forcing function (13-38). The nondimensional coefficients for these constants are r = 2.828, m = 4 and n = 5.32. The profiles of h along the x-axis for (13-37) or its modified form for *ln h* are plotted in Figure 13-15. The logarithmic modification yields slightly smaller values of h, but the difference is almost negligible. This figure indicates that the thickness h not only decreases in the wake of the storm but also that it oscillates with a wave length of about 1.3. The amplitude of oscillation is about 0.1. This oscillation is due to the internal wave of the inertial period since the dimensional wave length equals about 260 km, yielding a period of about 18 hours for the traveling speed of 4 m/sec of the storm. The value of h also decreases by about 0.2 uniformly in the wake of the storm. This corresponds to 20 m, which is of the same order as the upward displacement of the thermocline observed by Leipper (1967) along the track of the center of the hurricane.

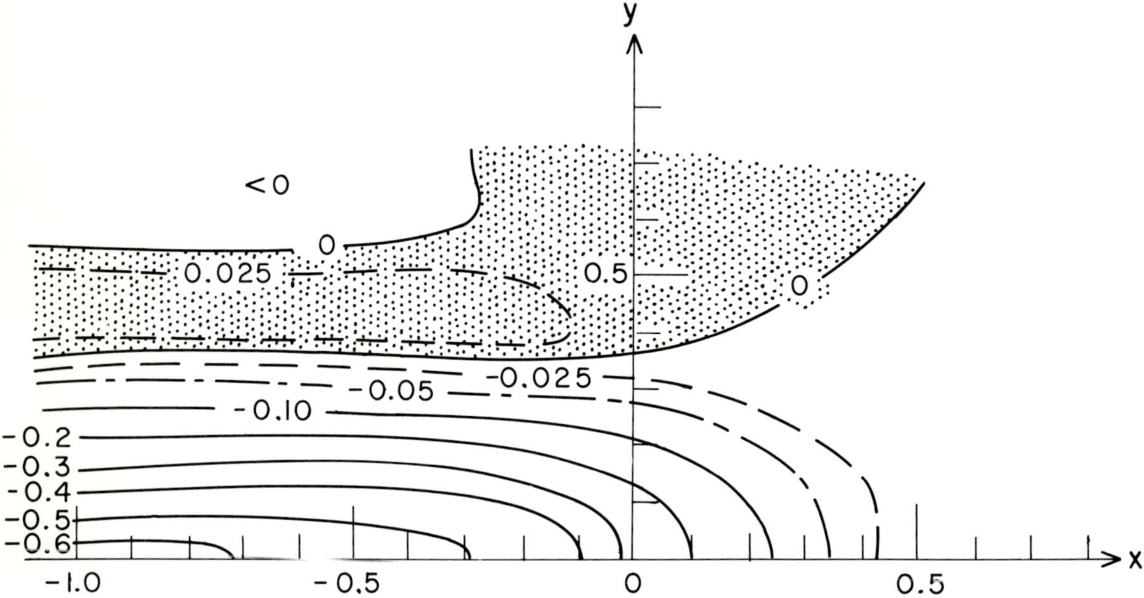

Figure 13.14. Forcing function for the curl of the wind stresses with a = *0.2,* R_o = *1 and* tan θ = *0.75.*

Response of an Ocean with a Coast

The effect of the coast on the ocean's response to a hurricane should be studied in order to interpret the observations of Leipper (1967) and Franceschini and El-Sayed (1968), since these observations were made in the semi-enclosed Gulf of Mexico. This effect is completely ignored in the theoretical models discussed heretofore. When the response of the ocean after the passage of a hurricane is considered, the nonlinear terms may be neglected. A simple geometry of the coastal boundary and shelf area can be assumed in order to determine general features of the response problem. The coast is straight and parallel to the y-axis, and the continental shelf of variable depth and constant width is flanked by a deep sea. Further, the deep sea consists of two layers with a motionless bottom layer, whereas the water is uniform on the shelf, consisting of only the upper layer of the deep sea.

In the shelf, the bottom friction becomes important. For storm surge research, some authors (Freeman, Baer and Jung, 1957; Reid and Bodine, 1968) used the bottom friction proportional to the square of velocity, which makes the equations of motion nonlinear. However, values of frictional coefficient are empirical and may be determined usually for natural or artificial channel flows and, thus, their validity for currents caused by a hurricane is questionable. The bottom friction acts only as a dissipational force for the response problem and its functional form is not critical. Therefore, the bottom friction is assumed to be proportional to the linear velocity.

The equations of motion and continuity on the shelf are given by

$$\partial \mathbf{u}/\partial t + f\mathbf{u} \times \mathbf{k} = -g\nabla\zeta + \tau/h_s - \mu\mathbf{u}/h_s \quad (13\text{-}39)$$

and

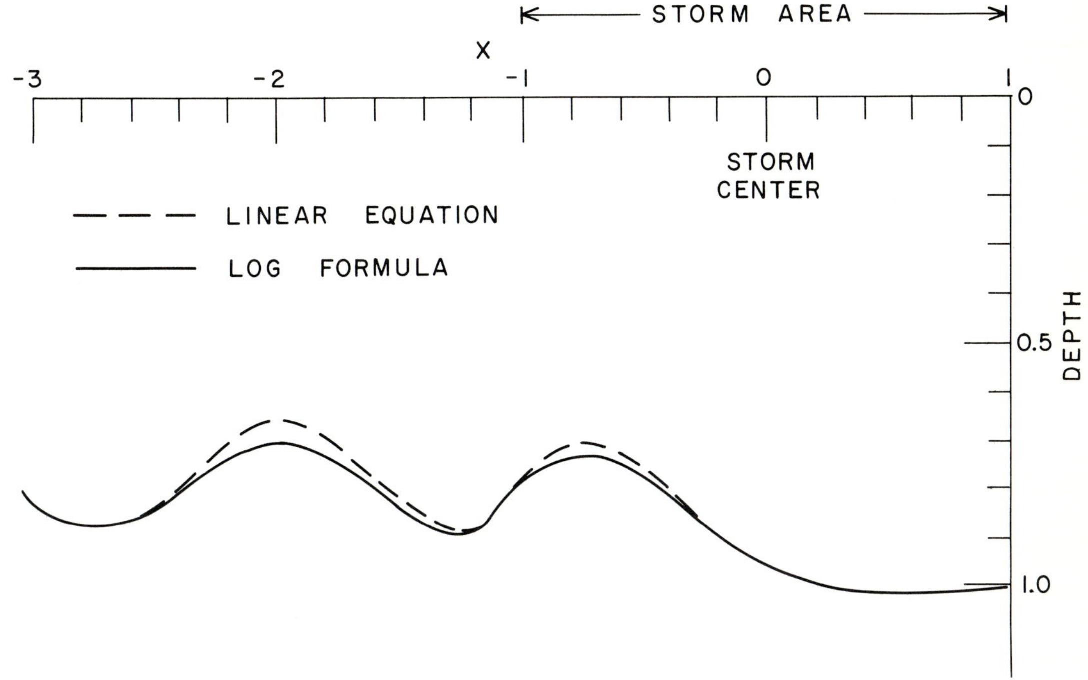

Figure 13-15. Depth of the upper layer h *along the* x*-axis for the linearized continuity equation (dashed line) and for the nonlinear continuity equation (full line).*

$$\partial\zeta/\partial t + \nabla\cdot(\mathbf{u}h_s) = 0 \qquad (13\text{-}40)$$

where ζ is the elevation of the free surface, μ is a coefficient of the bottom friction and h_s is the shelf depth. The moving coordinate X which was previously introduced for treating a traveling storm cannot be used if there is a coastal boundary crossing the course of the storm.

If the bottom contours of the shelf are almost parallel to the coast, the following conditions are satisfied on the shelf (Freeman et al., 1957): $|v| \gg |u|$, $|g\partial\zeta/\partial y| \ll |\partial v/\partial t|$. Further, when the response of the sea with time scale longer than the inertia period, the velocity v and elevation z can be approximately expressed by

$$v = \int_{-\infty}^{t} \tau_y h_s^{-1} \exp\left\{\mu(\lambda - t)h_s^{-1}\right\} d\lambda \qquad (13\text{-}41)$$

and

$$\zeta = \int_{\infty}^{x} (\tau_x + fvh_s)(gh_s)^{-1}\, dx \qquad (13\text{-}42)$$

where the lower limit of integration of (13-42) is taken at an infinite distance from the coast, since the elevation almost vanishes at the outer edge of the shelf (Ichiye, 1962).

The integral of equation (13-41) indicates that the velocity component parallel to the coast decreases exponentially with time after the passage of a hurricane because of the bottom friction. The solutions (13-41) and 13-42) do not represent the resurgence of storm surges which are often observed along the Atlantic coast nor damping of surges due to radiational dissipation of long waves into the deep sea. These phenomena are represented with the second order approximation to the solutions of (13-39) and (13-40). Analytical models indicate that such damping is stronger for a storm moving toward the coast than for one moving parallel to the coast (Ichiye, 1962).

The system of equations (13-39) and (13-40) for any wind stress function can be treated more generally by introducing the Laplace transform operator p for the time derivative $\partial/\partial t$. By solving equation (13-39) about u and v, the velocity components can be expressed with ζ and τ as

$$(\bar{p}^2 + f^2)\, uh = -gh\bar{p}\nabla\zeta + \bar{p}\tau + ghfk \times \nabla\zeta - fk \times \tau \tag{13-43}$$

where

$$\bar{p} = p + \mu h^{-1} \tag{13-44}$$

Substitution of (13-43) into equation (13-40) yields

$$\left\{p(\bar{p}^2 + f^2) - gh_s\bar{p}\ \nabla^2\right\}\zeta - g\bar{p}\ \nabla\zeta\cdot\nabla h_s - gf\left(\frac{\partial\zeta}{\partial y}\frac{\partial h_s}{\partial x} - \frac{\partial\zeta}{\partial x}\frac{\partial h_s}{\partial y}\right) = -(f\,\mathrm{curl}_z\,\tau + \bar{p}\ \nabla\cdot\tau). \tag{13-45}$$

When we consider only the response of the sea with a time scale longer than an inertia period, p can be neglected compared with f. Further, if μh_s^{-1} is smaller by orders of magnitude than f, only the last term of the left-hand side of equation (13-45) may be important for a time scale longer than an inertia period. Since it is assumed that the bottom contours of the shelf are almost parallel to the coast, $\partial h_s/\partial y = 0$. Thus, equation (13-45) can be simplified into

$$(\partial h_s/\partial x)(\partial\zeta/\partial y) = g^{-1}\,\mathrm{curl}_z\tau. \tag{13-46}$$

This equation indicates that the sea level perturbation is restored to the original state after the passage of a hurricane because the right-hand side of equation (13-46) vanishes at that time and also because the disturbances caused during the storm are dissipated into the deep sea as long waves as discussed by Ichiye (1962).

Since the order of magnitude of μh_s^{-1} is 10^{-5} sec $^{-1}$ for tides or seiches in a basin with depth of 50 to 150 m (Defant, 1961), the assumption that $f \gg \mu$ is valid for a normal condition. However, the strong current due to a hurricane may cause stronger bottom friction. Then, for a time scale longer than an inertial period, equation (13-45) becomes

$$\mu\nabla^2\zeta + (\mu h_s^{-1}\partial\zeta/\partial x + f\partial\zeta/\partial y)\, dh_s/dx = g^{-1}\,(\mu h_s^{-1}\,\nabla\cdot\tau + \mathrm{curl}_z\,\tau. \tag{13-47}$$

This equation also indicates that the disturbance of the sea level vanishes after the passage of the storm. This conclusion from equation (13-46) and (13-47) will be applied as a boundary condition for disturbance caused by a storm in the deep sea adjoining the shelf.

For the two-layer deep sea, linearized equations of motion and continuity become

$$\partial\mathbf{u}/\partial t + f\mathbf{u}\times\mathbf{k} = -g\epsilon\nabla h + \tau h^{-1} \tag{13-48}$$

and

$$\partial h/\partial t + h_o\,\nabla\cdot\mathbf{u} = 0 \tag{13-49}$$

where, in this case, h is the thickness of the upper layer and h_o is its undisturbed value. The friction term is dropped since it is small due to a large depth here. Introducing p as $\partial/\partial t$ and eliminating u and v from equations (13-48) and (13-49), we have

$$p\left\{c_h^2\,\nabla^2 - (p^2 + f^2)\right\}h = p\,\mathrm{div}\,\tau + f\,\mathrm{curl}_z\tau \tag{13-50}$$

where

$$c_h^2 = g\epsilon h_o.$$

When we consider only response with a time scale longer than an inertia period, equation (13-50) becomes

$$(c_h^2 \nabla^2 - f^2)\, h = \text{div}\, \tau + p^{-1} f\, \text{curl}_z\, \tau. \tag{13-51}$$

The operator p^{-1} represents integration with time. Equation (13-51) indicates that the disturbance of the upper layer thickness continues after the passage of a storm since the p^{-1} term does not vanish at such a stage.

If the shelf is adjacent to the deep sea, the disturbance of the surface elevation and of the current due to a storm vanishes after passage of the storm as discussed before. Therefore, the effect of the shelf is the same as the vertical coastal barrier after the passage of the storm. If the outer edge of the shelf is along the y-axis, the boundary condition at $x = 0$ for equation(13-51) is given by

$$h = h_o. \tag{13-52}$$

If the scale of a storm is about 100 km, it is much larger than the distance s defined by

$$s = (g\epsilon h_o)^{1/2} f^{-1}.$$

Rossby (1936) calls this distance the radius of deformation, equaling about 14 km for $h_o = 100$ m. Therefore, with the boundary layer technique, a solution of equation (13-51) with the boundary condition (13-52) can be expressed approximately by

$$h = h_o + M(t, o, y)e^{-x/s} - M(t, x, y) \tag{13-53}$$

where

$$M(t, x, y) = -f^{-2}\, \text{div}\, \tau - f^{-1} \int_{-\infty}^{t} \text{curl}_z\, \tau dt.$$

The solution (13-53) explicitly indicates that the effect of curl of the wind stress on the thickness of the upper layer lasts after the passage of the storm. Also, it shows that the effect of the coast or shelf is limited within 10 or 20 km from the coast or outer boundary, respectively.

On the other hand, the data of Franceschini and El-Sayed (1968) indicate that the thickness of the upper layer after the hurricane is not constant along the outer edge of the shelf but deeper on the left-hand side of the track. This may be explained by an effect of asymmetry of the wind stress over the shelf. When we consider the response of the sea level over the shelf for a period longer than an inertial period, equation (13-46) is valid. Substituting equation (13-46) into (13-43) and neglecting p and $\bar{p}$ against f, we have

$$uh_s = -h_s f^{-1}\, \text{curl}_z\, \tau (\partial h_s / \partial x)^{-1} + \tau_y f^{-1}. \tag{13-54}$$

The first term of the right-hand side of this equation represents the effect of the change of the depth on the shelf and the second term represents the Ekman transport.

When a width of the shelf and the horizontal scale of the hurricane are similar, the two terms of the right-hand side of equation (13-54) are of the same order of magnitude. During Hurricane Inez in 1966, the distance from the center to the maximum wind stress is 20-40 km, whereas the width of the shelf in the western Gulf is at least 50 km. The first term is always negative within a hurricane because of positive curl $_z\tau$, with magnitude much larger on the right-hand side than on the left-hand side, whereas the second term is either negative or positive in the front or rear and positive on the right- or left-hand side respectively. Therefore, after the hurricane, the first term cancels the second term on the right-hand side, and more shelf water flows out into the slope on the left-hand side of the track, increasing thickness of the upper layer and the temperature in the mid-depth on the continental slope as observed by Franceschini and El-Sayed (1968).

Concluding Remarks

The response of the ocean to hurricanes or other meteorological disturbances is an important problem not only for physical oceanography but also for meteorology and other branches of oceanography. More work should be done both in observational and theoretical fields.

More data are needed on the ocean conditions after passage of a storm both in the shallow and deep areas. Quick serial observations should be made repeatedly on a few transects across the storm track in order to study a process in which the disturbed ocean is restored to the normal state. It will be necessary to reserve one or two research vessels for this special purpose during a hurricane season. Time series monitoring of the current and temperature at several stations will be crucial for understanding processes occurring during intense agitation by a storm. Subsurface buoy systems without any exposed parts will do the work. Bottom friction on the shelf and mixing of the water through a thermocline during a storm are not well known and can be measured only with the subsurface buoy systems.

In theoretical fields discussed in this chapter several points are left out. One example is the distribution of the current and thermocline depth within and near a storm moving with a speed close to the internal wave velocity. This problem is essentially nonlinear and can be treated only numerically. Another example is the response of the shelf water and the deep basin with a shelf or a coast during passage of a storm. This problem can be treated as linear but time-dependent; and, thus, a numerical method may be more appropriate than an analytical solution.

Two-layer models of the deep sea and homogeneous models of the shelf are extreme simplifications of the actual state and, at best, useful only for predicting change of the thermocline depth and of surface storm surges, respectively. In order to understand effects of a storm on the distribution of water properties, we should know mixing processes through the thermocline and in the horizontal direction. However, these processes can be mathematically incorporated into hydrodynamic equations only after their physics is fully understood. This may be accomplished only through more observational and experimental works.

Acknowledgment

This work is supported by the Oceanography Section, National Science Foundation, NSF Grants GA-26498 and by the Office of Naval Research N00014-68-0308.

References

Courant R. 1962. *Methods of mathematical physics, Vol. 2, partial differential equations*. New York: Interscience Pub., 407-490.

Defant, A. 1961. *Physical oceanography*, Vol. 2. New York: Pergamon Press, 155-160.

Dunn, G.E. and Miller, B.I., 1964. *Atlantic hurricanes*. Baton Rouge: Louisiana State Univ. Press.

Fisher, E.L. 1958. Hurricanes and the sea-surface temperature field. *J. Meteorol.*, 328-333.

Franceschini, G.A. and El-Sayed, S.Z., 1968. Effect of Hurricane Inez (1966) on the hydrography and productivity of the western Gulf of Mexico. *Deutsche Hydrographische Zeits.*, 21:193-202.

Hidaka, K. and Akiba, Y., 1955. Upwelling induced by a circular wind system. *Records Oceanog. Works*, Japan, 2:7-18.

Freeman, J.C., Jr., Baer, L. and Jung, C.H., 1957. The bathystrophic storm tide, *J. Mar. Res.*, 16:12-22.

Ichiye, T. 1955. On the variation of oceanic circulation (5). *Geophys. Mag.*, Tokyo, 26:283-342.

———. 1962. *Storm surges on a continental shelf*. National Hurricane Research Project No. 50, Pt. 1, U.S. Weather Bureau: 255-266.

———. 1970. *Motion in a two-layer ocean caused by a moving disturbance*. Tech. Rept. to NSF GA-13430 and ONR 2119(04).

Jordan, C.L. and Frank, N.L. 1964. *On the influence of tropical cyclones on the sea surface temperature field*. Florida State University, Project Report for NSF Grant GP-621.

Jordan, C.L. 1965. *Evidence of surface cooling due to typhoons*. Technical Note 9-SAIL-1, ESSA Sea-Air Interaction Lab. Report No. 1, 187-190.

Kajiura, K. 1956. *A forced wave caused by atmospheric disturbances in deep water*. Tech. Rept. of Texas A&M, Dept. of Oceanog., Ref. 56-26T.

Leipper, D.F. 1967. Observed ocean conditions of Hurricane Hilda, 1964. *J. Atmos. Sc.*, 24:182-196.

Malkus, J.S. 1962. Large-scale interactions. *The sea*, vol. 1, M.N. Hill, ed. New York: Interscience Pub., 88-294.

Perlroth, I. 1962. Relationship of central pressure of hurricane Esther (1961) and the sea surface temperature field. *Tellus*, 14:403-408.

_____.1967. Hurricane behavior as related to oceanographic environmental conditions. *Tellus*, 19:258-268.

O'Brien, J.J. and Reid, R.O. 1967. The non-linear response of a two-layer, baroclinic ocean to a stationary, axially-symmetric hurricane, Pt. 1, Upwelling induced by momentum transfer. *J. Atmos. Sc.*, 24:197-207.

____ 1967. The non-linear response of a two-layer, baroclinic ocean to a stationary, axially-symmetric hurricane, Pt. 2, Upwelling and mixing induced by momentum transfer, *J. Atmos. Sc.* 24: 208-215.

Reid, R.O. and Gilbert, K. 1970. *Studies of meso-scale air-sea interaction*. Annual Report of the Themis Project, Texas A&M University (Sub Task G).

____and Bodine, B.R. 1968. Numerical models for storm surges in Galveston Bay. *J. of Waterways and Harbors Div., Proc. Am. Soc. Civ. Eng.*, 5805:33-56.

Rossby, C.G. 1936. Dynamics of steady ocean currents in the light of experimental fluid mechanics. *Pap. Phys. Oceanogr. Meteor*, 5(1): 1-43.

Stevenson, R.E. and Armstrong, R.S. 1965. Heat loss from the waters of the waters of the Northwest Gulf of Mexico during hurricane Carla. *Geofis. Internat.* Mexico City, 5:49-57.

Sugg, A.L. and Carrodus, R.L. 1969. *Memorable hurricanes of the United States since 1873*. Technical Memorandum WBTM-SR-42, ESSA.

14

Bathythermograph Sections across the Path of Hurricane Celia

Robert L. Molinari and Guy A. Franceschini

Abstracts

Three vertical sections constructed from expendable bathythermograph (XBT) profiles taken in the Gulf of Mexico across the track of Hurricane Celia, are presented. The pre-hurricane and one of the post-hurricane sections, which nearly coincide geographically, indicate that the storm may have caused upwelling of colder sub-surface waters. Comparison of the two post-hurricane sections suggests that persistence of this effect of upwelling may be dependent on the length of time the strong winds existed over each area of concern.

Using the expendable bathythermograph, the *Alaminos* took three vertical sections of observations across the path of Hurricane Celia, a storm which traversed the northwestern Gulf on 1 and 2 August 1970. *Alaminos* Cruise 70-A-10 provided a pre-hurricane section, and Cruise 70-A-11 provided two post-hurricane sections. The track of Celia and the positions of the XBT observations are given in Figure 14-1.

Since the 70-A-10 and the second 70-A-11 crossings were nearly coincident, comparison of pre- and post-hurricane conditions was possible. The data of the earlier cruise were taken on 4 and 5 July, the latter on 4 September. The sections drawn from the profiles of both sets of data are given in Figure 4-2. Comparing the patterns of upper layer isotherms (22.5-27.5°C) indicates that one month after the storm's passage the near-surface thermal structure is very similar to the pre-storm pattern. Also, the thermal patterns given by 70-A-10 XBTs 1-3 and 70-A-11 XBTs 104-107 show that the deep structure to the right of the path is relatively unchanged. However, the deeper

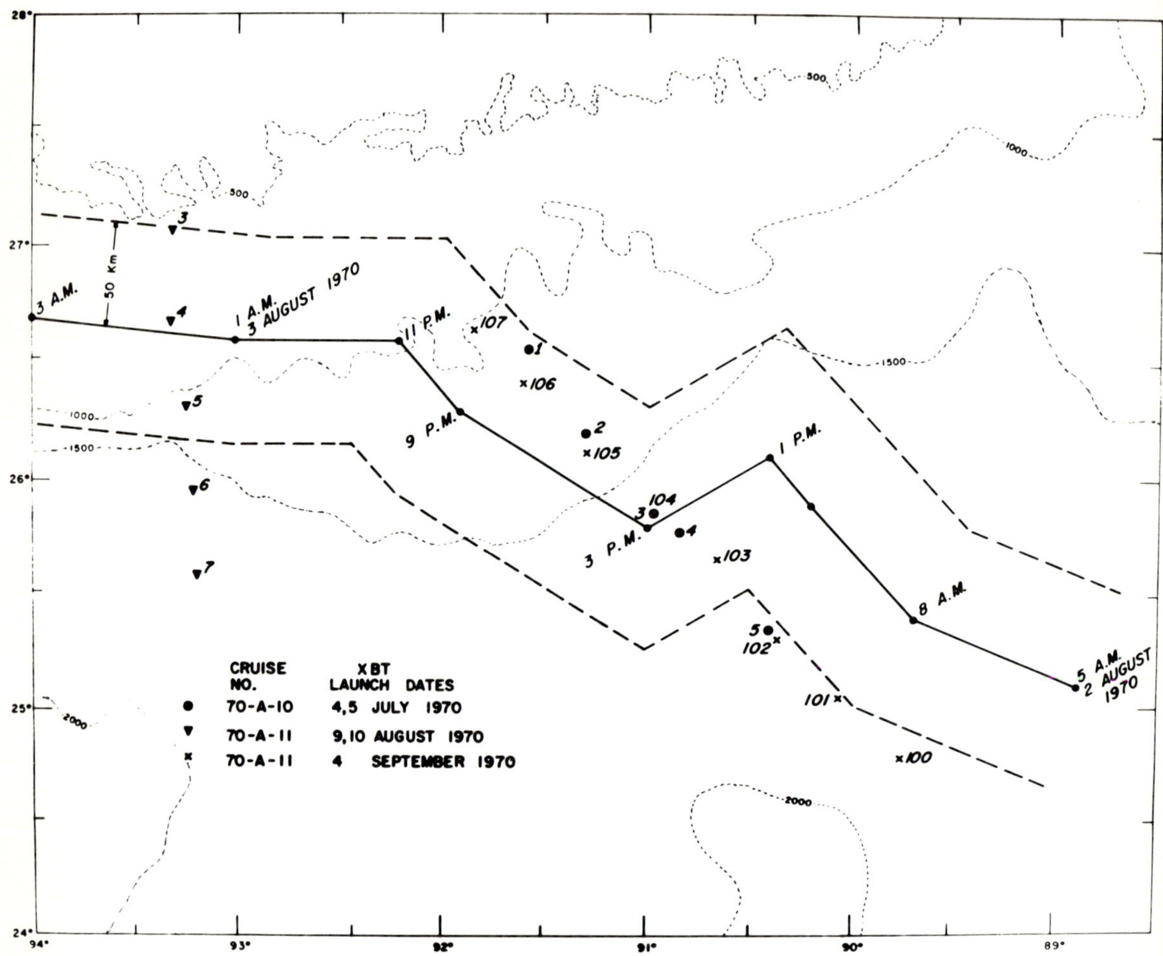

Figure 14-1. The path of Hurricane Celia (solid line) constructed from Weather Bureau bulletins and the XBT positions of Cruises 70-A-10 and 70-A-11. The dashed lines represent a distance of 50 km from the storm center. The contours, in fathoms, are from the U.S. Naval Oceanographic Office Chart HO 126.

post-hurricane isotherms in the vicinity of, and in particular to the left of the storm's path, are at shallower depths than those found before the storm. Bathythermographs 101-104 appear to be in the region of hurricane influence, since the isotherms at XBT 100 exhibit depths similar to those north of XBT 104. Contrasting pre- and post-hurricane isotherm depths in the area of XBTs 101-103 indicates that the deeper isolines have a greater vertical displacement than those nearer the surface. For instance, the 20°C isotherm is 30 m above the depth indicated by the 70-A-10 data, while the 12.5°C isotherm appears to have risen 70 m.

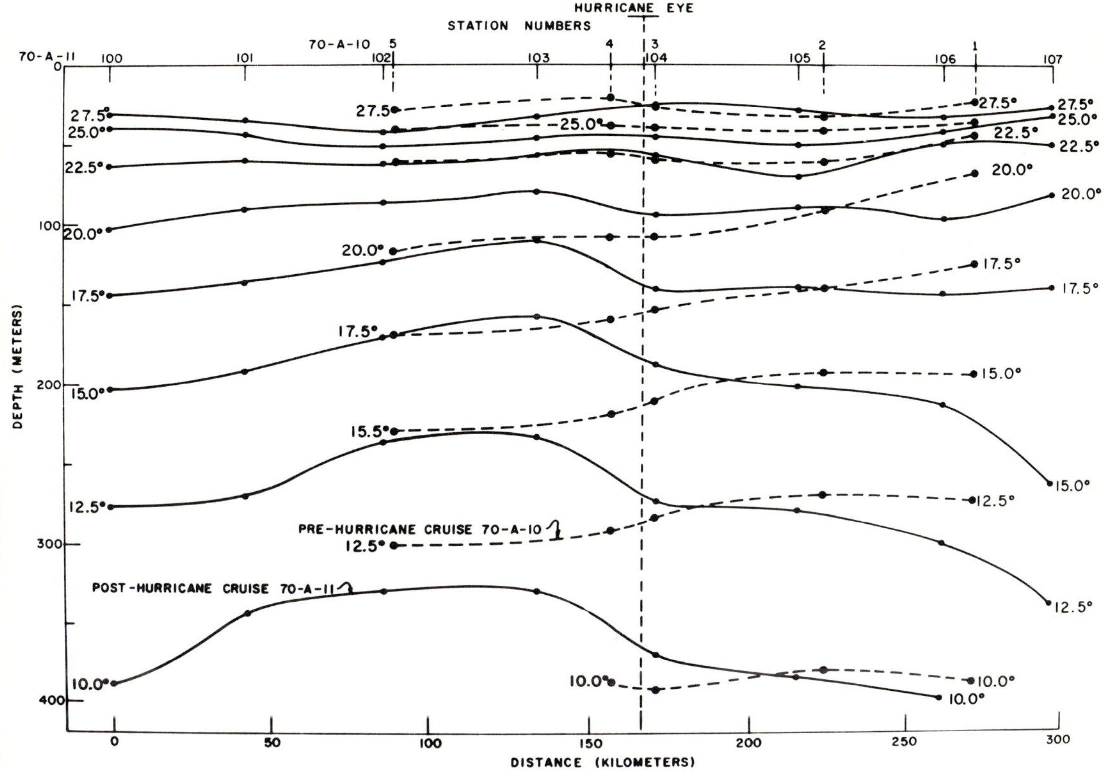

Figure 14-2. The vertical sections constructed from XBT profiles of Cruises 70-A-10 (dashed lines) and 70-A-11 (solid lines).

Figure 14-3, which presents the data from the first crossing of the hurricane's path, made on 9 and 10 August, shows no striking isotherm rise as depicted in Figure 14-2. There is a 10-20 m rise in the 17.5-27.5°C isotherms at XBT 3, but the 10-15°C isolines show a marked drop at this position. As there are no pre-hurricane data for this position across the storm's path it is difficult to ascertain the possible influence of the storm.

An apparent discrepancy arises when the two post-hurricane sections are compared. The second crossing, made one month after the storm's passage, shows a greater isotherm displacement than the first section made only one week after the storm. A possible explanation for this observed feature is the path of Celia (Figure 14-1) in the vicinity of the XBT lines.

Hurricane bulletins from the Weather Bureau indicate that Celia's intensity and speed remained relatively constant in this area, with hurricane force winds extending 100 km from the storm center. The dashed lines to the left and right of the track in Figure 14-1 are 50 km from the path—well within the high-wind region. Because of the bend in the storm's path near the second 70-A-11 section, the hurricane force winds remained over this

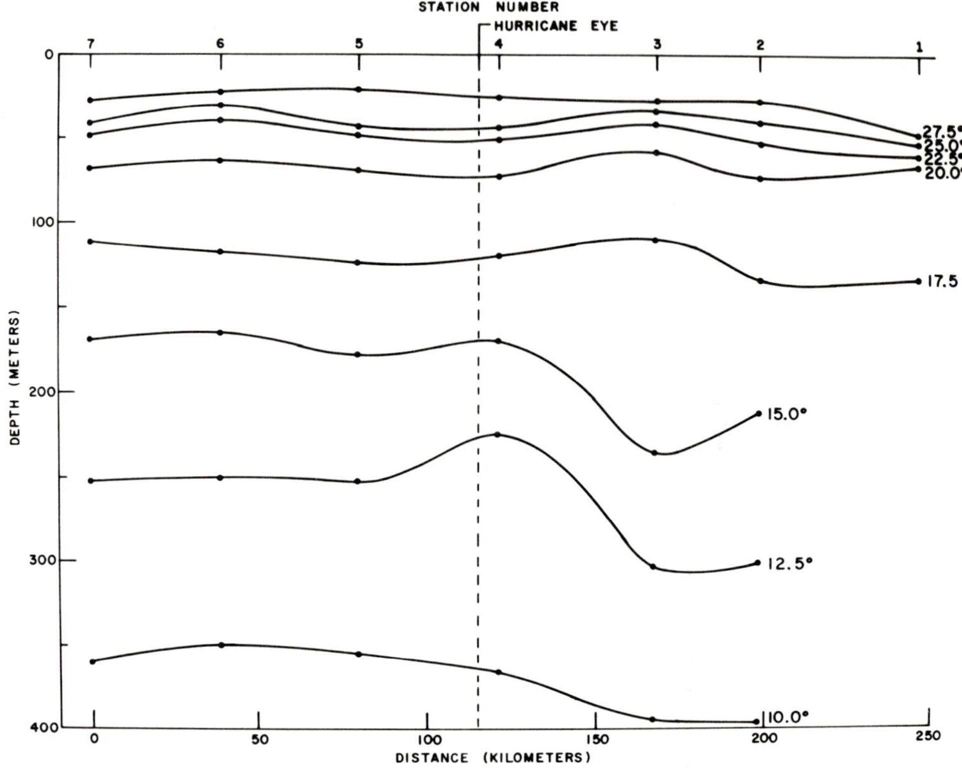

Figure 14-3. The vertical section constructed from XBT profiles of the earlier 70-A-11 crossing of the storm's path.

area for almost 12 hours. However, the storm traversed a straight path through the area of the first XBT line, and the high winds, as given by the 50 km lines, remained in this only region only 2 hours. Thus the hurricane had a longer time to exert an influence on the thermal structure in the region of XBTs 100-107 and may have caused the more pronounced upwelling pattern.

These bathythermograph sections suggest that although the surface thermal structure may return to pre-hurricane conditions, the deeper isotherms may remain displaced as long as one month after the storm's passage. The data also suggest that the residence time of a storm in a particular area is one parameter which may determine the persistence of the upwelled pattern.

Acknowledgments

The authors would like to thank Dr. D. Fahlquist (Office of Naval Research, Contract Nonr. 2119[04]) for allowing Dr. Molinari to participate on Cruise 70-A-11, and Mr. H. Cornelio for taking the XBT observations on Cruise 70-A-10. This research was partially supported by the National Science Foundation grant to the Texas A&M Research Foundation, Project 707.

Section 6
Tides

15
Tides in the Gulf of Mexico

B.D. Zetler and D.V. Hansen

Abstract

A hypothesis is proposed to explain the observed diurnal tide in the Gulf of Mexico. The tide in the Gulf is believed to be co-oscillating with the tide in the nearby Atlantic Ocean with amphidromic points in the Florida Strait near Miami and in Yucatan Channel. Harmonic constants for the tide and tidal current principal diurnal constituents in the overall area of co-oscillation support this theory. If the diurnal tidal current in Yucatan Channel is presumed to match that in the Florida Strait in both amplitude and phase, then calculations of the tidal amplitude inside the Gulf based on volume continuity are in good agreement with the observed value of about 15 cm for K_1.

The semidiurnal tides appear to be somewhat less controversial, with an amphidromic point apparently located in the Gulf roughly midway on a line between the Mississippi delta and the Yucatan Peninsula. Nevertheless, there are some controversial aspects that cannot be resolved with available tide and tidal current data.

The tide and tidal current amplitudes and phases in the entrances to the Gulf are used to estimate a tidal energy dissipation of 2 X 10^{17} ergs/sec. An associated calculation of the mean amplitude of the tidal currents on the shelf, about one-half knot, is consistent with the somewhat larger amplitudes that have been observed in the passes of the estuaries bordering the Gulf.

Existing Theory

The tide at many places in the Gulf of Mexico is diurnal (one high tide and one low tide each lunar day of 24.84 hours) or mixed (a large inequality between the heights of the two high waters and/or between

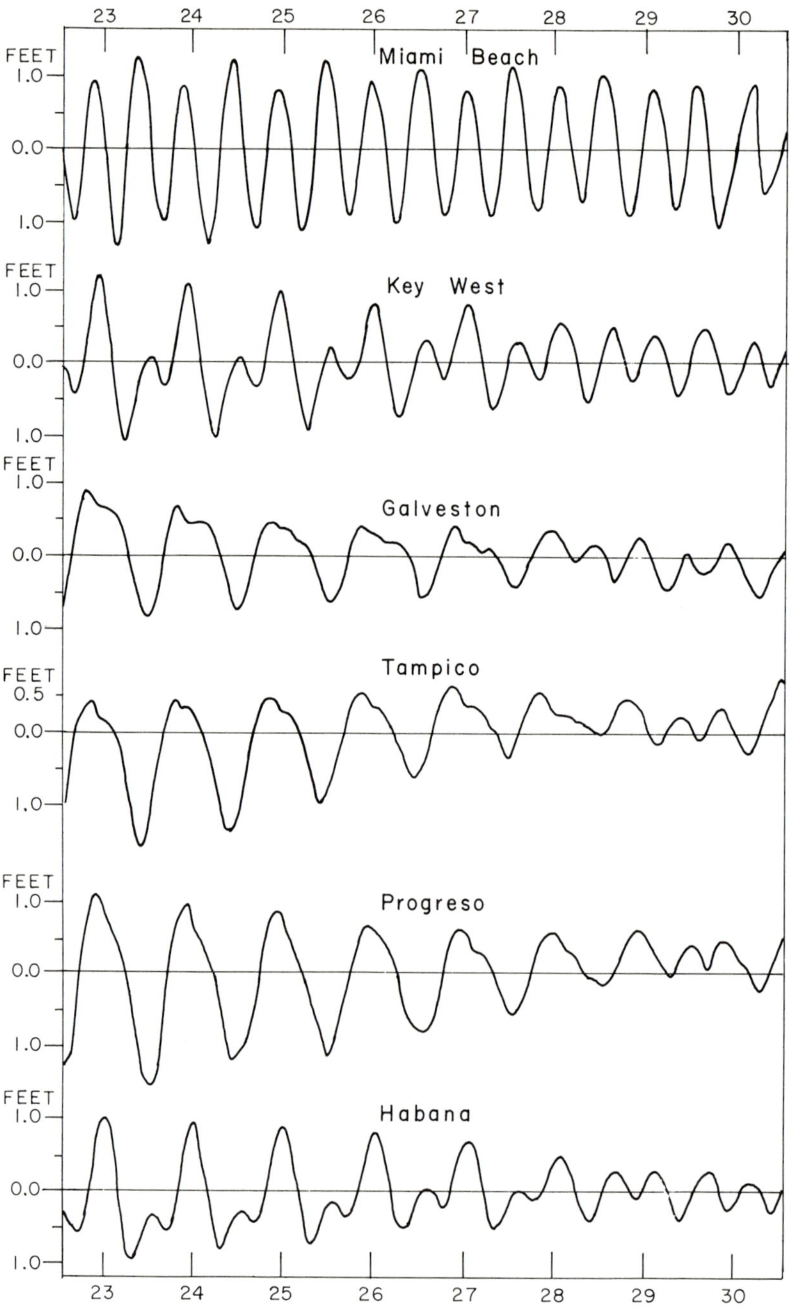

Figure 15-1. Tide curves, at various places in the Gulf of Mexico, June 23-30, 1948, from Marmer (1954).

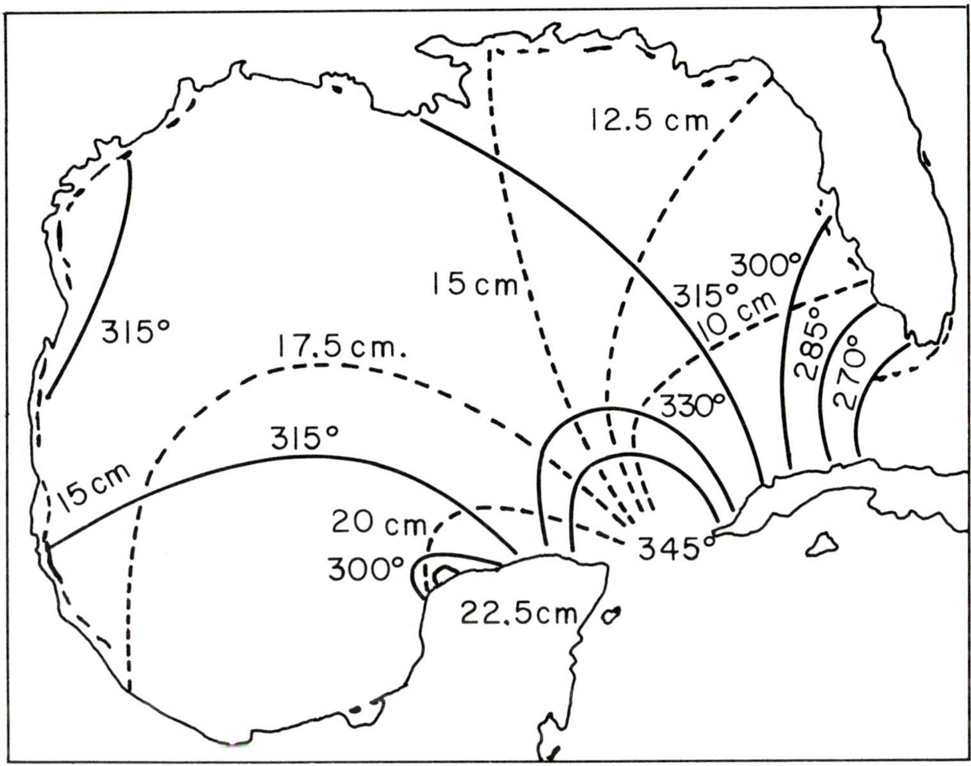

Figure 15-2. Lines of equal phase (——) and lines of equal amplitude (– – –) for K_1 *in the Gulf of Mexico, from Grace (1932).*

the heights of the two low waters in a lunar day). This is in sharp contrast with the tide on the east coast of the United States, where the tide is semidaily with little inequality between the heights of either the two high waters or the two low waters in a lunar day. Figure 15-1 shows typical tide curves at some representative places in the area. The diurnal regime of the Gulf has been the subject of numerous studies and there have been important differences in the conclusions.

The Gulf of Mexico is a basin connected to the Atlantic Ocean by the Florida Strait and to the Caribbean by the Yucatan Channel. The opening between the Bahamas and Cuba contributes negligibly to volume transport because of its shallow depth.

Early researchers (Harris, 1907; Endros, 1908; and Sterneck, 1921) suggested that the diurnal tide in the Gulf is some form of co-oscillation with the tide in the nearby Atlantic, but essentially opposite in phase. This implies a standing wave entering through one or more connecting channels, the Florida Strait or the Yucatan Channel, or both.

Grace (1932, 1933) studied both the diurnal and semidiurnal tides in the Gulf of Mexico in terms of the Gulf's response to astronomic tide-generating forces and of a co-oscillation with the tide in the Atlantic, excited in the Gulf by periodic flow in the channels. Using observations at various places in the Gulf and representing bathymetry of the Gulf by a small number of rectangular boxes, he concluded that the diurnal tide is primarily co-oscillating, entering the Gulf through the Florida Strait and exiting through the Yucatan Channel about 5 or 6 hours later (Figure 15-2).

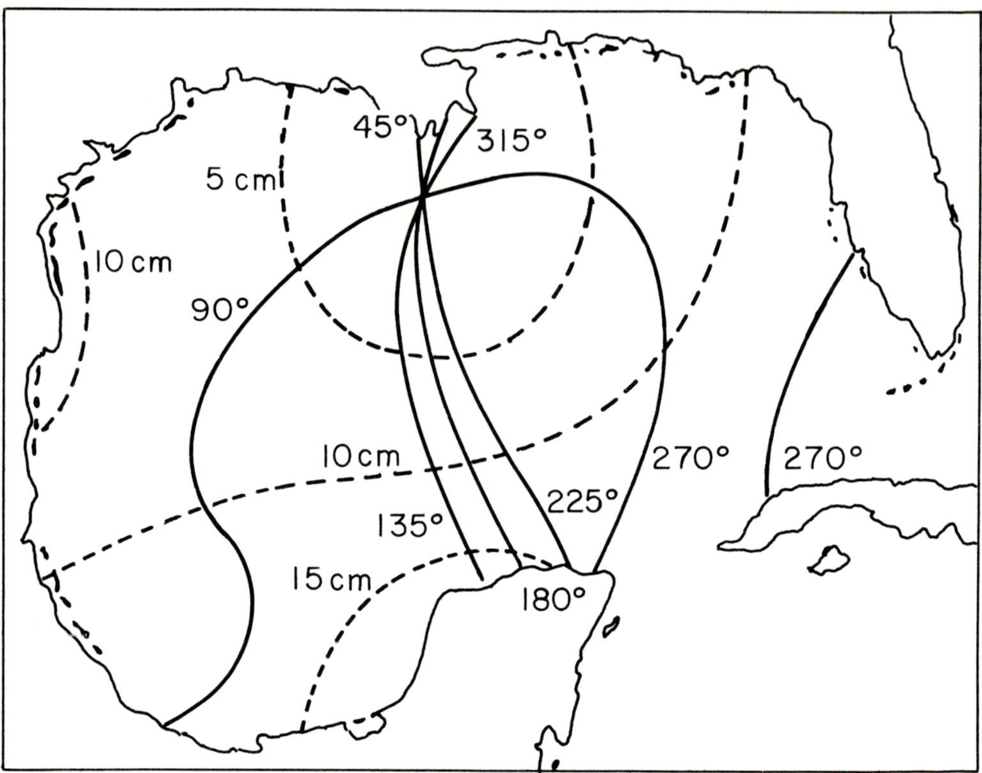

Figure 15-3. Lines of equal phase (——) and lines of equal amplitude (– – –) for M_2 *in the Gulf of Mexico, from Grace (1932).*

Grace's model is therefore consistent with the ideas of the earlier workers, although he based his work entirely on tidal data from within the Gulf.

Marmer (1954) attributed the diurnal tide in the Gulf to the resonance period of the basin. He summed up his analysis along these lines: "Qualitatively, the simplest explanation of the relatively large diurnal component in the Gulf is that the length and depth of its basin are such that its free period of oscillation approximates 24 hours; that is, it approximates the period of the diurnal tide-producing forces and therefore responds better to the diurnal forces than to the semidiurnal forces."

Von Arx et al. (1955) calculated the diurnal tidal current in the Florida Strait, using the cross section of the strait and the tidal prism in the Gulf and assuming no tidal exchange through the Yucatan Channel. However, this calculation was done as part of a possible interpretation of results from another study, and the constant rate of inflow through the Yucatan Channel was used merely to simplify the calculation.

The semidiurnal tides in the Gulf of Mexico are generally small, but the tidal regime appears to be more complex, so much so that Dietrich (1963) showed a question mark in the Gulf in his world cotidal chart. Harris (1904) explained the M_2 tides observed along the periphery of the Gulf as due to a progressive wave entering from the Florida Strait; the later lunitidal intervals in the northeast corner of the Gulf he attributed to the slow movement of a progressive wave over a long shallow shelf. Grace (1933) found that his mean error in amplitude was halved and the mean error in phase significantly reduced by omitting Cedar Keys and St. Marks from his calculations. He justified the

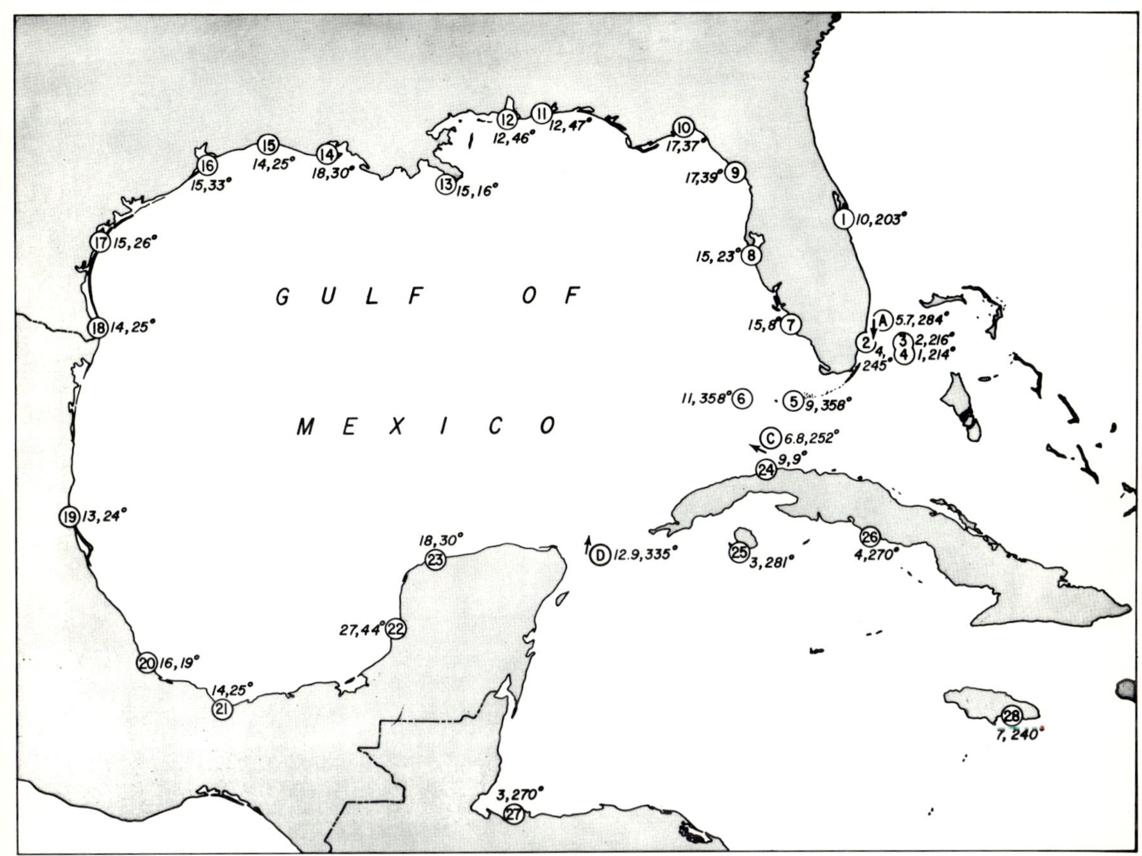

Figure 15-4. K_1 *harmonic constants. Circled numbers correspond to numbers of tidal stations in Table 15-1; amplitude is given in cm and phase in degrees referred to Greenwich transit. Circled letters with arrows correspond to letters of tidal-current stations in Table 15-2; amplitude is given in cm/sec and phase of indicated direction is in degrees referred to Greenwich transit.*

omission "because of local conditions which have not been allowed for in the discussion," such as the shallowing in the areas of these stations. His cotidal chart (Figure 15-3) shows a counterclockwise rotation about an amphidromic point south of the Mississippi River delta with very rapid changes in phase along the delta and along the north side of the Yucatan Peninsula. Sterneck (1920) also found an amphidromic point near the center of the Gulf, but he postulated a clockwise rotation of the cotidal lines. Although Michaelov et al. (1969) indicate that their study confirms Sterneck's interpretation of the semidiurnal tide in the Gulf, their cotidal lines rotate counterclockwise, like Grace's. The Michaelov et al. paper considers the diurnal regime in the Caribbean but does not extend this aspect of their numerical computations into the Gulf of Mexico.

Discussion

Marmer's hypothesis for a resonant diurnal tide does not seem to match the tidal harmonic constants. In a resonant basin, one expects to find an amphidromic point somewhere near the middle and the phase angle to be opposite on either side of the basin. It is clear from Figure 15.4 and Table 15-1 (representing all available K_1 harmonic con-

Table 15-1
Tidal Harmonic Constants, Gulf of Mexico

Stations	Latitude North	Longitude West	K_1 Ampl. (cm)	K_1 G (degrees)	O_1 Ampl. (cm)	O_1 G (degrees)	M_2 Ampl. (cm)	M_2 G (degrees)	S_2 Ampl. (cm)	S_2 G (degrees)	Observation year	Observation length
FLORIDA												
1. Patrick Air Force Base	28°14′	80°36′	10	203	7	207	51	6	9	25	1960	1 year
2. Miami Beach	25°46′	80°08′	4	245	3	267	37	20	7	44	1933	1 year
3. North Bimini	25°44′	79°19′	2	216	1	265	35	19	7	49	1966	326 days
4. Cat Key	25°33′	79°17′	1	214	1	331	34	20	7	49	1938-39	1 year
5. Key West (west coast)	24°33′	81°48′	9	358	9	354	17	71	5	93	1939	1 year
6. Garden Key, Dry Tortugas	24°38′	82°52′	11	358	11	354	15	84	4	97	1860	1 year
7. Naples	26°08′	81°48′	15	8	13	2	21	143	9	155	1966	1 year
8. Anna Marie	27°32′	82°44′	15	23	14	12	16	134	6	146	1933-34	1 year
9. Cedar Keys	29°08′	83°02′	17	39	14	29	33	190	12	216	1939	1 year
10. St. Marks Light	30°04′	84°11′	17	37	16	30	32	205	13	231	1933-34	1 year
11. Warrington	30°21′	87°16′	12	47	12	37	2	132	1	130	1859	1 year
ALABAMA												
12. Mobile Point Light	30°14′	88°01′	12	46	11	36	2	117	1	129	1850-51	1 year
LOUISIANA												
13. Southwest Pass	28°56′	89°26′	15	16	13	15	2	131	1	97	1959	29 days
14. Lighthouse Point	29°31′	92°03′	18	30	15	19	12	244	4	249	1960	29 days
15. Calcasieu Pass	29°47′	93°21′	14	25	13	14	16	259	5	254	1933-34	1 year
TEXAS												
16. Galveston, bay entrance, South Jetty	29°20′	94°42′	15	33	15	24	14	276	5	272	1936-37	1 year
17. Port Aransas	27°50′	97°04′	15	26	15	20	8	262	2	265	1958-59	1 year
18. Padre Island	26°04′	97°09′	14	25	15	20	7	256	1	274	1958	29 days
MEXICO												
19. Tampico	22°15′	97°51′	13	24	13	21	7	250	2	257	1942-43	1 year
20. Veracruz	19°11′	96°07′	16	19	15	12	9	243	3	251	1954	1 year
21. Coatzacoalcos	18°09′	94°25′	14	25	14	19	8	244	2	246	1946-47	1 year
22. Campeche	19°50′	90°32′	27	44	18	30	22	271	7	298	1900	105 days
23. Progreso	21°17′	89°40′	18	30	17	21	6	271	2	280	1946-47	1 year
CUBA												
24. Havana	23°09′	82°20′	9	9	10	5	13	46	4	70	1947-48	1 year
25. Carapachibey*	21°27′	82°55′	3	281	3	321	9	74	3	57	1953	29 days
26. Casilda*	21°45′	79°59′	4	270	2	292	11	74	4	48	1949-50	1 year
HONDURAS												
27. Puerto Cortes*	15°50′	87°57′	3	270	2	329	6	87	2	29	1949-50	1 year
JAMAICA												
28. Port Royal*	17°56′	76°51′	7	240	5	239	5	92	2	41	1963	1 year

*Stations 25-28 are in the Caribbean Sea.

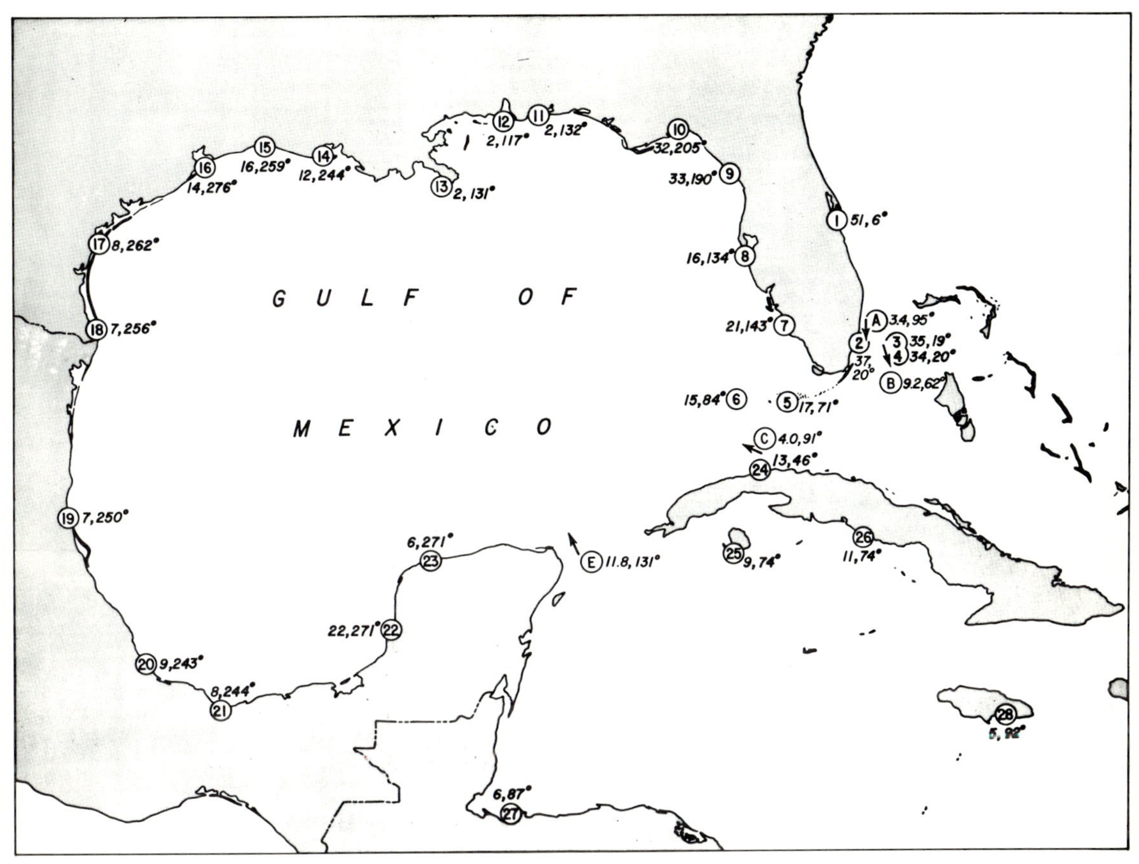

Figure 15-5. M_2 *harmonic constants. Same method of notation as in Figure 15-4.*

stants on or near the coast) that the diurnal tide is essentially in phase throughout the basin. Furthermore, Royer and Reid (1966) estimated the fundamental period of the Gulf at less than 8 hours, contrary to Marmer's hypothesis. Grace's calculation also showed that the direct tide contributes only a very minor part of the K_1 tide in the Gulf.

Although Grace's method appears to be fundamentally sound, his representation of Gulf geometry was quite approximate, and the phase progression *southward* through the Yucatan Channel cannot be reconciled with the available K_1 tidal constants in the Caribbean. Many of the latter were not available when Grace made his study.

Dietrich's (1963) K_1 world cotidal chart shows the tide progressing northward through Yucatan Channel and arriving in the Gulf essentially in phase with the K_1 wave coming through the Florida Strait. His cotidal lines match reasonably well the phases at additional tide stations in the Caribbean, analyzed subsequent to his paper.

Since an understanding of the tidal regime in the Gulf requires information on the tidal currents in the Florida Strait and Yucatan Channel, an intensive search was made for records of tidal-current observations in the two channels. Only some old observations by Pillsbury and some surface observations in 1965 off Hollywood, Florida, by the General Dynamics Monster Buoy were

Table 15-2
Harmonic Constants of Tidal Currents at Entrances to Gulf of Mexico

Stations	Latitude North	Longitude West	M2 Ampl. (cm/sec)	M2 G (degrees)	M2 Dir. (degrees)	K1 Ampl. (cm/sec)	K1 G (degrees)	K1 Dir (degrees)	Observation year	Observation length
A. Florida Current, off Hollywood	26°01′	79°51′	3.4	95	180*	5.7	284	180*	1965	15 days
B. Florida Current, off Fowey Rocks	25°33′	79°55′	9.2	62	160	–	–	–	1885	7 days
C. Florida Current, between Rebecca Shoals & Cuba	23°38′	82°34′	4.0	91	285	6.8	252	285	1887	2 days
D. Yucatan Channel	21°36′	86°16′	17.0	-	-	12.9	335	5*	1887	4 days
E. Yucatan Channel	21°36′	86°32′	11.8	131	310	–	–	–	1887	1 day

*Approximate direction.

found. General Dynamics furnished these data; the first 15 days appeared to be most suitable for analysis (Smith et al., 1969). The available K_1 and M_2 tidal harmonic constants for the Florida Current and the Yucatan Channel are listed in Table 15-2.

Computation of Tidal Amplitude

For an irregular basin with surface area, S, and incoming currents, u_i normal to the basin boundaries at cross sections, a_i, volume continuity requires the rate of change of surface elevation, h, with time, t, to be $\iint_S \partial h/\partial t \, dS = \iint_a u_i \, da_i$.

Since currents enter the Gulf only through the Florida Strait and the Yucatan Channel, u is nonzero only at cross sections a_1 and a_2 of these channels, and the mean current speeds in these channels are denoted by $\overline{u}_1$ and $\overline{u}_2$, the bar representing an average over the pertinent area. Then, $S \, \partial \overline{h}/\partial t = \overline{u}_1 a_1 + \overline{u}_2 a_2$.

At this point, the current speeds include both permanent flow and tidal current but, in a consideration of an oscillating tidal current, the permanent currents flowing in through the Yucatan Channel and out through the Florida Strait more or less cancel each other. The relatively small difference is related to river drainage, rainfall and evaporation in the Gulf.

Considering now the harmonic tidal current through the two channels, with w the angular speed of K_1, ϕ_1 and ϕ_2 the phase lags in the two channels, and $\widehat{u}_1$ and $\widehat{u}_2$ representing the mean amplitudes of the tidal currents, the height becomes

$$\begin{aligned}\overline{h}(t) = {} & [(\widehat{u}_1 a_1 \sin \phi_1 + \widehat{u}_2 a_2 \sin \phi_2)^2 \\ & + (\widehat{u}_1 a_1 \cos \phi_1 + \widehat{u}_2 a_2 \cos \phi_2)^2]^{1/2} \\ & \times (Sw)^{-1} \sin [\, wt + \tan^{-1} (\widehat{u}_1 a_1 \sin \phi_1 \\ & + \widehat{u}_2 a_2 \sin \phi_2)/(u_1 a_1 \cos \phi_1 \\ & + \widehat{u}_2 a_2 \cos \phi_2)] \qquad (15\text{-}1)\end{aligned}$$

Sufficient data are not available to evaluate this expression for comparison with tides observed on the coasts of the Gulf, but Grace's theory is necessarily consistent with it. Grace computed the tidal currents in the channels by fitting his dynamical model to the available tidal constants from around the Gulf. He noted that his K_1 results do not match well the current observations by Pillsbury in either amplitude or phase. In addition to the question of how reliable harmonic constants from such short series may be, Grace noted that the observations may not be representative of the average tidal currents in the channels.

If the K_1 tidal current in the Yucatan Channel is assumed to have the same phase (as suggested by the tidal constants) and amplitude as in the Florida Strait, where some current measurements are available, equation (15-1) simplifies considerably to

$$\overline{h}(t) = \hat{h} \sin (wt + \phi_1),$$

where

$$\hat{h} = (a_1 + a_2)\hat{u}_1 / wS.$$

The area of the Gulf of Mexico is estimated to be 1.7 X 10^{12} m^2 and the cross-sectional areas of the Florida Strait and Yucatan Channel are 4.8 X 10^7 m^2 and 25.8 X 10^7 m^2, respectively. The computed amplitude of the K_1 constituent in the Gulf is therefore $\hat{h}$ = 15 cm, which is in good agreement with the amplitudes listed in Table 15-1 for the periphery of the Gulf. The K_1 tidal current amplitude near Yucatan Channel computed by Michaelov el al. is considerably less than the above hypothesis, but the computed tidal current amplitudes in the latter study are quite small in general. It is not clear from their presentation that the computed values are valid in the channel.

Standing-Wave Hypothesis

Figure 15-4 and Table 15-1 show that the amplitudes for the diurnal constituents are very small in the Florida Strait at Miami and Bimini. In many areas, the associated diurnal tidal current is even less significant under such circumstances, as may be understood from the amplitude portion of equation (15-2), $u_1 = whS/(a_1 + a_2)$. This indicates that the tidal amplitudes are weighted by their relative frequency in regard to currents in such a closed region. Thus, the semidiurnal tide has approximately twice the associated current of the diurnal tide.

However, the diurnal tidal current is definitely not small in the Florida Strait, which suggests the possible existence of a standing-wave node in this area. Schmitz and Richardson (1968) found K_1 and 0_1 each comparable to M_2 in transport measurements for the Florida Current. Steinberg and Birdsall (1966) and Clark and Yarnall (1967) reported a large diurnal variation in their spectral analysis of phase changes for an acoustic signal continuously monitored across the strait. The analysis of the Monster Buoy surface-current data showed K_1 and 0_1 somewhat larger than M_2. The K_1 and 0_1 amplitudes and phases for the first four tide stations in Table 15-1 confirm that there is a standing-wave oscillation in the strait, with a node close to the latitude of Miami. The tidal harmonic constants for the western part of the Caribbean also show small amplitudes and rapid changes in phase, thereby indicating a comparable regime in that area. It is not difficult to visualize the two regimes as one, with Cuba's mass as a barrier to connecting the cotidal lines.

The K_1 tidal regime in the Florida Strait can be documented more adequately by analyses of tidal observations at additional points along the east coast of Florida between Vero Beach and Key West. The rapidly changing K_1 phase along this coast has not been a problem in terms of supplying tidal predictions for navigational purposes because the amplitude is so small. However, these harmonic constants are important to studies of variability of the Florida Current, as well as the nature of the tide in the Gulf of Mexico.

With reference to the observed phases of the current relative to the tide, an approximate Greenwich phase of the K_1 tide for the entire Gulf is 20°. The maximum flood at the node should precede this by 90°, or a phase of about 290°. The phase of 284° by Smith et al. (1969) fits this very

Table 15-3

Estimate of Energy Flux in Entrances to Gulf of Mexico (Units are 10^{16} ergs/sec)

	Yucatan Channel*	Florida Strait	
		Hollywood	Key West
M_2	8.5	.8	4.8
K_1	5.0	.1	−2.0

*Use has been made of tidal constants for Cozumel Island, Mexico, obtained subsequent to preparation of Table 15-1. At Cozumel, for M_2: 7 cm, 77° and K_1: 2 cm, 324°.

well. Harris' analysis of the Pillsbury data off Rebecca Shoal and Yucatan Channel gives Greenwich phases of 252° and 335°, respectively, approximately 40° from the anticipated value in opposite directions. Variations of this amount are reasonable for harmonic constants from 2- and 4-day series of current observations. Grace's computed phases of maximum current do not fit Pillsbury's observations as well, the variations being 141° and 75°, respectively.

Semidiurnal Tides

Grace compared the available observations of tidal currents in the entrance to the Florida Strait and in the Yucatan Channel with currents calculated from his theoretical development. The fit was found to be poor, and Grace pointed out that even the two sets of observed data in the Yucatan Channel "appear to be very different, unless the current data are unreliable." If Harris' hypothesis of a progressive wave entering the Gulf through the Florida Strait is correct, the west current should be in phase with the high tide. At current station B (off Fowey Rocks) and at C (between Rebecca Shoal and Cuba), the tidal current is about an hour later than the tide. This is a reasonably good fit in phase from short records, particularly when allowance is made for some contribution by a stationary wave oscillating east-west in the basin. It appears that the tidal current to the north at station E (Yucatan Channel) may also be about an hour later than the high tide, but the latter must be inferred from data for surrounding points and is not well determined.

Tidal Energy Dissipation

It is of some interest to evaluate the mean rate of tidal energy dissipation in the Gulf, which can be done conveniently by means of the data presented here. The barotropic energy flux of a single constituent through a tidal channel (cf. Proudman, 1953) is given by $\frac{1}{2}\rho g a \hat{h} \hat{u} \cos\phi$ where ρ is density, g is gravitational acceleration, a is the cross-section area of the channel, $\hat{h}$ and $\hat{u}$ are the amplitude of tide and tidal current at the section and ϕ is their phase difference. Evaluating this relation for the pairs of constituents available for the Yucatan Channel and the Florida Strait yields the set of results presented in Table 15.3. The substantially larger and more variable values at Key West relative to those at Hollywood arises primarily from greater and more variable differences in phase at the former section, which is not surprising in view of the expected influence of internal tides on short current records. Taking the Hollywood section as probably being more accurate, the energy dissipation associated with these two constituents in the Gulf is 1.4 X 10^{17} erg/sec. Assuming comparable phase relationships for the smaller diurnal and semidiurnal constituents as for K_1 and M_2, respectively, the total energy dissipation is about 2 X 10^{17} erg/sec, which may be compared to the Miller (1966) value of 3 X 10^{17} for the entire Gulf-Caribbean region.

Using $(4/3\pi)\gamma\rho u_O^3 S'$ as the total dissipation of energy due to frictional stress, where γ is taken as 2 X 10^{-3}, ρ=1 and S′ as that portion of the Gulf with a depth less than 200 m (estimated as 4 X 10^{11} m^2), u_O is calculated to be 0.8 knot, the root mean cube amplitude of the tidal current in shallow water in the Gulf. If allowance is made for the larger values in the cubed term increasing the mean, a mean amplitude of about one-half knot for tidal currents on the shelf is obtained. This is reasonably consistent with the somewhat larger amplitudes listed for the passes in the tidal current tables for the area.

Acknowledgment

This is a revised version of the article which appeared in *Bulletin of Marine Science*, Vol. 20, No. 1, March 1970, pp. 57-69.

References

Clark, J.G. and Yarnall, J.R. 1967. Long range ocean acoustics and synoptic oceanography, Straits of Florida results. *Proc. 4th U.S. Navy Symposium on Military Oceanography,* vol. 1: 309-365.

Dietrich, G. 1963. *General oceanography, an introduction.* New York: John Wiley and Sons.

Endros, A. 1908. Vergleichende Zusammenstellung der Hauptseichesperioden der bis jetzt untersuchten Seen mit Anwendung auf verwandte Probleme. *Petermanns Mitt.,* 54:86-88.

Grace, S.F. 1932. The principal diurnal constituent of tidal motion in the Gulf of Mexico. *Mon. Not. R. Astr. Soc. Geophys. Suppl.,* 3(2):70-83.

_____.1933. The principal semi-diurnal constituent of tidal motion in the Gulf of Mexico. *Mon. Not. R. Astr. Soc. Geophys. Suppl.*, 3(3): 156-162.

Harris, R.A. 1904. *Manual of tides,* 4 B. Rep. U.S. Cst Geod. Surv. for 1904, Appendix 5:313-400.

_____.1907. *Manual of tides,* 5. Rep. U.S. Cst Geod. Surv. for 1907, Appendix 6:231-545.

Marmer, H.A. 1954. *Tides and sea level in the Gulf of Mexico. Fishery Bull.* Fish Wildl. Serv. U.S., 55(89): 101-118.

Michaelov, V.D., Melishko, V.P. and Shchereleva, G.I. 1969. Estimation of the tides and tidal currents in the Gulf of Mexico and Caribbean Sea. *Transactions of G.O.I.N.*, 96:146-173.

Miller, G.R. 1966. The flux of tidal energy out of the deep oceans. *Jour. Geophy. Res.,* 71(10):2485-2489.

Pillsbury, J.E. 1891. *The Gulf Stream.* Rep. U.S. Cst Geod. Surv. for 1890, Appendix 10:461-620.

Proudman, J. 1953. *Dynamical oceanography.* New York: John Wiley and Sons.

Royer, T.C. and Reid, R.O. 1966. *Gravity waves in a rotating basin-normal modes.* Dept. of Oceanography, Texas A&M Univ., Proj. 471, Ref. 66-27T.

Schmitz, W.J. and Richardson W.S. 1968. On the transport of the Florida Current. *Deep-Sea Res.,* 15(6):679-693.

Smith, J.A., Zetler, B.D. and Broida, S. 1969. Tidal modulation of the Florida Current surface flow. *Jour. Mar. Tech. Soc.*, 3(3):41-46.

Steinberg, J.C. and Birdsall, T.G. 1966. Underwater sound propagation in the Straits of Florida. *J. Acoust. Soc. Am.,* 39(2):301-315.

Sterneck, R. 1920. Die Gezeiten der Ozeane. *Sber. Akad. Wiss. Wien,* 129:131-150.

_____.1921. Die Gezeiten der Ozeane. *Sber. Akad. Wiss. Wien,* 130:363-371.

Von Arx, W.S., Bumpus, D.F. and Richardson, W.S. 1955. On the fine-structure of the Gulf Stream front. *Deep-Sea Res.,* 3:46-65.

Authors

Jerald W. Caruthers

Jerald Caruthers was born in Port Arthur, Texas. He received his B.S. (1960), M.S. (1966) and Ph.D. (1968) degrees in physics at Texas A&M University. Since 1968 he has been an assistant professor in the Department of Oceanography at Texas A&M. He also served a NASA traineeship in the Physics Department at Texas A&M.

John D. Cochrane

John Cochrane received his B.A. in 1943 from the University of California in Los Angeles and his M.S. from Scripps Institution of Oceanography in 1948. At Scripps, he held positions of research assistant, associate in oceanography, junior research oceanographer and head of the Bathythermograph Division. He has taught in the Department of Oceanography at The Johns Hopkins University.

Now an associate professor of oceanography at Texas A&M, Mr. Cochrane's special interests include currents and waters of the Gulf of Mexico; western tropical and subtropical Atlantic Ocean and Caribbean; western tropical Atlantic and eastern tropical Pacific; reflection of surface gravity waves; internal waves; and temperature structure of the upper layers of the ocean.

Guy A. Franceschini

Guy Franceschini was born in North Adams, Massachusetts. He received a B.S. degree from the University of Massachusetts in 1950, an M.S. from the University of Chicago in 1952 and a Ph.D. from Texas A&M University in 1961. He is now a professor in the Department of Oceanography at Texas A&M, a position he has held since 1952. Dr. Franceschini's special field of research is air-sea interaction.

Andrew Garcia

Andrew Garcia grew up in Tampa, Florida, where he attended the University of South Florida. He received a B.S. in physics from that institution in 1963. In 1967 he received an M.S. degree in physical oceanography from Texas A&M University. After working as a civilian oceanographer for the U.S. Coast Guard in Washington, D.C., he returned to Texas A&M, where he is now continuing his studies toward the doctorate degree.

Donald V. Hansen

Donald V. Hansen is director of the Physical Oceanography Laboratory of the Atlantic Oceanographic and Meteorological Laboratories of the National Oceanic and Atmospheric Administration in Miami, Florida. He received a B.S. degree in physics from the University of Washington (1954), and M.S. (1961) and Ph.D. (1964) degrees in oceanography from that institution.

John F. Hewitt

John F. Hewitt was born in Belle Vernon, Pennsylvania. He attended the U.S. Naval Academy, where he received a B.S. in 1962, and the Naval Postgraduate School, where he received an M.S. in 1970. Lieutenant Commander Hewitt is an active duty Naval officer attached to the Navy's submarine service.

Jon M. Hubertz

After receiving a B.S. degree in physics from the University of Florida in 1963, Jon Hubertz was employed at Lamont Geological Observatory, where he was engaged in measurements of the earth's magnetic and gravitational fields at sea. In 1967 he received an M.S. degree in physical oceanography from Texas A&M University. After two years with the U.S. Naval Oceanographic Office in Washington, D.C., Mr. Hubertz returned to Texas A&M to work on a Ph.D. in physical oceanography. He is presently with the ocean dynamics branch of the Naval Oceanographic Office.

Takashi Ichiye

Takashi Ichiye is a native of Kobe, Japan. He received his B.S. (1944) and D.Sc. (1953) degrees from the University of Tokyo, Institute of Geophysics. Before coming to Texas A&M University as a professor in the Department of Oceanography in 1968, Dr. Ichiye was senior research associate at Lamont Geological Observatory. He has taught at Florida State University in Tallahassee and at the Japanese Naval Academy. He has been research oceanographer at Kobe Marine Observatory in Kobe, Japan, and at the Japan Meteorological Agency in Tokyo. He also served a year as visiting scientist at Woods Hole Oceanographic Institution.

Dale F. Leipper

Dale F. Leipper has been chairman of the Department of Oceanography at the Naval Postgraduate School in Monterey, California, since it was formed in 1968. From 1949-1964 he was head of the Department of Oceanography at Texas A&M University. While at Texas A&M, he also served as chairman of the Faculty Advisory Council of the College of Geosciences and as chairman of the Committee on Graduate Instruction.

Dr. Leipper received a B.S. from Wittenberg University (1937), and M.A. in mathematics from Ohio State University (1939) and a Ph.D. from the University of California, Scripps Institution of Oceanography (1950). His primary research interests are synoptic oceanography and large scale sea-air interaction as related to forecasting (coastal fog and stratus, effects of a hurricane on the sea, sea temperature, and the changing pattern of ocean currents).

Robert L. Molinari

Robert Molinari received his B.S. in meteorology at City College of New York in 1965 and his M.S. (1968) and Ph.D. (1970) degrees in oceanography from Texas A&M University. He is now doing postdoctoral research at the National Research Council, National Oceanographic and Meteorological Administration, Atlantic Oceanographic and Meteorological Laboratories in Miami, Florida. Dr. Molinari's fields of interest are numerical modeling of ocean currents and descriptive oceanography.

Worth D. Nowlin, Jr.

Worth Nowlin is an associate professor in the Department of Oceanography at Texas A&M University, where he received his B.A. (1958), M.S. (1960) and Ph.D. (1966) degrees. Dr. Nowlin has also served as program director, Physical Oceanography, for the Ocean Science and Technology Division of the Office of Naval Research in Washington, D.C.

Dr. Nowlin's areas of principal interest are descriptive studies, involving a field program of data collection on currents, water masses and distribution of physical properties in the Gulf of Mexico; theoretical studies of steady, wind-driven ocean currents; teaching descriptive oceanography, wave theory and theories of ocean circulation; and oceanographic instrumentation.

David F. Paskausky

David F. Paskausky is an assistant professor of oceanography in the Department of Geology and a member of the Marine Sciences Institute of the University of Connecticut. He received a B.S. in physics from the University of Chicago in 1960, an M.S. in physics from DePaul University in 1964 and a Ph.D. in physical oceanography from Texas A&M University in 1969. Dr. Paskausky has participated in three deep water cruises in the Gulf of Mexico and Caribbean on the R/V *Alaminos* of Texas A&M and in numerous cruises on Block Island Sound and Long Island Sound on the R/V *Uconn* and the R/V *T441* of the University of Connecticut. He has taught at Aquinas College in Grand Rapids, Michigan, the Milwaukee School of Engineering in Milwaukee, Wisoncin, and at Texas A&M University.

Willis E. Pequegnat

Willis Pequegnat is professor of biological oceanography in the Department of Oceanography of Texas A&M University. He was born and educated in California, receiving his Ph.D. in biological science from the University of California at Los Angeles.

Dr. Pequegnat has been associated with marine science for many years, beginning in 1940 with summer teaching of marine ecology at the Laguna Marine Laboratory and the Kerkhoff Marine Laboratory in California and continuing up to his present position at Texas A&M. In addition, he has visited, studied, and lectured at important marine science centers in Denmark, England, France, Germany and Italy while holding fellowships awarded by the Ford Foundation and the National Science Foundation. Since joining Texas A&M, he has led 10 oceanographic cruises into the Gulf and Caribbean.

Dr. Pequegnat has taught at several colleges and universities, including Pomona College, the California Institute of Technology, the University of Chicago and the University of California at Santa Barbara. He joined the staff at Texas A&M in 1963 after a three-year association with the National Science Foundation in Washington, D.C. His principal research interests are in the ecology of the deep sea and of rocky coastal regions.

Robert O. Reid

Robert O. Reid is a professor in the Department of Oceanography at Texas A&M University. Before coming to Texas A&M in 1951, he served as assistant oceanographer and supervisor of the program in physical oceanography at Scripps Institution of Oceanography's Marine Life Program.

Mr. Reid, a native of Milford, Connecticut, received a B.E. degree in 1946 from the University of Southern California and an M.S. in 1948 from Scripps. His special fields of research are wave theory, air-sea interaction and ocean current theory.

Richard T. Wert

Richard Wert received his B.S. in electrical engineering at the University of California at Berkeley in 1961, and his M.S. (1968) and Ph.D. (1970) degrees in physical oceanography from Texas A&M University. He is presently employed as a scientific officer (oceanographer) in the Office of Naval Research in Washington, D.C., doing contract research in program management for physical oceanography.

Dr. Wert's special areas of interest are theoretical physical oceanography and application of the computer to data acquisition in physical oceanography.

Bernard David Zetler

Bernard Zetler is a senior research oceanographer at the Atlantic Oceanographic and Meteorological Laboratories of the National Oceanic and Atmospheric Administration in Miami, Florida. He received his B.S. degree from Brooklyn College in 1936.

Mr. Zetler has served as director of the Physical Oceanography Laboratory at AOML and as associate professorial lecturer at George Washington University in Washington, D.C., and as adjunct professor at the University of Miami. He spent one year on special assignment doing tide research at the University of California.

Index